IMPERIAL NATURE

IMPERIAL NATURE

Joseph Hooker and the Practices of Victorian Science

JIM ENDERSBY

THE UNIVERSITY OF CHICAGO PRESS
CHICAGO AND LONDON

The University of Chicago Press, Chicago 60637
The University of Chicago Press, Ltd., London

Paperback edition 2010
Printed in the United States of America

19 18 17 16 15 14 13 12 11 10   3 4 5 6 7

ISBN-13: 978-0-226-20791-9 (cloth)
ISBN-13: 978-0-226-20792-6 (paper)
ISBN-10: 0-226-20791-9 (cloth)
ISBN-10: 0-226-20792-7 (paper)

Library of Congress Cataloging-in-Publication Data

Endersby, Jim.
Imperial nature: Joseph Hooker and the practices of Victorian science/Jim Endersby.
p. cm.
Includes bibliographical references and index.
ISBN-13: 978-0-226-20791-9 (cloth : alk. paper)
ISBN-10: 0-226-20791-9 (cloth : alk. paper)
1. Hooker, Joseph Dalton, Sir, 1817–1911. 2. Botanists—Great Britain—Biography. 3. Naturalists—Great Britain—Biography. I. Title.

QK31.H75E53 2008
580.92—dc22
[B]

2007034919

♾ The paper used in this publication meets the minimum requirements of the American National Standard for Information Sciences—Permanence of Paper for Printed Library Materials, ANSI Z39.48-1992.

FOR JOHN AND ELISABETH

TO WHOM I OWE MY LOVE OF

GARDENS AND LEARNING

CONTENTS

ILLUSTRATIONS

ACKNOWLEDGMENTS

My largest debt is to my former supervisor, James Secord, without whose wisdom, patience, and unfailing enthusiasm I would never have been able to complete this book or the PhD dissertation that preceded it.

I am grateful to everyone who listened to me talk about Hooker, read sections of the book, and provided various other forms of comment and encouragement. In particular, my thanks go to Sam Alberti, David Allen, Ruth Barton, Richard Bellon, Jim Betts, Nicola Bown, Janet Browne, Alex Buchanan, Anjan Chakravartty, John Christie, Adrian Desmond, Ray Desmond, Thomas Dixon, Gina Douglas, Richard Drayton, David Galloway, Cathy Gere, Geoff Gilbert, John Hodge, Shelley Innes, Nick Jardine, Robert Kohler, David Kohn, Martin Kusch, Sachiko Kusukawa, Roy MacLeod, Gordon McOuat, Ravi Mirchandani, Jim Moore, David Oldroyd, Alison Pearn, Clare Pettitt, Gail Pope, Duncan Porter, Greg Radick, Moira Rankin, Philip Rehbock, Catherine Rice, Simon Schaffer, Anne Secord, Sujit Sivasundaram, Peter Stevens, Rebecca Stott, Hugh and Charlotte Thurschwell, Pamela Thurschwell, Matthew Underwood, Dan Weinstock, Paul White, John van Wyhe, John Yaldwyn, and Richard Yeo. Audiences at seminars and conferences where I gave papers on Hooker were also most helpful.

An enormous number of librarians and archivists were essential to my work, and I would particularly like to thank Joanna Ball (Trinity College, Cambridge); John Flanagan, James Kay, Michele Losse, Kate Pickard, Leslie Price, Anna Saltmarsh, and Marilyn Ward (Royal Botanic Gardens, Kew); and the staff of the Rare Books Room, Cambridge University Library. Special thanks also to Dawn Moutrey, Lisa Newble, and all the other staff of the Whipple Library and Whipple Museum, Cambridge.

I am also indebted to the following libraries, archives, and individuals. In Australia: Australian Collections and Reader Services (National

Library of Australia); Judith Hollingsworth, Launceston Local Studies Collection (Launceston Library); Judy Nelson, Mitchell Library (State Library of New South Wales); Rod Home, Sara Maroske, and Monika Wells, Ferdinand von Mueller Correspondence Project (University of Melbourne/ Melbourne Botanic Gardens); Anna Hallett and Miguel Garcia, Library of the Royal Botanic Gardens, Sydney; Archives Office of Tasmania; University of Tasmania Library; Alex Buchanan, Tasmanian Herbarium; Australian Manuscripts Collection, State Library of Victoria. In New Zealand: Manuscripts and Archives Section, Alexander Turnbull Library (National Library of New Zealand); Donald Kerr and Georgia Prince, Auckland Central City Libraries; Lucy Marsden, Massey University Library/Te Putanga ki te Ao Matauranga; Nelson Provincial Museum; John Yaldwyn, Museum of New Zealand/Te Papa Tongarewa Archives; Hocken Library, University of Otago. In Britain: Cambridge University Library; Glasgow University Archive Service; Linnean Society of London; Lindley Library, Royal Horticultural Society; Library and Information Services, Royal Society of London; Archives, University of Strathclyde; Trinity College Library, Cambridge. In the United States: American Philosophical Society; Harvard University Herbaria.

I am grateful for permission to quote from unpublished manuscript sources; see the bibliography for a complete list of these.

Setting up the Joseph Hooker website (www.jdhooker.org.uk) has brought me many useful friends and contacts, but I would not have been able to establish it without the help of my brother, Rob Endersby, and his colleague John Banks; many thanks to both of them.

Primary funding for this research came from the Master and Fellows of Trinity College, Cambridge, and the Master and Fellows of Darwin College, Cambridge. I would also like to thank the American Philosophical Society, Philadelphia, for awarding me an Isaac Comly Martindale Fund library resident research fellowship; the Trustees of the Williamson Fund and the Newton Fund, Cambridge; and the Cambridge Commonwealth Trust for Australia. The publication was partly funded by a History of Science Publication Grant from the Royal Society.

INTRODUCTION

> In England, the higher departments of science are pursued by a few who possess independent fortune, by a few more who hope to make a moderate addition to an income itself but moderate, arising from a small private fortune, and by a few who occupy the very small number of official situations, dedicated to the abstract sciences such as the chairs at our universities; but in England the cultivation of science is not a profession.
>
> —Bulwer Lytton, *England and the English* (1836)

When Queen Victoria came to the throne, there were no scientists among her subjects; although the word itself was newly coined, it was rarely used before the last years of her reign, by which time, scientists were a highly visible, influential, and respectable group.[1] The newfound prominence of university-trained scientists—working in laboratories, offering expert advice to both government and industry—was one sign of the huge changes that had transformed Britain and its empire during the old queen's reign. Whether they celebrated or decried the fact, most Britons recognized that scientific expertise had helped to restructure their country, creating a decisive break with old certainties, undermining traditional social deference and religious faith, destroying old ways of earning a living and, with them, a whole way of life. In 1837, most Britons still lived on the land, often in the villages where their parents and grandparents had grown up, working with the same tools, often for the same masters. By the end of the century, steam, smoke, and noise had become the backdrop to the lives of most Victorians, who were now a nation of city-dwellers. Belief in tradition had been displaced by a commitment to endless, restless innovation: "Forward, forward let us range, Let the great world spin for ever down the ringing grooves of change." Tennyson's words, inspired by his first railway journey (and a misunderstanding about the nature of railway tracks), presaged a view that most of his countrymen came to share: that industry, technology, and science must inevitably drive the world onward. And, by the end of the century, most were confident that such changes would be gradual and evolutionary rather than revolutionary.

This summary of the role of science in the Victorian period is familiar (to the point of being clichéd); by staying within well-known themes—the professionalization of science, the growth of empire and the impact of science (especially of Darwinism) on religion—it obscures the richer, more complex stories that specialist historians have been uncovering in recent years. Far from being committed to a "drive towards the dignity of professional status," as one scholar has put it, the early Victorian scientific community was deeply divided.[2] Practitioners of the different sciences disagreed about the status and direction of their disciplines. Few, if any, were aiming for professional status, not least because they felt that being paid to do science was not entirely respectable. Insofar as one can generalize about such a diverse community, during the first few decades of Victoria's reign, British men of science still saw themselves as disinterested gentlemen, not as scientific tradesmen, much less as servants of centralized government, as were their French colleagues.

Concerns about the propriety of paid science particularly preoccupied the large numbers of scientific men whose careers relied on the work of unpaid enthusiasts, including botanical collectors, tide measurers, weather observers, and animal hunters. Many of these enthusiasts worked for love not lucre, and this sense of vocation sometimes meant they felt a little superior to those who—although they possessed greater scientific expertise—had chosen (or, in some cases, been compelled) to pursue science as a paid career.[3] Today we assume that an amateur will always defer to a professional, but that was certainly not the case during this period, when such categories were still being defined and negotiated. Elite practitioners of a science were still imagined to be modeled on the eighteenth-century ideal of a gentleman, someone like Sir Joseph Banks, who had put his vast personal fortune at the service of the search for knowledge, aiding nation and empire in the process but not enriching himself. Banks, who became a friend and adviser to the king, held no official position and took no salary. Ideally, the man of science should emulate such behavior, which meant that, during the early decades of the nineteenth century, being paid to do science put one, not among the elite, but in the same category as Banks's servants, the people he paid to collect, illustrate, and curate for him.[4] The association between receiving payment and low social standing lingered well into the second half of the century.

The question of what kind of career a man of science might aspire to pressed especially hard on naturalists, because their books, theories, and reputations depended on the vast mountains of specimens gathered from all corners of the empire by largely unpaid collectors. Metropolitan men of science relied on these networks of correspondents and collectors, yet while the specimens were always welcome, many of the people who supplied them would not have been, had they visited the metropolis. Nevertheless, this diverse group of colonials—from uneducated shepherds and traders, convict supervisors and seamen, to colonial administrators and missionaries with scientific pretensions—needed to be wooed and placated, flattered and thanked, if the vital flow of specimens was to continue. The complexity of the negotiations required to maintain such arrangements forces us to reevaluate our ideas about the relationships between those at the imperial center and those at its periphery. The idea that metropolis and colony create and transform each other is, of course, not a new one; what I have tried to do is analyze the details of how these complex associations worked in practice.[5] These relationships were far from being simply exploitative or one-sided. Worries about the relative social statuses of the participants and about the value of different kinds of expertise are among the factors that reveal a complex web of interdependence and mutual benefits, within which individuals bartered whatever they could in an effort to satisfy conflicting desires and competing agendas.[6]

In this book I focus on the career of Joseph Dalton Hooker (1817–1911) (fig. I.1) because the three themes that I have identified as dominating our understanding of much of Victorian science—the reception of Darwinism, the consequences of empire, and the emergence of a scientific profession—were all central to his life and work. Analyzing his life forces us to reconsider these central issues.

Hooker was born at Halesworth, Suffolk, on 30 June 1817. He was educated at Glasgow High School, later obtaining his MD at Glasgow University, where his father, William Jackson Hooker, was regius professor of botany. With the help of his father's influential friends, Hooker was appointed assistant surgeon aboard HMS *Erebus*, which—commanded by Sir James Clark Ross and accompanied by its sister ship, the *Terror*—was to spend four years (1839–43) exploring the southern oceans. The ships took shelter from Antarctica's winters in places such as New Zealand and Van

I.1 Joseph Hooker at his desk, microscope in hand (1886). From the original by T. B. Wirgman held at Kew. The image was published in the series "Celebrities of the Day" (*The Graphic*, 17 July 1886, p. 64). By kind permission of the Trustees of the Royal Botanic Gardens, Kew.

Diemen's Land (Tasmania), and these sojourns ashore allowed Hooker, considered by Ross to be the expedition's botanist, to collect plants in relatively unexplored regions.

Shortly after his return from the Antarctic, Hooker received a letter from Charles Darwin that congratulated him on his achievements and marked the beginning of a lifelong correspondence, through which the two became friends and collaborators and debated their many scientific interests. Meanwhile, Hooker was searching for a permanent paid position that would allow him to earn a living from his scientific work. After a brief period at the Geological Survey (founded in 1835), he traveled to the central and eastern Himalaya (1847–49) and then to eastern Bengal—thanks to financial support from the British government, the Royal Botanic Gardens, Kew, and his father. He returned to Britain in 1851 to complete his *Botany of the Antarctic Voyage*, which was published in serial form over many years until it eventually formed six large volumes: two each for the *Flora*

Antarctica (1844–47), the *Flora Novae-Zelandiae* (1851–53), and the *Flora Tasmaniae* (1853–60). It was a detailed description not merely of the plants he had collected but—thanks to the thousands of specimens collected by his network of colonial correspondents—a broad survey of the floras of the southern oceans. The book's broad scope enhanced his reputation, and with further help from his father's patronage network, Joseph Hooker was appointed deputy to his father, who had become director of the Royal Botanic Gardens, Kew, in 1841.

When Darwin published *On the Origin of Species* in 1859, Hooker—who had long known of his friend's theory—was the first man of science to defend natural selection in print. His continued support for Darwin's ideas helped induce many of their contemporaries to accept evolution.

In 1865 William Hooker died and Joseph succeeded him as director of Kew, a position he held until his retirement in 1885. Among his major later works were *Genera Plantarum* (with George Bentham, 1860–83) and *Flora of British India* (1855–97). He also continued to travel, visiting Syria, Morocco, and the Rocky Mountains. Hooker received numerous honorary degrees, including ones from Oxford and Cambridge, and was created CB (Companion of the Bath), then KCSI (Knight Commander of the Star of India) and GCSI (Grand Commander of the Star of India), eventually receiving the Order of Merit. The Royal Society gave him their Royal, Copley, and Darwin Medals, and Hooker was also elected president of the Royal Society.

As this brief biographical sketch makes clear, Hooker was one of the people whose career helped define the key issues concerning the status of nineteenth-century science: he was a close friend of Darwin's and one of his first and most important supporters; he was an internationally renowned botanist, whose work centered on utilizing and analyzing the natural resources of empire; and he was one of the first British men of science to turn a full-time, paid scientific position into a prestigious role, as director of Kew. Hooker was one of the people who created the modern scientist, not least because he showed how it was possible to earn a living from science without sacrificing one's respectability. He helped transmute the gentlemanly ideals of someone like Banks into modern scientific codes of conduct.

Nevertheless, once we examine the details of Hooker's career and compare them with those of his contemporaries, it becomes clear that there was nothing inevitable about the changes he participated in. The possibility that paid science might seem disreputable is one crucial reason why the established narrative of professionalization needs to be reexamined, but I shall also show that there was nothing predictable about Hooker's embrace of Darwinism, which was supposedly the common, secularizing ideology of the scientific professionalizers. Indeed, I shall argue that Hooker's acceptance of Darwinism was more complex and ambiguous than has hitherto been recognized; Hooker strove to reshape natural selection into something working naturalists could use but, far from embracing his friend's theory wholeheartedly, was anxious to distance himself from some of its implications.[7] His concerns were not unique and did not arise primarily from religious or political worries but from the practical difficulties of earning a living from science.

I shall return to the reception of Darwinism in the book's conclusion, because the practical problems that natural selection created for working naturalists make sense only once we understand naturalists' daily scientific work. Despite the fact that natural history was arguably the most important nineteenth-century science (certainly the most widely practiced), whose implications ranged from the economic to the theological, the daily activities of its practitioners have yet to be studied closely. Instead, historians have tended to concentrate on a handful of publications that deal with theoretical issues, ignoring the vast bulk of Victorian naturalists' libraries—notably the endless books on collecting, preserving, storing, and classifying specimens.[8] As a result, we have a rich history of scientific ideas but almost nothing on the scientific practices that made those ideas possible. By building my study around a detailed examination of what might initially seem the most mundane of practices—botanical collecting and classification—I hope to show how these activities shaped even the most sophisticated theoretical speculations.

What do I mean by "practice"? The *Oxford English Dictionary* defines the term as "the action of *doing* something," with subsidiary senses such as "working" and "method of working." For my purposes, I want to add the stipulation that this work or doing must involve tangible, material objects.[9] Thus, I have chosen to emphasize "writing" as a practice rather than the more abstract notion of "theorizing." Naturally, all scientific practices involve both manual and intellectual work—"classifying" being a good example—but I begin with the material aspects of sorting sheets of specimens into labeled drawers rather than with more abstract matters such as the existence (or not) of natural kinds. These decisions are, in part, polemical, intended to sharpen the contrast with a history of ideas in which disembodied concepts wander vaguely across an intellectual landscape, "influencing" people as and when the argument requires. However, my intention is not to replace this with a crude determinism, in which a scientific mode of production determines the content of scientific concepts; instead, I will show that focusing on what naturalists *did* gives us a better understanding of the nature of science because studying practice illuminates and connects everything, from the constraints of earning a living to the content of scientific theories. The surviving traces of scientists' activities encompass many written forms (including letters, diaries, notebooks, and publications), but I shall treat these initially as material artifacts—objects that were made, exchanged, bought, and sold—rather than merely as the bearers of ideas. In this way, I hope to relate these written traces more fully to other objects—such as instruments, specimens, and drawings. Analyzing the full range of this material is essential if we are to understand what science was and how it was transformed during this crucial period.

However, an account of scientific practices needs to be connected to the broader context of gentlemanly natural history. At a time when social and scientific acceptance and authority were closely linked, naturalists like Hooker—those who sought to raise their science to a position where it would be more highly regarded—sometimes struggled to be acknowledged as social equals by their fellow men of science. One route to such acceptance was joining the clubs, societies, and associations where the gentlemen of science met. But joining required money, and for those who were not lucky enough to be independently wealthy, that meant ensuring they were well paid for their scientific work; yet the genteel status to which they aspired was supposedly incompatible with earning a salary. Hooker epitomized the naturalists' dilemma, the often-contradictory combination of social and scientific aspirations that beset them as they struggled to establish science as a respectable way for a gentleman to earn a living, or—to put the problem another way—to show that one could work for a living and still claim genteel status; for with such status came a claim to the public's support and trust, and thus to the government's money.[10]

The thematic structure of this book ranges across Hooker's entire career, but in order to focus on his effort to establish himself and his discipline, I have concentrated my account on its early decades, prior to his becoming director of Kew in 1865. The crucial years in Hooker's career were the middle decades of the century, which were also those during which the role of the man of science was most in flux; hence, Hooker's search for a livelihood exemplifies issues of much wider concern. In the final chapter of the book, I examine Hooker's later career, as director of Kew and of Britain's botanical empire. Hooker's later governance of Kew frequently reflected the difficulties of his early career, which left him extremely sensitive to real or perceived slights to his status or that of botany—hence, Hooker's stiff-necked opposition to allowing picnics and "bands of music" to sully Kew and his public brawl with First Commissioner of Works Acton Smee Ayrton over the garden's scientific status.[11] From banning early opening on bank holidays to defending his precious herbarium (a library of dried plant specimens) from amalgamation, Hooker's directorship is best understood in the context of his earlier struggle to establish himself, scientifically and financially.

Gentlemanly Science

Was science to be a vocation or a profession? The ways in which this question was answered need to be understood within much wider Victorian debates over whether gentlemen were born or made—defined by character or by ancestry. Was nobility a kind of conduct that anyone, from any class of society, might learn to exhibit—perhaps from conduct or etiquette manuals—or was it ultimately something "in the blood"? These questions preoccupied the Victorians, as is evident by the regularity with which they appear in popular novels of the time: John Halifax, the eponymous hero of Dinah Craik's best-seller, working hard to prove that poverty and trade could not disbar him from the title "gentleman"; John Thornton, the mill owner in Elizabeth Gaskell's *North and South*, who—despite his humble birth—is shown as morally superior to his social betters; or Dickens's Pip, being turned into a gentleman by his initially unknown benefactor, Magwitch, who was trying to prove by proxy that, given enough money, it took only carefully taught manners to make a gentleman.

Despite the mid-Victorian cult of the self-made man, exemplified by the tediously enumerated heroes of Samuel Smiles's *Self-Help*, a pervasive unease surrounded these questions.[12] Smiles himself was conscious of his humble background, unsure of his own status as a man of letters (a new career which shared many characteristics with that of the man of science), and concerned by some of the problems his gospel of self-improvement created: If birth and wealth did not make a gentleman, what did? How was one to know a true gentleman from a persuasive fraud? These concerns were also apparent among those who joined the scientific debates over the role of heredity—from the breeders of dogs and horses to the gentlemen of the Royal Society—all of whom shared an interest in untangling the respective contributions of nature and nurture, that "convenient jingle of words" coined by the eminent Victorian scientific gentleman Sir Francis

Galton to summarize these arguments.[13] The ways in which men of science imagined and planned their working lives were shaped by these shifting notions of gentility and by the accompanying discussions of breeding, conduct, character, and etiquette.

To understand some of the concerns that shaped Joseph Hooker's career, it is useful to contrast the ways in which he sought a livelihood with the options his father faced. Joseph was the second son of William Jackson Hooker (1775–1858), who had inherited land but had sold it to buy a share in the local brewery, which was owned by his patron and later father-in-law, the banker and botanist Dawson Turner. Unfortunately, Hooker's move from landowner to brewer coincided with the economic depression that followed the end of the Napoleonic Wars, which hit the brewery's profits hard. In 1820, concerned about his ability to support his family, William decided to turn his passion for natural history into a paid occupation; as Joseph explained, some eighty years later, "reduced circumstances" obliged William to "turn his botanical attainments to material account."[14] Despite never having heard, much less given, an academic lecture, he drew on the support of influential friends, notably Banks, to become the regius professor of botany at Glasgow University.[15]

Patronage was crucial to William Hooker's career; in addition to Turner and Banks, Sir James E. Smith, one of the founders of the Linnean Society of London, provided valuable support.[16] While Hooker's enthusiasm was exemplary and his published work impressive, he could not have hoped to secure the Glasgow professorship at this time without such important supporters. By contrast, when Joseph Hooker attempted to win the Edinburgh chair of botany a quarter of a century later, he failed—despite being able to count the prime minister among the patrons his father had helped him secure. Even a thick file of testimonials from some of Europe's leading men of science failed to overcome the Edinburgh City Council's perception that they needed an experienced lecturer with a firm grasp on the medical uses of plants, not an expert taxonomist, however well connected.[17] In addition to crucial local factors, times had changed: a growing emphasis on individual expertise and intellectual merit was slowly beginning to displace the power of patronage (which, nonetheless, remained important throughout the century).[18] Joseph Hooker emerged from his ego-bruising failure determined to succeed on his own terms, to shape a career for himself that would be independent of patronage. His view was one that was becoming increasingly commonplace during the period: for middle-class Victorian men, "independence" was becoming highly prized. Its meaning was vague, but true manliness was established through self-reliance and hard work, not by relying on inherited wealth or aristocratic patronage. Victorian manliness was also about establishing a home and providing for a family, by being seen to "act" and above all to work. As we shall see, Hooker was typical of thousands of young men who reached manhood in the 1830s and turned to the empire to provide the opportunities they needed to establish their independence.[19]

The contrast between being comfortable with patronage and aspiring toward independence was only one of the differences between the Hookers. For William Hooker the move from brewery owner to university professor would almost certainly have been perceived as a step *down* the social scale.[20] He had sold the little land he had inherited to invest in the brewery at a time when income from industry was less solidly respectable than income from rents—but at least he did not labor in the brewery himself.[21] By contrast, the move to Glasgow was from proprietor to employee.[22] Not only did William Hooker work for a living, but his status was further compromised by the fact that he was paid to teach medical students at a time when medicine was the least prestigious of the learned professions.[23] A tiny minority of medical men became celebrated, highly paid physicians; most earned a modest living in a profession tainted by its association with trade. Yet during the century's early decades, it was almost impossible to earn a living from botany other than in a medical context, especially after botany became a compulsory part of English medical education following the passing of the 1815 Apothecaries Act.[24] The act ensured a growth in botanical teaching positions but also ensured that plants were only studied as part of the *materia medica,* the materials from which drugs were made. This branch of medicine was primarily associated with apothecaries, regarded as the lowest of the lowly medical men because they engaged directly in trade.[25] By contrast, physicians avoided manual work and regarded dispensing medicines as deleterious to their social and professional standing.[26] In George Eliot's *Middlemarch,* set during the 1830s, Dr. Lydgate displays his pretensions to higher status by deciding to "simply prescribe, without dispensing drugs or taking percentage from druggists."[27] Although dispensing drugs brought in a secure income, ambitious doctors like Lydgate sought to climb the social and professional scale by breaking their link with shopkeepers.[28]

William Hooker worked directly for Glasgow's Faculty of Physicians and Surgeons rather than for the university, and the faculty had no interest in those aspects of botany that might give it independent standing (such as the principles of plant classification or geographical distribution).[29] Instead, as with other ancillary subjects such as chemistry, the faculty expected students to acquire practical, economically useful skills, which reflected the ethos of a merchant city, built on trade and industry. Just as William Hooker was happy to rely on patronage to advance his career, he seems to have been largely sympathetic to this utilitarian approach: the Museum of Economic Botany he founded at Kew in 1848 was a direct result of the twenty years he spent at Glasgow, absorbing its industrial and commercial ethos.[30]

As we shall see, Joseph Hooker was less happy to see botany taught purely for its medical or other uses; during the contest for the Edinburgh chair, he bemoaned the fact that he would be teaching "that lowest of all classes of students, the medical."[31] Unfortunately for him, there were few other kinds of university posts in botany for many decades; a dozen years later, the *Athenaeum* was still complaining: "Of all the natural sciences Botany is perhaps worse treated in this country than any other" because it was "tacked on as an appendix to a course of medical study, and gets little or no consideration in any other direction."[32] It is not surprising, therefore, to find that Joseph Hooker—having been rebuffed by Edinburgh—made no further attempt to win a university post; the universities simply did not provide an opportunity to study plants for their own sake and thus no opportunity to pursue a scientific vocation.

Not only did university botanical teaching associate one with the lowest rungs of the medical profession, but it was not even well paid. Like most academic positions, such professorships were assumed to be held by gentlemen of independent means, and so their "salaries" were little more than honoraria. As a result, professors like William Hooker had to stand at the classroom door, collecting fees from their students in order to ensure their income.[33] Robert Graham, the professor of botany at Edinburgh (whom Joseph Hooker hoped to replace), found he had to continue practicing medicine because the income from the botanical chair was only one hundred pounds per year in the 1820s. And although William Hooker had sought the Glasgow professorship because of his financial problems, he found he still had to supplement his income by writing and publishing for the broadest possible audience.[34]

William Hooker's friend and former protégé John Lindley found himself forced to pursue a similar strategy. He supplemented his salary as professor of botany at University College London by producing popular publications, notably the weekly *Gardeners' Chronicle*, which was aimed at the same audience as Hooker's *Botanical Magazine:* middle-class gardeners and plant enthusiasts, male and female. Such works were often lavishly illustrated and focused on popular plants such as ferns and orchids.[35] The contrast with Joseph Hooker's publications was striking; as we shall see, Joseph Hooker's books were generally dry and highly technical, and he rather resented the need to illustrate them.

"More Complete & Philosophical"

Although Joseph Hooker was not interested in producing the same kinds of popular, illustrated works as his father, publishing was still crucial to his career and he was ambitious in his goals. In 1868, he mentioned to his friend James Hector in New Zealand that

> I am hard at work at a British Flora!—to be a more complete & philosophical through plain & simple descriptions of B. Plants than any of the 3 now in vogue—with notes on anatomical physiological etc points.
>
> It is an awful labour but is much wanted for the classes in England & Scotland—& *I want money* in this expensive post for a family man.[36]

This letter nicely juxtaposes Hooker's aim to be "more philosophical" with his open admission that he "wants money" to support his growing family—evidently still a concern even though he had become director of Kew in 1865. In some ways the proposed book was aimed at the same market his father and Lindley had served, the medical and botanical "classes in England & Scotland"; beginning in 1854, Hooker examined assistant surgeons for the Indian army for twelve years, and beginning in 1855, he examined candidates for the Apothecaries Company's botany medal. His attempts to improve botanical and medical teaching left him conscious of the need for better textbooks.[37] As Hooker noted, other authors sought to reach the same market, but he felt that the leading titles had their faults, which his own work would overcome in part by being more "philosophical" than its competitors.

What did Hooker mean by "philosophical"? He and his contemporaries used the term frequently and it had different meanings for different people. An audience of, say, London medical students would have given a very different answer from that offered by the elite gentlemen of science whose acceptance Hooker wanted. Yet, even for that latter audience, the term's meaning was much broader and its implications were much more significant than have been previously recognized.[38] Understanding the numerous implications of the term is the key, not merely to understanding Hooker's goals, but to understanding the Victorian scientific world in which he lived.

"Philosophical" was immediately derived, of course, from "natural philosophy," the study of causes in nature, which was traditionally contrasted with natural history's main task, the description of natural phenomena.[39] It was precisely that association with the prestigious physical sciences, especially physics and astronomy, that made so many naturalists aspire to be described as philosophical. Hooker's letter to Hector explained that "Bentham is not scientific enough for a classbook, nor is Babington."[40] His criticism suggests that "philosophical" and "scientific" were synonymous, and to some extent they were, but defining "philosophical" as "scientific" begs more questions than it answers—we still need to know what "scientific" meant in the mid-nineteenth century.

An initial sense of Hooker's definition of "philosophical" begins to emerge if we compare the "more complete & philosophical" book he eventually wrote—the *Student's Flora of the British Islands* (1870)—with those of his rivals: George Bentham's *Handbook of the British Flora* and Charles Cardale Babington's *Manual of British Botany*.[41] Hooker did not specify the third book "now in vogue"; but its identity can be deduced with some certainty: John Hutton Balfour's *A Manual of Botany* (3d ed., 1863) would be an obvious contender but for the fact that Hooker mentioned in the letter previously quoted that "Balfour urged me to this" (i.e., to write the new book). Assuming that Balfour was unlikely to have urged Hooker to write a rival to his own book, it seems most likely that Hooker was thinking of John Lindley's introductory *School Botany*, then in its twelfth edition.[42] Detailed comparisons will be made in later chapters, but for the moment there are three contrasts between Hooker's work and those of his competitors that merit comment. First, his book was more comprehensive than its rivals and endeavored to list every indigenous British plant species; as a result, Hooker had to omit illustrations (which were a major feature of both Bentham's and Lindley's books), as well as hints and tips on how to botanize. The result was a slightly forbidding volume, which appeared more sober and serious than its rivals.

This impression of seriousness was reinforced by Hooker's approach to classification, which is the second key difference between these books. In the *Student's Flora* he explained that the descriptive characters used to define the classificatory groups "have been rewritten, and are to a great extent original, and drawn from living or dried specimens or both," and although he had included keys (analytical tables used to identify plants;

see fig. 6.2) to the genera, the species themselves were simply given "curt diagnoses" because the use of keys for species "promote[s] very superficial habits amongst students."[43] (The same superficial habits were presumably fostered by the use of illustrations for identification, which may well have been another reason Hooker omitted them from his book.) As figure 6.2 shows, using either the key or the diagnoses required considerable expertise, so—despite being aimed at students—the *Student's Flora* discouraged superficiality. All this creates the impression that Hooker wished to emphasize that the study of botany was not to be entered into lightly; making botany more serious was an important aspect of Hooker's desire to raise its standing among the sciences.

Raising botany's status was also a motivation for another aspect of Hooker's classification: although the *Student's Flora* (like all Hooker's books) used the same classificatory system (known simply as the natural system) as his rivals, he "lumped" plants together into fewer species, genera, and orders than, for example, Babington, who was inclined to split apart large variable groups into numerous separate ones, each with its own name. For Hooker, the fact that the *Student's Flora* was more comprehensive in scope but contained fewer species was proof of his book's philosophical status. As shown in chapter 5, part of Hooker's conception of "philosophical" was "reducing a great number of dissimilar ideas under a few successively higher general conceptions."[44]

The third aspect of the *Student's Flora* that distinguished it from its rivals was that it gave details of the geographical distribution of the various species, a topic I will discuss below.[45] These three aspects of the *Student's Flora*—its seriousness, its "lumping" classification, and its emphasis on distribution—were important aspects of Hooker's definition of "philosophical."

Nevertheless, the *Student's Flora* also demonstrates that we should be cautious about identifying "philosophical" with such terms as "speculative" or "theoretical," for one obvious reason: it is a small, octavo book, twelve × eighteen centimeters (five × seven inches), weighing only six hundred grams (twenty-one ounces), and, thanks to the lack of illustrations, it sold for just ten shillings, sixpence.[46] This was clearly a book for the field, intended—as its title suggests—to be slipped into students' pockets and used on field trips and in the classroom. The meanings of "philosophical" began (and, in at least one important sense, ended) in the field; specimens had to be collected before they could be classified, but any classification would be meaningless until used to classify the plants themselves. A book like the *Student's Flora* was a point in a cycle that ran from field to herbarium and back to field again; Hooker used the expertise he had gained in both field and herbarium to write it, and then he and his publisher had to wait and see if other botanists would buy it, read it, and then collect and classify according to its precepts.[47]

Fieldwork—traveling, walking, observing, and above all collecting—was nineteenth-century natural history's primary practice. Every naturalist's education began in the field, and introductory books invariably stressed the need for practical collecting before the novice even contemplated the more abstruse branches of the subject.[48] Hooker prefaced his discussion of the definitions of species in the *Flora Indica* with the following observation: "Long and patient observation *in the field*, and much practice in sifting and examining the comparative value of characters, can alone give the experience which will warrant the expression of a decided opinion on a question of so much difficulty."[49]

As will be discussed in later chapters, these strictures were partly designed to encourage deference to the metropolitan expert while discouraging premature speculation among local botanists, creating a division of labor which, among other benefits, was essential to establishing botany's inductive credentials. However, the requirement for firsthand field experience applied to the metropolitan gentlemen as well as colonial collectors, as is clear from letters between Hooker and Darwin concerning Frédéric Gérard's work on species. Hooker—not long returned from his first voyage aboard HMS *Erebus*—disparaged Gérard's work because he was "neither a specific naturalist, nor a collector, nor a traveller" and therefore merely "a distorter of facts." To be qualified to speculate, Hooker wrote, "one must have handled hundreds of species with a view to distinguishing them & that over a great part,—or brought from a great many parts,—of the globe."[50] Hooker had no doubt that Darwin fulfilled these requirements, but his friend was less certain, and Hooker's strongly expressed view prompted Darwin to take up his firsthand study of living and fossil barnacles from all over the world (including some he had collected aboard the *Beagle*). The eight years spent poring over barnacles were, in part, meant to demonstrate that Darwin was indeed qualified to discuss species.[51]

However, while worldwide collecting (combined with the examination of other people's collections, "brought from a great many parts,—of the globe") was a necessary part of a naturalist's experience, it was not sufficient to earn one the prized adjective "philosophical." Much depended on *how* one collected. When Darwin asked Hooker for some of the *Erebus* grass specimens, Hooker replied that he was doubtful as to their usefulness, telling Darwin that "I did not collect with any idea of having the specimens made such a philosophical use of."[52] The "philosophical use" in question was to help answer a central question about plant distribution: what role the wind played in spreading the seeds of plants. Darwin had asked for the specimens on behalf of the German naturalist Christian Gottfried Ehrenberg, with whom he had been corresponding. Ehrenberg wanted to compare samples of grasses from Ascension Island with those he had from Malta, in order to help decide whether the wind could transport plants so far. He had specified that "there must be dependable names on them," and Darwin passed this request on to Hooker, who had visited Ascension during his voyage aboard the *Erebus*, telling him that the specimens "must be *named* or else they will be useless."[53] Hooker was probably concerned that the names he had given his specimens might be wrong, given that he would have identified these unfamiliar plants in the field, with only limited access to books or dried specimens for comparison. His anxiety would have been further exacerbated by the fact that, despite an extensive botanical education under his father, the quality of Joseph's early specimens had been poor, largely because he lacked practical experience of collecting in warm climates (see chapter 2). As we shall see, specimens were carefully crafted artifacts and considerable dexterity was needed to make them; imperfect, badly selected, or poorly preserved specimens were all but worthless.

Collecting for philosophical purposes depended on the collector's talents, which included recording the appropriate details of when and where the specimens had been gathered (either too little or too much detail caused problems) and using the appropriate techniques to preserve each specimen (which varied considerably from plant to plant). However, such craft skills were worthless unless collectors could successfully identify the plants, not least because they needed to distinguish well-known and common plants from potential novelties. As we shall see, identification and classification were inseparable, which meant that the practical and theoretical could not be separated. A good collector needed to name plants, and naming required considerable familiarity with Hooker's approach to classification and thus with what he called "the philosophy of system" that underlay it.[54] Philosophical and practical matters were as inextricably intertwined in the field as they were in the herbarium.

Attending to the often-neglected craft aspects of natural history collecting reveals that far from being a simple business of picking and pressing flowers, collecting was complex, highly skilled, and difficult to learn. Although learning these skills was hard work, acquiring them would gradually change the collector's relationship with a gentleman like Hooker.

The metropolitan gents needed expert collectors in order to avoid being bombarded with misidentified plants, worthless duplicates, and poorly preserved rubbish. One way to improve your collectors was to send them gifts, especially of botanical books, journals, and other tools. But, as the gifts' recipients painstakingly acquired the expertise they needed to collect well, they began to realize that possessing such expertise made them more useful to their distant correspondents—and thus harder to replace. That recognition opened up the possibility that collectors could drive a harder bargain in their negotiations over the value of their specimens and abilities; money was seldom the object, however. Goals varied, and as we shall see, Hooker could not always give his correspondents what they wanted, especially when they wanted to name their species themselves. Hooker's refusal to allow such naming created friction that had to be overcome to keep the network running. Attending to the apparently trivial craft aspects of collecting reveals that, far from being passive providers of specimens or inert recipients of metropolitan knowledge, the colonial naturalists were active participants in the making of scientific knowledge. Understanding the full range of opportunities (and restrictions) they faced is possible only once their practices are fully understood.

A Botanical Empire

As Darwin's request for Hooker's grass specimens illustrates, understanding the distribution of plants and animals was a key concern for nineteenth-century naturalists, which brings us to another sense of "philosophical." Global plant collections were vital for investigating the distribution of plants, a more prestigious study than mere classification since it offered to shed light on the natural laws that shaped the plant kingdom. Traveling European naturalists had long been aware that there was a connection between a region's physical geography—such as its climate, elevation, rainfall, and soil type—and its characteristic vegetation. However, there were puzzling exceptions to these apparent rules: the same species were sometimes found on tiny islands, separated by thousands of miles of ocean—how had they got there? In other cases, regions with very similar climates were found to be populated by physically similar but distinct species.

Understanding these complex patterns of vegetation had two attractions for a botanist like Hooker. As we have seen, uncovering the laws that generated the patterns would raise botany from the merely descriptive.Unraveling these mysteries had a more practical, economic benefit, however. Much of the wealth of Britain's empire rested on plants: from the timber and hemp from which her navy was built, to the indigo, spices, opium, tea, cotton, and thousands of other plant-based products that the ships carried.[55] Grasping the laws that shaped vegetation might allow valuable, new plants to be discovered, and it would certainly allow existing crops to be successfully transplanted from their original locations to British colonies, where they could be cultivated profitably—and a grateful government might reasonably be expected to reward the science that had added new crops to the empire. It is therefore not surprising that Hooker described the "great problems of distribution and variation" as "prominent branches of inquiry with every philosophical naturalist."[56] This was a widely held view: in 1833, the *Edinburgh Review* defined a "philosophical botanist" as one "who invents new principles of classification, who studies the structure and organs of plants, who develops the laws of their geographical distribution."[57] Including information about geographical distribution in his *Student's Flora* was part of Hooker's wider campaign to establish botanical geography as a fundamental element of the science.

Hooker would later comment that geographical distribution was seldom discussed when he had been a student; at that time, British natural history—and botany in particular—was largely concerned with the identification and naming of species. Classification was useful but unglamorous work, often thought to be intellectually undemanding and thus a relatively low-status activity. The botanist's colleagues, the zoologists, improved the standing of their studies by attending increasingly to matters such as comparative anatomy and physiology, and gradually botanists began to follow suit (hence the *Edinburgh Review*'s reference to studying "the structure and organs of plants," topics Hooker had initially planned to include in the *Student's Flora*).[58]

However, classification could not be discarded in favor of distribution studies. No one could attempt an explanation of the global patterns of vegetation until taxonomists had established exactly how many species each country held—hence Hooker's long struggles both to revise the principles of classification and to demonstrate its fundamental importance to all the other branches of botany. Perhaps the biggest problem he faced was establishing consensus over names and classificatory methods. Hooker's publications and his reputation would eventually confer the prestige and authority that, combined with the resources of Kew, gave him some ability to settle disputes, but this power was only acquired slowly. And even at the height of his fame and influence, Hooker could not simply overrule those who disagreed with him, not least because he wished to claim that botany was a mature science, whose principles were founded on empirical evidence, not on idiosyncratic opinion. He relied instead on his herbarium collections, physical evidence of his encyclopedic, global knowledge of plants, to provide the material to settle arguments.

Although many of the specimens in the Kew herbarium had been gathered by the Hookers themselves, many more had come from their correspondents and collectors around the world. Even a well-traveled naturalist like Joseph Hooker could not hope to see more than a fraction of the world's plants for himself if he was also going to have time to classify, analyze, and write about them. The sheer size and scope of his herbarium increased his reliance on his scattered collectors, many of whom had their own views about classification; their opinions, ironically, having been shaped and strengthened by Hooker's numerous gifts. As a result, Hooker constantly found himself struggling with their tendency to be taxonomic "splitters," meaning that they defined species too narrowly for Hooker's taste (who was, as we have seen, a "lumper"). In his *Flora Novae-Zelandiae* he told his readers, "I do not think that those who argue for narrow limits to the distribution and variation of species, can have considered a garden in a philosophical spirit."[59] He argued that the sheer variety of plants from all over the world that grew outdoors at Kew and flourished despite having been transplanted to Britain's climate was proof that most species were not naturally restricted to one small corner of the world. One of Hooker's preferred strategies for abolishing names that had been conferred by colonial or provincial naturalists was to compare the specimens the local naturalists had relied on with the much wider range he had available at Kew. Hooker argued that a global comparison demonstrated that the apparently distinct forms to which the local botanist had given names were in fact linked by intermediate forms from around the world. The varieties of the plant could then be lumped together as a single species, a process that, not surprisingly, many local collectors objected to, since it removed "their" names from the botanical record.

Hooker often found it hard to persuade collectors to adopt his broader definition of a species, which was only one aspect of his attempt "to develop the principles of classification" in such a way that species and other groups would be defined on "philosophical grounds" and "with a proper degree of precision."[60] Precision was, of course, the hallmark of the physical sciences, and many early Victorian men of science felt that one of the reasons that natural history compared unfavorably with more exalted studies like astronomy and physics was that they lacked mathematical precision. As part of his campaign to elevate botany's status, Hooker and some of his colleagues used a technique called botanical arithmetic to study the phenomena of plant distribution. Bringing mathematical tools to bear gave precision to the botanist's generalizations and also helped to distinguish botany from its closest scientific relative, zoology. Although zoologists were spending more of their time in laboratories, which they claimed gave greater precision to their work than was possible in the field, there was no comparable science of "zoological arithmetic," a lack that gave botanists a clear advantage in distribution studies. As Darwin commented in a letter to the American botanist Asa Gray, "Botany has been followed in so much more a philosophical spirit than Zoology."[61]

Hooker's views on distribution and classification give us some sense of what he meant by "philosophical," but the term's full importance becomes apparent only when we consider how he made a living from being philosophical.

"An Amateur Only"

In 1843, during the *Erebus* voyage, Hooker faced the problem of what he would live on when he returned. His commander, Ross, had offered to recommend Hooker's promotion to full surgeon and to support an application for half-pay while ashore in order to allow him to write up the record of his travels. Hooker wrote to tell his father that he would accept Ross's offer, but only if he found himself unable to obtain a paid position at Kew. Hooker explained, "As you know, I am not independent, and must not be too proud; if I cannot be a naturalist with a fortune, I must not be too vain to take honourable compensation for my trouble."[62]

The desire to be a "naturalist with a fortune," like Banks, was an understandable one, but Hooker acknowledged that, "having no competency [a sufficient income or estate] of my own," he had no choice but to seek a paid position; as the novelist Henry Fielding had put it, "There is no happiness in this world without a competency."[63] As we have seen, Hooker's father had sold the land he had inherited, spending his money on making "my Herbarium the richest of any private one in Europe."[64] This collection was, as we shall see, to be Joseph's main inheritance, but while he waited to come into his inheritance, governmental parsimony meant there was to be no post for him at Kew until 1855, twelve years after the *Erebus* returned to Britain. As a result, Hooker would spend more than a decade trying to find the "honourable compensation" he felt entitled to; and the issue of what kind of income would be regarded as "honourable" by the gentlemen of British science was a significant concern.

A decade after he had decried his lack of independence, Hooker wrote to his friend and collaborator George Bentham to explain his fury at those he saw as retarding the progress of the science. "I am a *rara avis,*" Hooker explained, "a man who makes his bread by specific Botany, and I feel the obstacles to my progress as obstacles on my way to the butcher's and baker's. What is all very pretty play to amateur Botanists is death to me."[65] Bentham, of course, was just such an amateur, "a naturalist with a fortune," who found no obstacles between him and his baker. Hooker's complaint might seem tactless, but of course Bentham could console himself with the thought that he enjoyed a higher social standing than Hooker, precisely because he did not need to make his bread from botany.

Five years later, in 1858, Hooker wrote to Asa Gray, who earned his botanical living at Harvard University, to discuss Bentham's suitability to be appointed a foreign member of the American Academy of Science. Hooker described Bentham as "essentially an amateur," albeit "a stunning one." Bentham's independence gave him the freedom to do only what he enjoyed, regardless of the needs of botany as a whole; Hooker was mildly critical of this, commenting that Bentham "says—& thinks—I [i.e., Bentham] am an amateur, I am not grounded in Philosophical Botany & do not care to be so now. I . . . do not feel myself to be a servant of Science."[66]

Given that Hooker contrasted dedicated servants of science like himself, who had put themselves out to master "philosophical botany," with amateurs, who pursued the subject for their own enjoyment, it would seem reasonable to equate "philosophical" with "professional"—which we would now regard as the obvious alternative to "amateur." Yet in 1857, Hooker had written to Darwin about the relief he felt in "turning from the drudgery of my 'professional Botany' to your 'philosophical Botany.'" Far from resenting Darwin's questions on geographical distribution, Hooker explained that he looked forward to them.[67] Hooker's "drudgery" was a mixture of helping administer Kew (of which he was now, finally, assistant director) while trying to finish his *Botany of the Antarctic Voyage.* He was in the midst of clarifying the names and identities of roughly eight thousand species before he could finally turn to the introductory essay, in which he would discuss the philosophical question of their distribution.[68]

By the time this letter was written, Hooker had obtained the honorable compensation he felt he had earned: a full-time, paid position at Kew, where he was poised to eventually succeed his father. He was, as he told Darwin, now a professional botanist. Yet, far from being a status he rejoiced in, he seems to have been anxious to escape it. It is noticeable that, although his comments on amateurs exhibit slight feelings of superiority, it is his *philosophical,* not his *professional,* standing that he is proud of; the latter is mostly "drudgery." In an 1860 letter to Darwin, Hooker described their mutual friend William Henry Harvey as someone who "calls himself an Amateur only,—lays no claim to be philosophical or scientific"—despite the fact that Harvey was the professor of botany at Trinity College, Dublin.[69] Once again, the contrast being made is not between amateur and professional but, as in the case of Bentham, between amateur and philosophical (and Hooker seems to distinguish the latter term from "scientific").[70]

It was important for men like Hooker not to exclude people like Bentham from the scientific community. In fact, Hooker supported Bentham for the presidency of the Linnean Society in 1861 because, as he told his friend Thomas Henry Huxley, he had a "prejudice against professional Scientifics being Presidents of these heterogeneous bodies; & in favour of independent men who make a bond of union between science as represented by the society & the outer world—& who if really scientific, are so as amateurs. Bentham is one such."[71]

This letter and the 1857 one to Darwin are among the very few instances where Hooker calls himself a professional. Historians have generally been less reticent and have not only applied the term to him but described the attempts he and his contemporaries made to earn a living from science as part of a conscious strategy of "professionalizing" their discipline.[72] My objection to this approach is that it obscures the crucial fact that for Hooker and his contemporaries the word "professional" had rather negative connotations, which is why they rarely used it. I propose to follow Hooker's lead and avoid the term, not least because close attention to the terms that *were* in regular use illuminates the important changes that were under way in Victorian science and society more generally.[73]

Hooker usually distinguished those who pursued botany seriously as "professed," rather than professional, botanists. For example, in the preface to his *Flora Novae-Zelandiae,* after stating that the book gave him an opportunity "of promoting a love and knowledge of the Science of Botany in those English Colonies which it has been my good fortune to visit," he noted that, "though it was called for by professed Botanists, and is therefore more scientific than a popular Flora should be, I have added to the technical characters such English descriptions as will enable the resident to name his plants, and I have written these in the simplest language that can be applied to Botany."[74] The contrast here is essentially between different degrees of expertise.[75] The same distinction was made, using the same term, in the *Flora Indica,* where he noted that too many medically and commercially valuable plants were "known only to the professed botanist" rather than to those who could profit from them.[76]

At this time, "professed" was most often used to describe a public declaration (or profession) of religious or other belief (in George Eliot's *Felix Holt*, the dissenting minister refers to his congregation as "professors") or to make a claim of expertise or qualification.[77] In both these senses, it implied a vocation rather than the modern sense of profession; it described an individual's character, not a paid socioeconomic role.[78] Claiming a vocation associated naturalists with disinterested study rather than commercial gain; widely used terms such as "votary of science" had similar quasi-religious connotations, implying a commitment that transcended narrow self-interest.[79] Unsurprisingly, "professed" remained the common term during the first half of the century: as late as 1860, the *Edinburgh*'s otherwise-hostile review of the *Origin of Species* began by praising Darwin's "pleasing style," which made it readable, "not only by the professed naturalist, but by that far wider intellectual class which now takes interest in the higher generalisations of all the sciences."[80]

Naturalists were not alone in wanting to be seen as professing a vocation rather than profiting from a profession. "Profession" was an ambiguous and not always complimentary term: prostitution was commonly referred to as a profession, and an 1830s etiquette manual waspishly commented, "We do not attempt to deny that 'Esquire,' in common with the terms 'Professor,' or 'Professional,' is occasionally abused, since a 'Professorship' rewards alike the saltatory labours of a dancing master, and gilds the graceful avocations of an 'Arcadian' [i.e., rustic] hair-dresser."[81] Men of science would have been in no hurry to identify themselves with either prostitutes or hairdressers, but more importantly many laypeople feared the power and self-interest of the professions. Lawyers, for example, were routinely satirized as needlessly argumentative and as prolonging cases to enrich themselves (a tradition Dickens drew heavily on in *Bleak House*), while doctors were mocked for being ignorant and greedy—and thus interested in promoting, rather than curing, disease. Defining science as a profession would have made its practitioners into another potentially self-serving interest group, with a clear commercial stake in their advancement.

Antiprofessional satires are evidence of the complex negotiation that accompanied the acceptance of a profession's claims to high social status.[82] Even as the modern sense of "learned profession" emerged gradually in the eighteenth century, groups such as lawyers and doctors clung to their claim of a vocation, which remained implicitly part of what distinguished a respectable profession from a mere occupation. Professing an expertise could be the first step toward higher status, but only once such assertions were accepted by the public; declaring that one pursued a "calling" (another term with religious overtones to describe an occupation), rather than mere lucre, was a way of advancing one's claims to respectability.[83] The moral overtones of terms like "professed," "cultivator," and "votary" all helped to instill trust in men of science by implying that they were disinterested seekers after truth, which—among other things—made them suitable advisers for government. Such terms were also inclusive and served to make these admirable moral qualities the common property of all men of science; by contrast, a sharp distinction between amateurs and professionals would exclude many of the people, such as colonial collectors, on whose goodwill and hard work the metropolitan experts relied.[84]

While they shared the concerns of other men of science, botanists had an additional reason for describing themselves as "professed"; the word helped distinguish their studies from those of the slightly unsavory profession—medicine—with which they were usually associated. "Professional" was often used to distinguish the merely medical study of plants from properly scientific work. Edward Forbes, professor of botany at London's King's College (fig. 9.2), argued that botany, because of its medical importance, "forms a connecting link between professional and purely scientific studies."[85] Although it was the connection with "professional studies" (training doctors) which allowed Forbes to earn his living from botany, he was keen to assert the "purely scientific" nature of his studies, thus raising them from their status as a mere "appendix" to the medical course.[86]

The consistent use of "professed" rather than "professional" expressed a desire to distance botany from its medical context, as well as a wish to be judged by such factors as one's seriousness, expertise, and commitment—rather than on how one earned a living. As we shall see in later chapters, the same is true of the term "philosophical," which—among other meanings—denoted approval of methods and ideas without reference to how philosophical botanists financed their researches. And of course, the relatively small number of badly paid botanical positions would have made botanists reluctant to embrace a term which implied that only those who held such positions were to be taken seriously.

As we have seen, some of these concerns were common to all men of science, whereas others were specific to naturalists or just to botanists; the significance of diverse disciplinary affiliations has sometimes been neglected because common goals have been assumed to exist. For example, Frank Turner quotes the following letter from Huxley to Hooker about a proposed research fund: "If there is to be any fund raised at all, I am quite of your mind that it should be a scientific fund and not a mere naturalists' fund. . . . For the word 'Naturalist' unfortunately includes a far lower order of men than chemist, physicist, or mathematician. You don't call a man a mathematician because he has spent his life in getting as far as quadratics; but every fool who can make bad species and worse genera is a 'Naturalist.'"[87] Turner describes Huxley's remarks as embodying the "cutting edge of the professionalizing spirit," which was rapidly excluding "amateurs" from science.[88] However, while Huxley and Hooker agreed that only an expert (such as a professed naturalist) should qualify for funding, it is not apparent that they therefore assumed such expertise was the exclusive province of professionals. What is clear is that Huxley thinks naturalists as a group rated considerably lower than practitioners of the mathematical and physical sciences, so from Hooker's perspective, Huxley's letter about the research fund was simply an episode in the long struggle to raise naturalists to the status of physicists by excluding the makers of "bad species and worse genera" from their ranks. What distinguished him from such "fools" was that he was philosophical and they were not; for Hooker, philosophical status was tested by such matters as how one defined species, not by how you earned your living. Moreover, botany's status problems were closely connected to its unstable classificatory practices, and Hooker, being a botanist, was more sensitive to these issues than Huxley.

Conflating Hooker's interests and attitudes with those of his contemporaries is one problem with existing accounts of professionalization, but a more serious one is the teleological assumption that the eventual shape of a career must have been its original goal. In recent years, numerous historians have argued that the idea of professionalization carries with it the anachronistic assumption that the career of a modern professional scientist was the implicit goal of their nineteenth-century predecessors.[89] By contrast, David Allen has divided those who earned a living from natural history into three groups: "refugee professionals" (who abandoned a recognized profession, such as law or medicine, to try and earn a living from natural history); those who switched from another calling for nonfinancial reasons; and *"rentiers manqués"* (who expected to be wealthy gentlemen amateurs but who were disappointed). He notes that this third group did not turn professional voluntarily, nor in many cases did the first two: ill health or (as in William Hooker's case) financial hardship frequently forced them to try earning a living from natural history.[90] This provides a useful reminder that few of those who pursued careers in natural history wanted to work for a living; we should therefore abandon the idea that they had a conscious program for achieving this rather-undesirable goal. Joseph Hooker's career was typical only in the sense that it was improvised under highly contingent (and often adverse) conditions.

In an effort to avoid the teleology implicit in existing professionalization models, Richard Bellon has recently argued that Hooker was a professional but "only in the very singular sense that he and his contemporaries would have understood." Hooker's ideal of the man of science focused on personal characteristics, like honesty, and Bellon argues persuasively that only such men could engage in a disinterested pursuit of truth; appropriate codes of gentlemanly behavior were essential to this pursuit, as was government funding, but the latter was to be regarded as a reward for service—not a fee.[91] While I find much to agree with in Bellon's account, his description of Hooker's attempts to use his institutional positions to reform science in accordance with these ideals concerns the post-1860 period, when the British state was both more able and more willing to fund science. Yet even during these later decades of the century, the question of who was a professional was, if anything, even less clear than it had been previously. In 1877, William Carruthers, keeper of the botanical department of the British Museum, testified to the Devonshire Royal Commission about the pros and cons of merging the herbarium collections at the museum with those at Kew. If such a national herbarium were to be based at Kew, visiting it would require a long train journey from London, and Carruthers argued that this would place it "out of the reach of the busy men who frequently use it to the advantage of science": "Of course the working botanist who devotes himself exclusively to the science would follow the collections wherever they went; but the active professional man, and the man of business, who devote their spare hours to botany, would be deprived of the assistance necessary to their work which they now obtain at the British Museum."

It is not the devoted "working botanists" but the part-time devotees who are described here as "professional men," since they were often lawyers or doctors. However, by this time, it was no longer clear who the professionals were, since Carruthers went on to note that of the nineteen botanical articles in recent volumes of the Linnean Society's *Transactions*, "four are produced by professional botanists, and fifteen by others."[92] Not only are professionals now on both sides of the full-time/part-time divide, but it is also entirely unclear as to where Carruthers would have placed leisured gentlemen like Bentham, who was then devoting his whole time to writing his *Flora Australiensis* (1863–78) while collaborating with Hooker on the *Genera Plantarum* (1862–83).

It seems impossible to impose a workable definition of "professional" on Victorian men of science, however hard we might try to stabilize it by specifying its precise scope or sense. I therefore propose to give up the attempt and—to reiterate a point I made earlier—focus instead on analyzing the terms that were in use at the time, to see what they tell us about such matters as the unease that working for payment caused men like Hooker and the ways they attempted to resolve these tensions. This preoccupation with what are sometimes called "actor's categories" is not just a matter of academic pedantry; language is not a neutral medium for conveying facts but a complex method of persuasion. The terms men of science used to describe themselves not only shaped the public's perception but also helped created the community's sense of itself. Aspiring new members of the community received unconscious guidance as to how they should behave as they learned to describe themselves as votaries or servants of science. The same is true of such categories as "professed" and "philosophical"; they began as improvised responses to issues (especially monetary ones) that confronted all men of science, but these makeshift solutions were then carried forward into later periods to become aspects of the ideal modern scientist. Why, for example, are we so taken aback by scientific fraud? One answer is because it is so easy (peer review is an inefficient police force) yet remains rare. We assume that if a scientist makes a claim, he or she should be trusted, ultimately because a "gentleman's" word is his bond. However, we should not make the assumption that these modern ideals were the anticipated consequences of earlier practices.

One alternative to discussing Hooker in the context of professionalization is to consider the institutions he used to build his career. His strategy was a compromise between adapting himself to an institution and trying to remake that institution to serve his goals. This is exemplified by his time at the Geological Survey (1846–47), where he obtained the position through his father's patronage but quickly mastered fossil botany to justify his appointment. Until he was appointed, he had published almost nothing on fossil plants, but he quickly wrote three essays on plants from which coal was formed for the Survey's *Memoirs* (1848).[93] The salary of one hundred and fifty pounds per year, together with the fact that much of the work could be done at Kew, allowed him to continue work on the *Flora Antarctica*. Yet he was not able to use the Survey to fulfill all his needs; his proposal that it should finance the formation of a British herbarium of all extant plant species—ostensibly to provide comparisons for the fossil ones—was rejected as being peripheral to the Survey's goals, and Hooker soon left the Survey to travel in India.[94] By contrast, in the 1850s Kew offered a greater opportunity for herbarium building, but the government was not providing the garden with either clear policies or guaranteed funding during most of the 1841–65 period.[95] Indeed, the gardens had been threatened with closure in the late 1830s shortly before being brought under direct government control.[96] Although some saw Paris's Jardin des Plantes as an explicit model for what Kew might become, many Britons were still suspicious of anything French, so both Hookers had to improvise alternative models that made Kew into a place where a career as a philosophical botanist could be pursued.[97] Kew was not a stable institution within which to pursue an occupation; the scientific version of Kew that emerged late in the nineteenth century was as much a product of Joseph Hooker's career as a context for it. The structures of institutions provided resources for careers, but those who took advantage of the opportunities often struggled to reshape institutions to serve their vocational goals.

For Hooker, the herbarium was Kew's heart and he devoted much of his life to enlarging his father's collections. Nicholaas Rupke has noted that Richard Owen's career also focused on creating a national natural history museum, and despite their obvious differences, Hooker and Owen shared a practical concern with the fate of their institutions rather than with the more esoteric questions that preoccupied their independently wealthy contemporaries. Despite ending up on opposite sides of the Darwinian debate, they had surprisingly similar unease about Darwinism's impact on their institutions, their collections, and their networks of collectors.[98] Owen's colleague at the British Museum John Edward Gray was another museum-based naturalist whose institutional context shaped his career. Gordon McOuat has argued that his huge correspondence network and cataloging efforts were essential to the work of all the leading naturalists of the day, yet Gray's concern for his institution's continued funding and his desire to avoid fragmenting his networks led him to remain aloof from the natural selection controversy. As a result, he has been overlooked by historians because he cannot be fitted into the dominant framework of the Darwinian debates.[99] In a similar way, I will show in my conclusion that Hooker's occasionally equivocal response to Darwinism was shaped by his role as a herbarium builder. Owen, Gray, and Hooker took three very different positions on Darwinism, yet their attitudes can be fully understood only by shifting our attention from the evolutionary debates to their protagonists' respective disciplinary and institutional contexts.

"Nothing Is So Essential as a Character"

Victorian men of science professed allegiance to a range of disciplines and institutions, but despite these differences, they were all developing careers at a time of considerable social change. The rapid growth of British cities brought them many opportunities, but the cities also brought anonymity, creating uncertainty about whom to trust. As the Congregationalist minister Robert Vaughan noted in his influential work *Age of Great Cities* (1843), in a village, "every man is known" and "his movements are liable to observation"; these circumstances ensured that "a strong check is constantly laid upon the tendencies of the ill-disposed." By contrast, the "crowded capital is to such men as some large and intricate forest, into which they plunge, and find, for a season at least, the places of darkness and concealment convenient for them."[100] The city also made it hard to assess someone's place in society, and a whole range of sources testify to widespread concern at the ease with which city-dwellers could claim a social standing they had no right to.

Like so many of their fellow citizens, the men of science were pursuing new ideas and new ways of making a living. Their claims to expertise and credibility were as yet unverified, so good manners, courtesy, and an aura of respectability had to do the work that would eventually be done by formal qualifications and institutional or professional affiliation. To be a trusted member of the early Victorian scientific elite one had to be a gentleman, yet—as we have seen—there was considerable uncertainty as to who or what a gentleman was. Charles Dickens's early work *Sketches of Young Gentlemen* (published in 1838, the year Hooker obtained his MD) contains satirical observations on the varieties of young "gentlemen" that young ladies were likely to meet. Among them was a particular kind of young gentleman who "has so often a father possessed of vast property in some remote district of Ireland, that we look with some suspicion upon all young gentlemen who volunteer this description of themselves." This unverifiable wealth is as dubious as his claims to be "a universal genius; at walking, running, rowing, swimming, and skating, he is unrivalled; at all games of chance or skill, at hunting, shooting, fishing, riding, driving, or amateur theatricals."[101] Dickens's humor expresses a common concern, that the "large and intricate forest" of the crowded city allowed counterfeit gentlemen to pass themselves off as the real thing. Detecting these fakes was no easy matter, and many of the novels, advice books, and etiquette manuals that dealt with gentility were aimed at identifying the true gentlemen amid the anonymous throng.

For a man of science like Hooker, whose "vast estates" were a collection of dried plants and whose "universal genius" was in matters few understood, proving himself trustworthy and reliable was no simple matter. He and his contemporaries stressed that good manners were an essential prerequisite for joining the scientific community, which meant conducting disagreements with fellow men of science in measured, courteous tones—preferably in private. In the 1850s, Hooker and Darwin shared their concerns over Huxley's vehement tone and enthusiasm for public brawling; they initially decided not to nominate him for membership of the Athenaeum Club, for fear that his discourtesy might generate opposition that would harm the public image of all men of science.[102] A few years later, Hooker described to Darwin the advice he had given the entomologist Henry Bates: "to take care not to quarrell [*sic*] with or show contempt for his brother Entomologists; & to take their sneers & suspicions in perfect good part." As long as Bates avoided such ill-mannered altercations, Hooker was convinced that "if he only goes on quietly & goodnaturedly working hard & publishing such papers as he has in 3 years he will be the first living philosophical Entomologist"—but only as long as "he makes no enemies" among his fellow naturalists.[103] Hooker wrote to Bates himself, commiserating with him over the ungenerous treatment being meted out by his rivals in London's scientific societies, who Hooker believed were jealous of Bates's accomplishments. Hooker nevertheless urged him, "Do I entreat you smile at their sneers & [?tell] them good humouredly that 'time will show,'" adding, "To get employment especially, nothing is so essential as a character for never being offended, which after all is true dignity."[104] Hooker's advice to Bates makes a clear connection between being philosophical and being well mannered. Having and keeping "a character" was a key aspect of the developing mid-Victorian sense of what it meant to be a gentleman.[105]

Gentlemanly codes of courtesy became an important aspect of the definition of "philosophical" in part for the same reason that the term "professional" was avoided: labeling people as philosophical, rather than professional, defined them by their practices, ideas, and standards of behavior rather than by their source of income; it was a term that allowed members of the diverse scientific communities to negotiate their way through the uncertain and shifting relations between social and scientific status; it was a word with which naturalists attempted to bridge the uneasy gap between a prevailing ideal of practicing science for love, as a gentlemanly vocation, and their often urgent need for money. Above all, "philosophical" is the term that allows us to connect the daily work of science to the great issues that divided Victorian men of science and the society around them. Reexamining those connections gives us new ways of understanding the complex, shifting place of science in society.

CHAPTER ONE

Traveling

In the late summer of 1839, the twenty-two-year-old Joseph Hooker was walking through London, accompanied by Robert McCormick, nearly forty years old and an experienced British naval surgeon who had already been on several voyages. The two men were preparing for a long ocean voyage, and as they crossed Trafalgar Square, McCormick recognized an old shipmate, a man Hooker later recalled as tall and "rather broad-shouldered," with "an agreeable and animated expression when talking, beetle brows, and a hollow but mellow voice." As Hooker remembered, when they stopped to chat, the newcomer's "greeting of his old acquaintance was sailor-like—that is, delightfully frank and cordial."[1]

The frank and cordial man was Charles Darwin, and this chance encounter marked the start of one of the nineteenth century's most important scientific friendships, a friendship initially founded on the shared experience of traveling. Darwin could not have known that he was already one of Hooker's scientific heroes; although his *Journal of Researches* (the *Voyage of the Beagle*) was still unpublished, a family friend had managed to get Hooker a set of printer's proofs, knowing how much he enjoyed travelers' tales. While he was waiting to set sail on his own voyage, Hooker slept with Darwin's words under his pillow so that he could read them before he got up. Many years later he remembered that they had "impressed me profoundly, I might say despairingly, with the variety of acquirements, mental and physical, required in a naturalist who should follow in Darwin's footsteps." Nevertheless, that was what Hooker hoped to do and he was inspired by Darwin's book, which stimulated his enthusiasm "to travel and observe."[2]

In the informal world of early-nineteenth-century science, with few university degrees and even fewer paid positions available, Darwin had

1.1 In the wake of Cook: HMS *Erebus* and her sister ship, HMS *Terror*, anchored in Christmas Harbour, Kerguelen's Land. From J. C. Ross, *A voyage of discovery and research in the southern and Antarctic regions, 1847*. Reproduced by kind permission of the Syndics of Cambridge University Library.

chosen traveling as the best way to make a reputation for himself. His inspiration had been the renowned German naturalist Alexander von Humboldt, whose narrative of travels in South America inspired many young men to undertake similar, hazardous scientific adventures.[3] Among them was Hooker, whose first voyage, like Darwin's aboard the *Beagle*, was a British naval expedition mainly concerned with mapmaking. When Hooker and Darwin first met, Hooker had just been appointed to a post aboard HMS *Erebus*, which, accompanied by its sister ship, the *Terror*, was about to set sail for Antarctica as Britain's contribution to international efforts to map terrestrial magnetism (fig. 1.1). The expedition, which became known as the Magnetic Crusade, had been launched after a successful lobby of government by the British Association for the Advancement of Science (BAAS) and the Royal Society.[4] Some leading British men of science were concerned that their nation's scientific efforts were falling behind those of their European rivals and saw the expedition as an opportunity to reverse this decline. With this goal in mind, Humboldt himself had been persuaded to support the campaign, and the Royal Society had arranged for his letter urging the importance of the expedition to be printed and circulated among influential government figures.[5] To be even a minor figure in such a high-profile expedition was, Hooker realized, an excellent opportunity. As he later told Darwin, "from my earliest childhood I nourished & cherished the desire to make a creditable Journey in a new country, & with such a respectable account of its natural features, as should give me a niche amongst the scientific explorers of the globe."[6]

However, while Darwin and Hooker had similar goals, they traveled in very different styles: Darwin went as a gentleman companion to the *Beagle*'s captain, Robert FitzRoy, while Hooker was an assistant surgeon, subject to regular naval discipline. Darwin's father's wealth had allowed his son to spend long periods ashore, accompanied by his personal servant; by contrast, Hooker would have to fit his botanizing into whatever spare moments his medical and other duties permitted.

James Clark Ross, commander of the Magnetic Crusade, had met Hooker before they set sail. When Hooker made it clear that he hoped to become the expedition's official naturalist, Ross had told him that he was looking for someone "perfectly well acquainted with every branch of Nat. Hist., [who] must be well known in the world beforehand, such a person as Mr. Darwin." Ross had therefore decided to appoint the more experienced McCormick as naturalist. On receiving this unwelcome news, Hooker wrote to his father, complaining "what was Mr. D. before he went out? he, I daresay, knew his subject better than I now do, but did the world know him? the voyage with FitzRoy was the making of him (as I hoped this exped. would me)."[7]

Hooker confessed that he "must have looked very sorry and angry, which however [Ross] did not see, as he went on, speaking as kindly and almost as affectionately as ever, offering to write me letters of introduction to the surgeon and chief officers of the ship at Chatham, charging them to give me every opportunity of going ashore." Despite this magnanimity, Hooker immediately did the rounds of his father's influential friends, whose support had secured his place on the *Erebus*, asking their advice. He told his father that these advisers had "strongly disadvised my going except as the only Naturalist in the ship, the more especially as *Dr. McCormick was to be my superior*." This seniority would give McCormick first claim on all the collections made during the voyage; as Hooker wrote, "all my notes on Molluscs and sea animals will naturally revert, from the Admiralty, to the Zoologist, besides which he will have more time on shore than I can."[8] Without access to these collections, Hooker would have little chance of publishing a broad work on natural history to compare with Darwin's *Journal*.

Despite his disappointment, Hooker realized that the expedition was the best opportunity he had to make a name for himself; as he wrote, "No future Botanist will probably ever visit the countries whither I am going, and that is a great attraction."[9] After some further negotiations, he was able to tell his father "I am appointed from the Admiralty as Asst. Surgeon to the *Erebus,* and Capt. Ross considers me the Botanist to the Expedition and promises me every opportunity of collecting that he can grant."[10] Given that McCormick's expertise was primarily zoological, Hooker seems to have realized that he would be able to take charge of the botanical collections, publish floras of the countries the ships visited, and thus establish a scientific reputation. But even if this ambitious plan failed, Hooker realized that accepting the post would mean that "I can always fall back on the service [i.e., the Navy] as a livelihood."[11]

A few months later, on 30 September 1839, the *Erebus* and *Terror* set sail from England. It would be four years before they returned. Life aboard ship was uncomfortable and often dangerous (the ships came close to being wrecked on more than one occasion), and Hooker was often lonely. Yet, the voyage was indeed to be the making of him; he began his adult life as an underpaid naval surgeon but—thanks in large part to his travels—ended up as Sir Joseph Hooker, GCSI, KCSI, OM, president of the Royal Society, director of the Royal Botanic Gardens at Kew, and loaded with honorary degrees and medals from the world's scientific societies.

However, while Hooker's career illustrates some of the ways in which traveling provided the basis for a scientific career, his eventual success was also the product of botany's rise up the scientific hierarchy of the day. From a peripheral aspect of medical training, botany would eventually become one of the great imperial sciences, playing a key part in exploring, cataloging, and exploiting the natural wealth of the empire. Hooker was to play a central role in the transformation of botany's status—raising his science as he raised himself.

From Glasgow to Newcastle

In 1838, a year before the Ross expedition set sail, Hooker had attended his first meeting of the BAAS, in Newcastle; this was a crucial event for him, since this was where the lobby for the Magnetic Crusade was launched. These meetings brought together a rich diversity of practitioners of all the sciences from every class and background. In addition to the celebrated scientific dignitaries and their aristocratic patrons, the meetings were attended by self-improving mechanics, part-time meteorologists, astronomical enthusiasts, and artisan naturalists. The *Westminster Review* said of the latter group that "there are literally hundreds of such men scattered over the land—and they are a blessing to it" and went on to describe the sight of the "worthies of this class" attending "almost every meeting of the British Association for the Advancement of Science," where they could "be seen enjoying the happiest day of their lives, by listening to dry and seemingly abstruse discourses in the Natural History section."[12]

Part of the attraction for working-class "worthies" and other scientific newcomers was that the BAAS provided a rare opportunity for newcomers to catch a glimpse of the elite of British science. Hooker wrote to his grandfather Dawson Turner to describe a local bookshop that was "a sort of rendezvous for all the Newcastle strangers," where he had spent some of his mornings because he "was sure of seeing 8 or 10 of the gentlemen, on one forenoon in particular, while reading there, I saw Buckland, Lyell, Sedgwick, Herschell [sic], Richard Taylor, Hutton, & several others."[13] Yet, despite his excitement at meeting some of the association's celebrities, Hooker felt that, "with regard to the scientific department of the Association, it fell far behind the amusement & eating."[14] As he would soon realize, in the informal, sociable world of Victorian science, opportunities for amusement and eating, the chance to meet and make friends, were at least as important as formal meetings.

The astronomer John Herschel, whom Hooker glimpsed at Newcastle, had just returned from five years of significant astronomical work at the Cape Colony and was fêted by the BAAS. He was one of those who were convinced that British science was in decline, and he used his new celebrity to become a leading member of the "magnetic lobby," joining the effort to convince the government to finance the Antarctic expedition.[15] A couple of years later, while the Ross expedition was at sea, Herschel explained the importance of the magnetic survey to George Grey, then governor of South Australia. Herschel wrote that a full survey "of all the colonized and colonisable parts of Australia" and Britain's other colonies was becoming vital, because "Surveyors are too apt to *work by compass* . . . making undue and erroneous allowances for the Magnetic declination, or deviation of the needle from the true meridian." He claimed that most surveyors were unaware that the compass needle needed to be corrected in the light of local variations in the earth's magnetic field and that "it is hardly possible to over estimate the amount of confusion and litigation which must be caused to the next generation when land becomes more valuable" if the surveys on which land tenure was based were not done accurately. Measuring magnetism was at least as important to the carving up of new colonies as it was to fully understanding the nature of the terrestrial globe. Herschel went on to inform Grey that magnetic surveys were already under way in Canada and at the Cape, and that the governments of the Australian colonies had an ideal opportunity to make their own, since "a magnetic observatory now exists in full activity in Van Diemen's Land" which could provide "a centre of reference . . . upon which any extent of operation might be securely based." [16] That observatory had been set up by Ross's expedition (fig. 8.3) and Hooker had been present at its inauguration, so the economic value of the imperial scientific enterprise of which he was part would have been obvious to Hooker from the earliest stages of his career.

Meanwhile, the Newcastle meeting gave Hooker early news of the impending Antarctic expedition—which gave him a chance to fulfil his ambition of joining the scientific elite he saw around him—but it also offered a rather sharp reminder that his science, botany, was not held in anything like the same esteem as Herschel's astronomy. Whether one examines the money the BAAS disbursed, its published reports, or the comments of its leading lights, it is clear that during the 1830s and 1840s natural history was not highly regarded and that botany sat even lower than zoology. These attitudes directly affected Hooker's career prospects, because the association's implicit disciplinary hierarchy was closely linked to the funding it provided (and it was the largest source of scientific funding in Britain at the time). [17]

At Newcastle, Hooker was struck by the lack of attention paid to botany and natural history. He observed that "the sections were very variously attended, the Geological was always the most crowded, then the Mechanical & Physical, the Medical was farthest behind of any"—which not only confirmed its lowly status but tainted botany by association—and "next to [the Medical], I am sorry to say, the Natural History, which (from want of papers I believe) did not sit at all on the last day," adding "there was only one Botanical paper read." [18] If the dearth of papers led Hooker to suspect that natural history was not held in the highest esteem, his opinion would have been confirmed by the general committee's report on its expenditure: in 1838 the BAAS distributed a total of £3,742.10 to fund the research of its various sections, of which the natural history section (Section D) received a princely £6—for work on both botany *and* zoology. By contrast, the mathematical and physical sciences (Section A) got £2,263.10. The Newcastle meeting reflected BAAS priorities: in the first dozen years of the association's existence, the average expenditure on botany was 2% of the total.[19] By way of comparison, geology and geography (Section C) received over 8%, while Section A received 38.5%.[20] The priority accorded to the physical sci-

TABLE 1. BAAS, number of published reports, by section, 1835–1844

Section	1835	1836	1837	1838	1839	1840	1841	1842	1843	1844	Total	%
A (mathematical and physical sciences)	7	6	6	4	7	6	8	2	4	8	58	40
B (chemistry and mineralogy)	1	1	2	0	0	1	0	1	0	1	7	5
C (geology and geography)	0	0	0	0	1	1	2	4	4	5	17	12
D (natural history)	1	3	1	0	1	2	2	5	5	6	26	18
E (statistics)	4	4	3	0	0	2	1	0	1	0	15	10
F (medicine)	1	0	1	0	0	0	0	1	0	0	3	2
G (mechanical sciences)	1	0	3	2	0	1	3	3	3	2	18	13
Annual totals	15	14	16	6	9	13	16	16	17	22	144	100

Source: Morrell and Thackray 1981

ences relative to botany was also reflected in the government's funding of Ross's Antarctic voyage, which cost the Admiralty £109,768.6.6, making it the largest scientific endeavor that had been mounted in Britain.[21] By contrast, the expedition's botanist—Hooker—was paid just £114 a year and had to provide much of his own equipment.[22]

The hierarchy that was implicit in the association's expenditure was borne out by its published reports, whose number reflects the importance accorded to each specialist section (table 1). Although at first sight, Section D seems well represented, its reports were almost all zoological. For example, in the 1839 BAAS *Report* there were 2 lengthy zoological papers, but none on botany; in 1840, there was a 58-page paper on Ireland's fauna and a 2-page paper on methods for preserving both animal and vegetable substances, with no other botanical papers; in 1841, a committee on preserving the vegetative power of seeds got 2 pages to report its findings, while the comparative anatomist, Richard Owen, got 145 pages to report on fossil British reptiles. In 1843, zoology got 64 pages, botany 5; in 1844, zoology won again (163 pages to 5); and again in 1845 (146 pages to 9); and once more in 1846 (142 pages to 11). The pattern is too strong to ignore—not only was natural history poorly funded, but within Section D botany was clearly zoology's poor relation.[23]

This implicit hierarchy of sciences was not confined to the BAAS; it reflected widely held views, as is evident from the way the sciences were covered by general Victorian periodicals, especially those like the *Edinburgh Review* and *Athenaeum* that were pro-science and campaigned for progressive causes such as university reform.[24] Botanical works were popular and thus often reviewed, but the reviews regularly implied that botany, however pleasant a pastime it might be, was not really a serious study. For example, in 1833 the *Edinburgh Review* discussed the *Memoir and Correspondence of the Late Sir James Edward Smith* (cofounder of the Linnean Society).[25] In a series of distinctly double-edged compliments, the reviewer commented that "in ordinary society" botanists had "a great superiority over the other cultivators of physical science" because serious studies produced unduly serious students, whereas the botanist enjoyed "the simultaneous and healthful exercise of mind and body." His studies were also accessible: "the domestic party, from which the lover of plants has gone forth, feel a lively interest in the pursuits of his day; and minds of every cast who surround him in the evening circle, are capable of appreciating the discovery of a plant that no eye had before seen." After the new plant was inspected: "the female pencil is called forth to copy the new or rare plant; the ambition of the youthful aspirant is inflamed; the piety of more aged listeners is warmed; and the vanity of all is flattered when the labours of the day are recorded by learned societies, and when the new acquisition is perpetuated in coloured drawings, and transmitted in duplicates to grace the *herbaria* of foreign nations. Such is the ordinary day of the botanical traveller."[26] The reviewer's comments reveal some of the difficulties that beset botanists who wished to become more philosophical: botany was suitable for "minds of every cast" rather than intellectually demanding; it was a healthy exercise for the young and was capable of arousing "piety" through its traditional association with natural theology.[27] Lastly, it was attractive to women. The prevalence of "the female pencil" was a problem for the would-be philosophical botanist (I will come back to the gender aspects of botanical illustration in chapter 4).[28] The reviewer's double-edged praise implied that botanists were pleasant folk but that their studies perhaps lacked seriousness.

The *Edinburgh*'s review had been written anonymously by David Brewster, the Scottish astronomer and physicist, who was a key figure in the 1830s debate over the purported decline of British science.[29] Brewster was a "declinist," convinced by the mathematician Charles Babbage's claim that Britain was falling behind France because French savants, unlike their British counterparts, received direct government funding.[30] John Lindley, professor of botany at University College London, was the most vociferous botanical declinist; when Charles Daubeny took the Oxford botanical chair, Lindley used the *Athenaeum* to comment that Daubeny had accepted it only because the income from the chair of chemistry was inadequate: "If such be the case with chemistry, the state of other branches of natural history at Oxford must be even more deplorable, if indeed that be possible; and it must require no common share of zeal and disinterestedness, to induce a man to undertake the duties of the botanical chair."[31] Such comments were part of a campaign to raise botany's status which culminated in Lindley's attempt to modernize Kew. Through his anonymous reviewing, Lindley was able to co-opt the *Athenaeum* into his campaign for botanical reform. In 1831, he used its columns to comment: "Thirty years ago [botany] was considered in England of so low a character, as to be calculated for the amusement of women, or the intellect of children, rather than for man."[32] Lindley claimed to believe those days were gone—but Brewster's comments in the *Edinburgh Review,* written just two years later, suggest otherwise. Brewster professed disingenuous surprise that "it has been customary, principally among mathematicians and natural philosophers, to undervalue the study and collection of minute objects of natural history." Nevertheless, he had to concede that most botanists are "perhaps on a level with the astronomer who only observes and discovers new stars," because "like Mineralogy, Botany has hitherto been chiefly a science of observation; and a botanist who knows a plant only by its parts of fructification, has made as little progress as the mineralogist, who pronounces upon a mineral, by throwing its lustre upon his eye, and by shaking it knowingly in his hand."[33]

Such remarks highlight the dilemma that faced men of science, especially naturalists; in trying to draw attention to the parlous state of their sciences, they ran the risk of damaging their reputations still further. As Richard Drayton has shown, Lindley was one of several people who, during the 1830s, used anonymous reviews to attack the mismanagement of Kew, complaining of its shabby, mislabeled plants and dilapidated buildings and arguing that it contributed nothing to the nation's needs. Their goal was to persuade the government to take direct control over what was still a royal garden and transform it into a national garden whose managers could direct the efforts of Britain's ad hoc empire of colonial botanic gardens.[34] In pursuit of this agenda, the reviewers might have been tempted to exaggerate botany's low status and lack of achievements in order to dramatize the need for more resources. However, such tactics could harm botany's status further, since the diverse readership of the periodicals would have included many people who had no notion of the underlying political battles and would simply have concluded that botany was not to be taken seriously.[35]

Nevertheless, Lindley's campaign was partially successful in that Kew was indeed brought under government control in 1841, but it was William Hooker who became the first director of the "new Kew," primarily because he had requested a lower salary than Lindley.[36] Government money was scarce and "retrenchment" was in fashion, while at the same time some British men of science opposed accepting any government support, fearing it might lead to Continental-style jobbery and bureaucratic centralization.[37]

Paid positions for men of science remained scarce. In 1846, the *Athenaeum* reviewed some of the latest German work on advanced physiological matters and noted that "less of this kind of work is done by Englishmen than by any other people in the world" because "our men of science are not men of leisure. Few, indeed, are the places of profit which they hold,—and most of them are occupied with other pursuits, whereby they gain their bread."[38] A decade later, the situation had still not improved; when the *Athenaeum* reviewed *An Elementary Course in Botany* by Arthur Henfrey (who had succeeded Forbes as professor of botany at King's College, London, a few years earlier), it asserted that, until the English universities started conferring science degrees, "the botanists of this country will never have fair play, and our botanical literature will always partake of the popular and semi-scientific character which Prof. Henfrey laments the necessity of in his preface to the present volume."[39] Botany's "popular and semi-scientific character" was partly a product of its practitioners' financial need to reach a broad audience. The *Athenaeum* acknowledged that "there is no want of books on botany in our language. Everyone is fond of flowers, and books with a modicum of information on their structure abound for popular use." Popularity created an audience for lectures and books—and thus an income for its badly paid practitioners—but the need to address those who were merely "fond of flowers" forced botanists who needed to make money to adopt a tone that lowered their standing relative to practitioners of the physical sciences.

The problems confronting botanists throughout the middle decades of the century were similar—if a little more acute—than those facing other men of science. Hooker was one of many young men who needed to earn a living from his science while avoiding roles that would have compromised the status of either himself or his science. The Newcastle BAAS meeting provided Hooker with an opportunity both to earn a living and to establish a reputation, since it marked the launch of the Magnetic Crusade, but BAAS meetings also provide a glimpse of what "philosophical" meant to the country's leading men of science.[40]

David Brewster had been a founder of the BAAS, although he did not remain a leader for long because of scientific and political disagreements with its more prominent members.[41] His faint praise of botany in the *Edinburgh Review* ended with some encouraging words for botanists, noting that the mineralogists he had compared them to had risen through the scientific ranks to the extent that "Natural Philosophy . . . now claims Mineralogy as one of its most interesting branches, and Botany will, we doubt not, soon rise to the same dignity."[42] Botany had plainly failed to garner such dignity as yet, but botanists could still be hopeful because alongside the dedicated fact-gatherers there also worked "the philosophical botanist, who invents new principles of classification, who studies the structure and organs of plants, who develops the laws of their geographical distribution."[43] This de facto definition of "philosophical" is similar to one offered by the geologist Roderick Murchison when he presided over the BAAS more than a decade later. In 1846, Murchison used his presidential address to praise the naturalist Edward Forbes's attempt to explain plant and animal distribution using geological theorizing as "the first attempt to explain the *causes* of the zoological and botanical features of any region anciently in connexion."[44] This was in sharp contrast to natural history's usual preoccupation with cataloging species. As the polymathic philosopher William Whewell had commented in an earlier address to the BAAS, "the mere gathering of raw facts may be compared to the gatherings of the cotton from the tree. The separate filaments must be drawn into a connected thread, and the threads woven into an ample web, before it can form the drapery of science" (botanists in his audience must surely have noticed that he had chosen a botanical analogy to describe unphilosophical fact-gathering).[45] Murchison's praise of Forbes included the observation that the association's meetings had played an important role in helping Britain's naturalists shift their attention from merely cataloging phenomena to searching for their underlying causes; as a result, "British naturalists have annually become more philosophical, and have given to their inquiries a more physiological character, and have more and more studied the higher questions of structure, laws and distribution."[46]

Murchison's triad of "structure, laws and distribution" is strikingly similar to Brewster's definition of a philosophical botanist: one who is concerned with systematics (the principles and laws of classification), plant anatomy and physiology (structure), and plant distribution (particularly as a way of discovering the laws that governed vegetation). It is clear from the BAAS reports that by "naturalists," Murchison and others were mainly referring to zoologists; nevertheless, Murchison's own science—geology—had undergone a transformation in the early decades of the century that served as a model for many of the botanists. Influential theorists like Charles Lyell had provided a philosophical basis on which to speculate about the laws that governed the earth's history, while men like Murchison and Adam Sedgwick had conducted extensive fieldwork to map the strata of Britain. And Murchison's travels, which he referred to as his "campaigns," took the standard systems of classification—such as the Cambrian and Silurian systems—from Britain's rocks and spread them all over the world. The laws and standards that governed this implicit empire were of much more than academic interest: in 1844, Murchison had predicted that gold would be found in Australia's Silurian rocks, a prediction that was proved right in 1851, triumphantly confirming the imperial economic importance of his science.[47] Almost as importantly, the economic importance of geology led to the founding of the Geological Survey, which provided some of the first government-funded scientific positions in Britain.[48] Although there were tensions between the elite, independently wealthy gents of London's Geological Society and the paid government geologists at the Survey, the latter provided a reasonably respectable way for a few men of science to earn their living.[49]

The *Erebus* Voyage

Not only did Hooker's four years aboard the *Erebus* provide him with a way to earn a living (following the voyage, he received half his naval surgeon's pay while he wrote up his collections), but the Magnetic Crusade also gave him a tangible illustration of how the leaders of British science sought to combine science's practical applications with important theoretical work. The BAAS president, Vernon Harcourt, defined the crusade's purpose as being to construct "a theory based on a *legitimate representation of known facts.*" Just as Herschel had done in his letter to Grey, Harcourt emphasized the practical benefits for navigation of magnetic mapping, but he also shared the hope of discovering "new cosmical laws—a discovery of the nature and connexion of imponderable forces."[50] Whewell described it as "by far the greatest scientific undertaking which the world has ever seen," not least because the crusade embodied the BAAS goal of coordinating the work of science's part- and full-time devotees (regardless of whether or not they were paid) to ensure that the nation derived maximum benefit from their efforts: it collated data from a widespread network of nonexpert collectors (such as naval officers) by imposing standards to achieve uniform observations that could be collated and analyzed.[51] Hooker's botanical practices were to be strikingly similar to those he saw in use during the voyage: he set up a network of botanical collectors that—albeit on a smaller scale and a more informal basis—mirrored the crusade's magnetic observatories, and Hooker also struggled to standardize his collectors' work. The crusade provided both his first scientific experience and a model of prestigious, exacting science.

However, when Hooker joined his ship at Chatham, prestigious science was a long way off. His cabin was just six by four feet and, as he described it to his grandfather Dawson Turner, contained "a bed-place, a book shelf, a seat, table, etc.; below the bed are very large drawers for our things."[52] Most of his scientific equipment had to be supplied by his family: Turner sent him a traveling thermometer and William Hooker gave him "a beautiful Chronometer watch" as well as books and other equipment (fig. 1.2). Seventy years later, Joseph Hooker was still complaining about the government's meanness, telling a fellow naturalist that "except some drying paper for plants I had not a single instrument or book supplied to me as a naturalist—all were given to me by my father."[53]

When the ships set sail no one knew how long the voyage would last. The expedition failed in its main goal, which was to reach and map the south magnetic pole, because at the time, no one knew that Antarctica was a continent. Nevertheless, the *Erebus* and *Terror* were to sail farther south than any vessel had ever ventured before, a record they held for almost half a century.[54]

More than six months after leaving Britain, the ships arrived at the tiny island of Kerguelen's Land, part of an isolated archipelago that lies more than 3,000 miles southeast of the southernmost tip of Africa (when the French navigator Yves Joseph de Kerguélen-Trémarec first found it in 1772, he aptly named it Desolation Island). Kerguelen's Land had long held a special place in Hooker's imagination: as a child, he had loved travelers' tales and vividly remembered sitting on his grandfather's knee and looking at the pictures in Captain Cook's *Voyages*. He was particularly struck by the picture of Cook's sailors killing penguins on Kerguelen's Land, and Hooker later remembered thinking that "I should be the happiest boy alive if ever I would see that wonderful arched rock, and knock penguins on the head."[55] The Ross expedition arrived there on 12 May 1840 and stayed for three months while they established a magnetic observatory, giving Hooker ample time to explore the island's curious plant life.

1.2 The case of the chronometer used by Joseph Hooker aboard the *Erebus* (the chronometer itself is missing, and the case now contains a barometer, also thought to have been used by Hooker). By kind permission of the Trustees of the Royal Botanic Gardens, Kew.

Captain Cook had listed just 18 species clinging to the volcanic terrain—a total that Hooker felt indicated a serious lack of effort by Cook's naturalist; Hooker found 30 species on his first day ashore. Over the next few months, he was able to dramatically extend the island's flora—to 150 species. Unlike previous travelers, Hooker paid particular attention to the minute flowerless plants—mosses, lichens, and similar things (known collectively as cryptogams)—in which he was particularly interested.[56] As he noted, "the Lichens appear here to form a greater comparative portion of the vegetable world than in any other portion of the globe." Indeed, as he sailed south, he observed—as others had done before—that the flowering plants gradually disappeared (there were none in Antarctica itself), leaving the cryptogams as the dominant flora.[57] These seemingly predictable shifts in the nature of each island's flora were an intriguing hint that the distribution of plants might be governed by natural laws, which botany had hitherto lacked.

Kerguelen's Land offered many other fascinating puzzles to the prospective philosophical botanist. Its flora was clearly related to that of South America; Hooker would later find some of its characteristic species of plants on the Falkland Islands and in Tierra del Fuego. However, he also found Kerguelen plants on Lord Auckland's islands, off the southernmost tip of New Zealand. As he would later write in his *Flora Antarctica*, "It may appear paradoxical, at first sight, to associate the plants of Kerguelen's Land with those of Fuegia separated by 140 degrees of latitude, rather than with those of Lord Auckland's group, which is nearer by about 50 degrees." However, he had been forced to do so because Kerguelen's Land had so many species that were "similar to, and in many of the species identical with, those of the American continent."[58]

The fact that the Kerguelen plants seemed more closely related to those of America than Australasia was curious enough, but even more surprising was that they seemed entirely unrelated to those of Africa, even though the island was closer to that continent than to any other.[59] Even more puzzling was the existence of species that were not found anywhere else in the world, the most famous being the Kerguelen's Land cabbage (*Pringlea antiscorbutica*), "the most interesting plant procured during the whole voyage," in Hooker's view (fig. 1.3).[60] This curious plant was (as its scientific name indicates), an antiscorbutic—a vegetable that helped prevent scurvy. As Hooker noted, its presence "on a spot upwards of 1000 miles from any land where fresh vegetables can be obtained" was quite remarkable; it almost seemed to have been "planted by Nature's hand for the poor mariner."[61]

Like most of his contemporaries, Hooker assumed that each species of plant or animal had come into existence just once. Like other men of science, he maintained (in public, at least) a discreet silence on the question of whether living things had been created by divine or natural means, but in scientific circles, the balance of opinion was increasingly shifting toward the idea that such creation was—like other features of the natural world—governed by scientific laws that would eventually be discovered. If the distribution of the world's plants could be understood, it would bring botany closer to finding such laws—and bring men like Hooker a little closer to the scientific standing they craved.

Even if the hand of Providence had not created the Kerguelen's Land cabbage for the benefit of scurvy-racked sailors, it seemed that the world's

1.3 The mysterious Kerguelen's Land cabbage (*Pringlea antiscorbutica*). From J. D. Hooker, *Flora Antarctica* (1844–47). Reproduced by kind permission of the Syndics of Cambridge University Library.

islands had been created as a series of experiments to test theories about the distribution of living things. Among other things, their enigmatic inhabitants hinted at connections between lands that were now separated but perhaps had once been linked. Shortly after his return to Britain, Hooker received a letter from Darwin offering him access to his own collections of southern plants.[62] Inspired initially by their common experience of travel, the two men immediately struck up a friendly correspondence—within weeks they had dropped the Victorian formality of addressing each other as "dear sir," opting for "My dear Darwin" and "My dear Hooker." Within just a few months of their first letter, they were discussing the observations they had made on their respective voyages to see what light they might shed on the mysterious question of the distribution of species. Among the evidence they discussed was the Kerguelen's Land cabbage, which Hooker noted "is found only in that Island & is the most remarkable plant of its whole Nat Ord in the whole S. Hemisphere there is nothing at all like it any where else." Assuming, as he did, that "there was a beginning to the creation of plants on our globe," the question that inevitably arose was "why should not the said cabbage grow on lands we suppose older than Kerg Land."[63] This was written in response to Darwin's well-known revelation that "I am almost convinced (quite contrary to opinion I started with) that species are not (it is like confessing a murder) immutable."[64] Darwin's melodramatic (and slightly tongue-in-cheek tone) was certainly prompted by the possibility that his new friend might be shocked by the admission of a belief in the transmutation (evolution) of species, which was still considered an unorthodox attitude. However, Hooker's reply simply discussed the evidence, such as the puzzling cabbage, in a calm and philosophic manner, concluding that "there may in my opinion have been a series of productions on different spots, & also a gradual change of species. I shall be delighted to hear how you think that this change may have taken place, as no presently conceived opinions satisfy me on the subject."[65] Hooker's willingness to discuss such subjects helped cement their new friendship, but—as we shall see in later chapters—it would take many years of discussion before the two naturalists reached agreement.

Meanwhile, philosophical discussions of Hooker's plants lay far off in the future—he had a long voyage to survive. From Kerguelen's Land, the ships sailed to Van Diemen's Land (Tasmania), arriving almost a year after they had left Britain. During the long months at sea, when there had been only a few seaweeds to examine, Hooker kept himself busy by using a dredge to collect minute sea creatures from the ocean's depths. Since his medical duties were comparatively light, Hooker was able to spend many hours learning to identify and draw tiny marine invertebrates, sharpening his skills as a naturalist in the process. As he told his father, "Were it not for drawing, my sea life would not be half so pleasant to me as it is. In the Cabin, with every comfort around me, I can imagine myself at home."[66]

However, when the ships finally reached Tasmania, Hooker received a tragically sharp reminder of just how far from home he was. Among the waiting letters was a solemnly black-edged one from his father addressed to "my very dear and only son"—Joseph's older brother, William, had died in Jamaica, several months earlier, but there had been no way of informing Joseph until the ships landed.[67] As Joseph wrote in reply: "Nothing I can say on that subject can console you, but the oft-repeated promise that you shall prove me a faithful son & one that will make up your loss as well as he could. You have lost a son, but I a brother and companion of 20 yrs. standing, & now I have none, with whom on my return I can talk over my childhood's days." His father's letter also expressed his anxiety over rumors that he had heard that Joseph was also in ill health; Joseph wrote at length to reassure his father he was perfectly well, but added, "Knowing your affection for me I cannot help sorrowing that 4 weary months must yet pass before you will see this hand writing that lets you know how misinformed you are."[68]

Such long delays in sending and receiving letters added to the acute loneliness of the voyage. Although some of his shipmates, including McCormick, became good friends, strict naval discipline and the insistence on maintaining a rigid hierarchy among the officers (and an unbridgeable gulf between the officers and men) meant there were few people aboard whom Hooker could even speak to as equals, and fewer still who shared his scientific pursuits.[69] So it was with considerable delight that he found a friend in Tasmania, Ronald Campbell Gunn, a passionate botanical enthusiast.[70] When the celebrated Arctic explorer Sir John Franklin arrived in Van Diemen's Land as lieutenant governor in 1837, he had immediately begun promoting science in the colony, and Gunn's expertise on the country's natural history soon brought him to Franklin's notice. As a result of his scientific interests, Gunn had risen from being Hobart's assistant police magistrate to becoming Franklin's private secretary and secretary to the Tasmanian Society, which Franklin founded. These connections ensured that Gunn met all scientific visitors, but he would have met Hooker in any case, since he had been sending plants to William Hooker for many years.[71]

Gunn and Hooker soon became close friends, going on collecting trips together and spending evenings classifying and comparing their finds. As we shall see in later chapters, their friendship proved mutually beneficial—

although they had quite different goals, both their careers flourished as a result of their botanical correspondence. The same was true for a friend Hooker made soon afterward, when the *Erebus* arrived in New Zealand and he made contact with the Reverend William Colenso, a printer and missionary who had been sent to New Zealand by the Church Missionary Society and who also collected plants for William Hooker.[72] Although correspondence could allow a naturalist to make invaluable contacts and, through them, gain vital specimens, the experience of meeting and working together created an entirely different kind of friendship; Gunn and Colenso were to be two of Hooker's most important collectors, more loyal and harder working than any of the dozens of others who contributed materials to his books. Traveling provided many benefits beyond firsthand experience of the countries whose plants Hooker was to classify.

"Unremitting Attention to Botany"

Like many of his contemporaries, Hooker had acquired most of his medical education outside Glasgow University at the city's many extramural colleges, such as Anderson's College.[73] In 1832, Glasgow suffered particularly badly during Britain's first cholera epidemic. The city's rapid growth had left it overcrowded and unsanitary; malnutrition was common and disease swept through the slums regularly. To address these chronic public health problems, the generation of men who ran the Glasgow medical schools in the 1820s and 1830s shifted their research from individual case studies to large-scale statistical and epidemiological studies, partly modeling their work on that of the French, who had pioneered medical statistics a generation earlier.[74] Their students would have absorbed a medical philosophy that stressed the importance of practical experience, taking detailed notes on all cases, and keeping meticulous records.[75] Such an education was common for botanists. Edward Forbes stressed that "The mental process is the same at the bed-side of the patient and in the cabinet of the naturalist: its first element, correct observation, leading to correct diagnosis; the second, accurate discrimination, leading to sound methods of treatment in the one case and philosophical views of affinity in the other."[76] Hooker's later decision to conduct arithmetical analyses of his large-scale botanical studies was probably influenced by the teachers he met in Glasgow, many of whom were pioneers of statistical medicine.[77]

Joseph Hooker was a rather shy and studious young man who often felt isolated from his contemporaries. These feelings, together with the ambiguous place of botany's students and teachers within the university's social and academic hierarchy, left him feeling ambivalent about his university days. In 1841, Joseph heard that his father's friend George Walker-Arnott had failed to get the Glasgow botany chair after William moved to Kew. Joseph commented that his "lingering affection for Glasgow University" was now "cast off with scorn." He complained that, with a few exceptions, he had felt slighted by the other professors while a student because he had never been invited to their homes, telling his father: "I have felt, & not a little, the want of any kindly feeling on the part of the older, or indeed the majority of your fellow Professors."[78] These social slights seem to have been closely connected in Joseph's mind with the lowly standing of botany. He added: "if ever I live to be, by unremitting attention to Botany at home, a Botanist, (which is my ardent hope), & if ever my acquaintance be thought worth the seeking, I shall go to Glasgow, on purpose to cultivate the friendship of those few who cared for me in my obscurity, when a little attention is worth so much; & to demand from the others some little submission before I condescend to acknowledge them."[79] This fantasy of social revenge—leaving behind his obscurity and forcing "submission" from those who had once snubbed him—helps explain Joseph Hooker's lifelong sensitivity to perceived slights to his status and that of his studies.

Hooker's "unremitting attention to Botany" during and after the *Erebus* voyage would eventually prove to be the making of him, just as he had hoped—although transforming the experience of travel into a paying scientific career would prove to be a long, slow process. The central importance of travel for Hooker is evident if we compare his career with that of his friend Thomas Thomson. Thomson was also the son of a Glasgow professor (in his case, of chemistry), and the two sons took their medical degrees together at the university.

At the end of their university studies, Hooker and Thomson attended the BAAS Newcastle meeting but—despite the similarities in their lives up to that point—after Newcastle their careers diverged markedly. When Hooker left for the Antarctic, Thomson sailed for India, having received a commission as a surgeon in the East India Company. By the time they met up again in India, where Hooker traveled from 1847 to 1851, Hooker had already published more than thirty-five papers as well as his *Flora Antarctica* (1844–47; fig. 1.4); by contrast, Thomson published only twenty papers in his life, half of which were cowritten with Hooker.[80] Thomson's most substantial botanical publication, the first volume of the *Flora Indica* (1855), was produced in collaboration with Hooker, and its introductory essay was almost entirely Hooker's work, despite both names appearing on the title page.[81] Thomson's botanical reputation remained modest; after the *Flora*

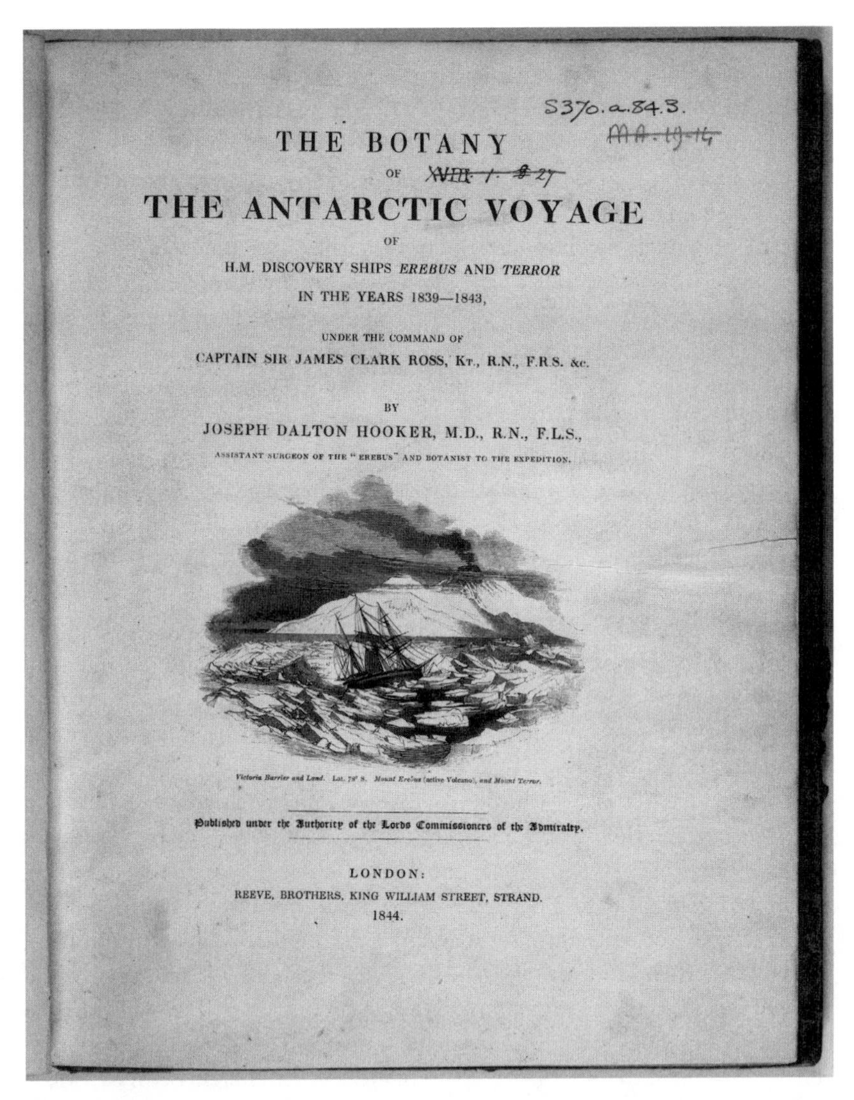

THE BOTANY

OF

THE ANTARCTIC VOYAGE

OF

H.M. DISCOVERY SHIPS *EREBUS* AND *TERROR*

IN THE YEARS 1839—1843,

UNDER THE COMMAND OF

CAPTAIN SIR JAMES CLARK ROSS, KT., R.N., F.R.S. &c.

BY

JOSEPH DALTON HOOKER, M.D., R.N., F.L.S.,

ASSISTANT SURGEON OF THE "EREBUS" AND BOTANIST TO THE EXPEDITION.

Victoria Barrier and Land. Lat. 78° S. Mount Erebus (active Volcano), and Mount Terror.

Published under the Authority of the Lords Commissioners of the Admiralty.

LONDON:

REEVE, BROTHERS, KING WILLIAM STREET, STRAND.

1844.

1.4 Title page of Hooker's *Flora Antarctica*, vol. 1 of *The Botany of the Antarctic Voyage* (first part, 1844). Reproduced by kind permission of the Syndics of Cambridge University Library.

Indica failed to earn him the East India Company's patronage and further installments had to be abandoned, he returned, deeply disappointed, to India, where he became superintendent of the company's botanic gardens in Calcutta and professor of botany at Calcutta Medical College. He held both posts until 1861, when ill health once more forced his return to England.

By that time, Hooker was one of Europe's most highly regarded and influential botanists. When Thomson died in 1878, Hooker had already been knighted, received the Royal Society's Royal Medal, and was finishing a five-year term as the society's president.[82]

The differences between Hooker and Thomson's careers had many sources. Thomson, for example, lacked William Hooker's patronage, but Joseph Hooker's decision to join the navy, rather than the East India Company, was also crucial; not only was he able to travel much more widely than Thomson, but his travels always brought him back to England, to the centers of scientific society and patronage, and back into his father's sphere of influence; meanwhile, Thomson was away in India, isolated from the world in which Hooker built his career.

However, while Hooker's father had excellent contacts in the natural history world, he did not—as we have seen—have wealth to bequeath to his son. Writing from the *Erebus* in 1841, Hooker told his father that he had been working on crustaceans because of the lack of opportunities to botanize while at sea: "All this renders me most anxious to see the termination of the voyage, for I have no wish but to continue at Plants." He added: "Could I with honour leave the expedition here, I would at once and send home my plants for sale as I collected them, but now my hope and earnest wish is to be able on my return home to devote my time solely to Botany and to that end the sooner we get back the better for me. My habits are not expensive, but should I not be able to live at home with you, I would have no objection to follow Gardner's steps and gain an honorable livelihood by the sale of specimens."[83] Hooker must have quickly realized that selling specimens was not in fact considered honorable, because he never contemplated such a course again. Yet, earning a living preoccupied him during the voyage, when, for example, he discussed whether he should stay in the navy or return to the colonies.[84] As late as 1851, he was still wondering if he should give up botany for a mineralogical position at the British Museum, telling his friend the botanist George Bentham: "It is £400 & an excellent house" and "I hate the idea of giving up Botany, but I am advised to try for it by Gray particularly & my Father proposes it."[85]

Until 1855, neither Hooker's father nor Kew could afford to employ him full-time; in fact (as we shall see in the next chapter), William had considerable difficulty in persuading the government to let him employ even a single botanical collector and ended up depending on volunteers. Many of these were former students, whose colonial postings allowed them to collect plants. Thomson was a typical correspondent: a colonial doctor, who traveled around India gathering plants while on active service in the East

India Company's army. There were many such men (and a much smaller number of women), some of whom were born in the colonies but most of whom had gone there to take advantage of the opportunities the rapidly expanding empire offered.

Since William Hooker's correspondence networks would eventually grow to become one of the main assets he was able to offer his son, Thomson's decision to go to India effectively turned him into a resource Joseph could draw on.[86] By contrast, Joseph's decision to go to the Antarctic eventually made it possible for him to utilize that resource. Nevertheless, it is important to understand that neither of them saw their relationship as exploitative; on the contrary, the friendship they forged at university was one that lasted all their lives and was vital to their collaboration.

Given the relationship between the status of a science and its financial rewards, Hooker knew he would have to address botany's higher questions. As the *Erebus* neared England, he abandoned his original plan to write a modest volume describing new plants and instead began planning the first full floras of Antarctica, New Zealand, and Tasmania.[87] The broad geographical scope of his work was largely prompted by his ambition to write on botanical geography, which he would later describe as "a philosophical study in the foremost ranks of science" concerned with "the laws which govern the development, progression, and distribution of forms and species."[88]

After returning from the voyage, Hooker spent almost two decades searching for a secure, paid botanical position; that search is central to understanding his lifelong concern with improving the status of botany. During the later 1840s, Hooker's need for an income became even more pressing after he became engaged to Frances Henslow, daughter of the Cambridge botany professor John Stevens Henslow.[89] Joseph and Frances had to delay their marriage for several years until he could afford to support them both. He could not take his place among the BAAS gentlemen if his discipline was poorly regarded, but nor could he earn an income that would allow him to mix socially with independently wealthy men like Darwin and Murchison. Newcastle was his first lesson in the important relationship between the abstract issues of disciplinary status and the immediate ones of paying his bills: it was vital to find a philosophical way of earning a living.

CHAPTER TWO

Collecting

Such Travellers and Residents in distant countries as may desire to increase the collections of vegetable treasures in the noble and Royal Gardens of Kew (which are daily thrown open to the inspection and use of the public) will have the goodness to address their packages to Sir William Jackson Hooker, K.H. Director of the Royal Botanic Gardens of Kew, London.
—W. J. Hooker, *A Few Plain Instructions for Collecting and Transporting Plants in Foreign Countries*

Collecting was a Victorian passion: from shells, seaweeds, flowers, and insects to coins, autographs, books, and bus tickets, Victorians of every class collected, classified, and arranged their treasures before exchanging unwanted finds with other enthusiasts. Natural history collecting was perhaps the most widespread form of this passion; today's museums contain millions of Victorian specimens, and our libraries are full of nineteenth-century works on how to collect, preserve, and display these collections, yet few historians have looked in any detail at the day-to-day work of naturalists: at what they collected, why they collected, and—most importantly of all—*how* they collected.[1] Perhaps the very popularity of scouring the beaches or hedgerows for shells or bird's eggs suggests that these were simple pastimes, about which nothing much can or needs to be said. While popularity was undoubtedly an important aspect of collecting (although its implications are more complex than they initially appear), I shall argue in this chapter that natural history collecting was anything but simple; it was a complex craft activity, requiring skills that took time and effort to master. Analyzing how those skills were acquired and transmitted reveals—among other things—that the apparently humble business of collecting was an essential tool for negotiating who had scientific and social authority and how they could use it.

Imperialism formed the context for much Victorian science; as William Hooker's request at the head of this chapter shows, the growth of empire—both formal and informal—meant there were many British "Travellers and Residents" to be found "in distant countries" who could supply Kew with specimens. Such people were once considered marginal to the development of the sciences; they were considered merely passive recipients of knowledge that was created in Europe. In recent decades, the inadequacy of such a view has become increasingly apparent, but there is still considerable debate about what might replace it.[2] Over the next few chapters, I hope to show how investigating scientific practices allows us to understand the full range of participation in the making of scientific knowledge. The seemingly mundane business of picking and pressing flowers proves to be a way of constructing imperial relationships: when a collector in some remote land accepted the authority of a distant metropolitan expert by following his guidance as to what and how to collect, he or she was simultaneously accepting a subordinate role and thus a colonized status. On the other hand, as some of these collectors worked to improve their expertise, they came to realize that it was their metropolitan correspondents who were becoming dependent—since accomplished collectors were hard to find. This realization could make collectors more demanding in their dealings with the metropolis, forcing a renegotiation of relations between those at the center and those on the imperial periphery.

The craft aspects of collecting remind us that natural history specimens—botanical, zoological, and geological—are not natural objects but artifacts, things that were made according to exacting standards.[3] Simply picking a plant did not produce a botanical specimen: in addition to knowing *which* plants to pick (a topic I will return to), collectors had to gather specific parts of the plant, label them correctly, and preserve and mount them properly (fig. 2.1). Learning to do all this was part of a collector's training, but even the most distinguished European specialist could not simply impose his view of collecting. Precisely because collecting was so popular, it was a largely voluntary activity. As a result, local collectors had to be encouraged to comply with expert judgments (a standard that no one complied with would be worthless), so teaching someone how to collect was part of a complex process in which botanical standards were simultaneously being set and enforced. Once the conventions were learned, they became part of what is sometimes called tacit knowledge, an expertise that is continuously drawn on without being consciously referred to;

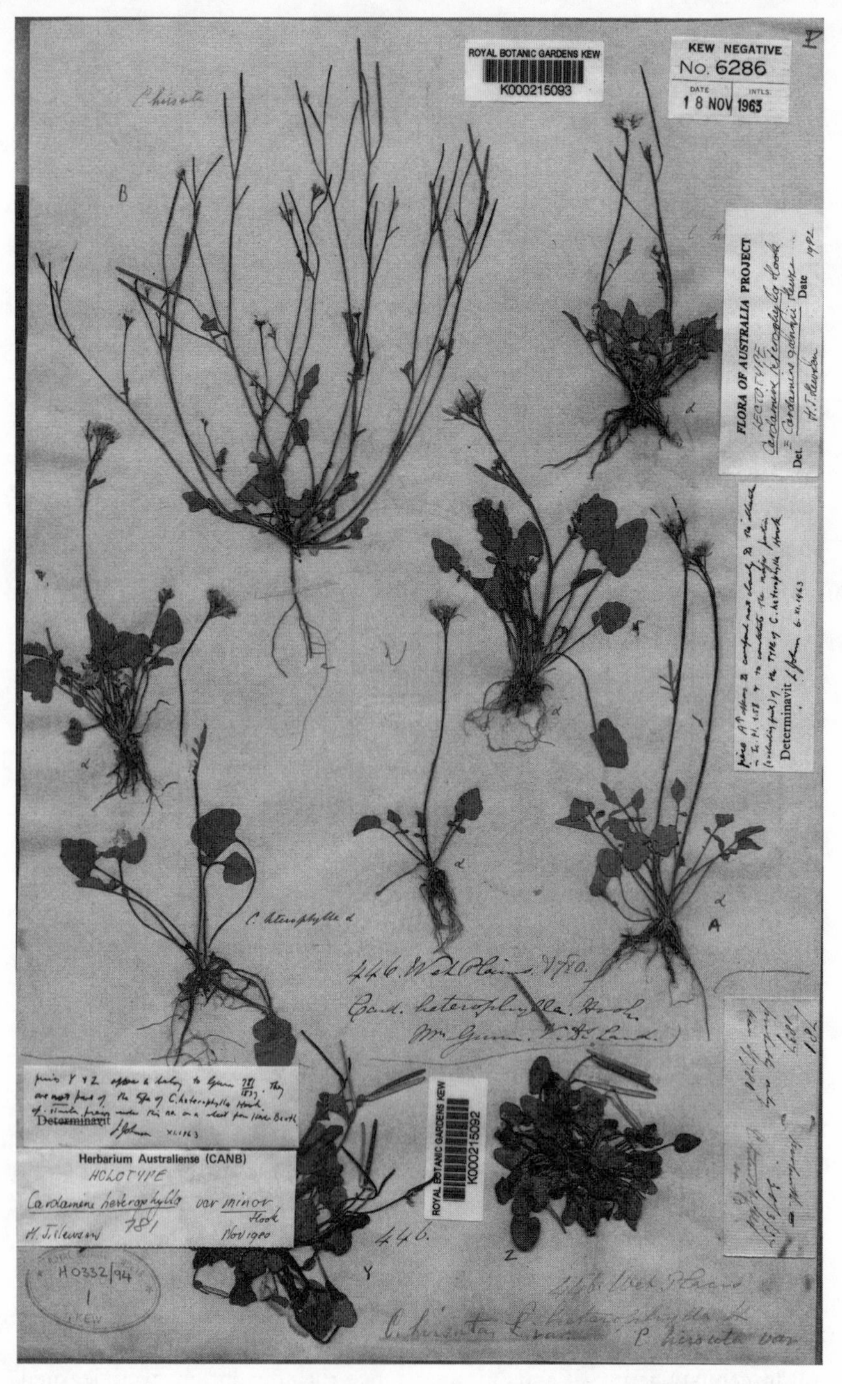

2.1 Specimen of *Cardamine heterophylla* collected by Robert Campbell Gunn. By kind permission of the Trustees of the Royal Botanic Gardens, Kew.

for example, we "know" a familiar face without being able to say exactly how we know it.[4] However, the acquisition of such tacit knowledge was complicated when collectors felt they knew their locality's flora and fauna better than these distant experts; this conviction could inspire them to devise alternatives to aspects of the metropolitan standards. As a result, everyone—from the most exalted metropolitan naturalist to the humblest colonial collector—was a participant in a cycle in which all kinds of objects and information about them moved back and forth, from periphery to center and back again. Scientific knowledge of plants—of their classification, distribution, and uses—emerged from this cycle, and those at the periphery were vital to its creation. Understanding what was added (and what was removed) from a plant when it was transformed into a specimen is the first step in analyzing that cycle, since it reveals the types of information that specimens incorporated and thus the uses to which they could—and could not—be put.

Learning the Collector's Trade

Some of the difficulties inherent in collecting are revealed in letters Hooker wrote aboard the *Erebus*. The ship's first port of call was Madeira, where Hooker got his first taste of overseas field collecting. Since the islands' flora was well known, his collections could be compared with those of other travelers—a comparison that did not always redound to his credit. William Hooker was unimpressed and urged his son to greater efforts. Joseph's collections were also shown to Robert Brown, underlibrarian at the British Museum and de facto curator of its botanical collections, who had circumnavigated Australia on the *Investigator* (1801–5) and pioneered the study of the southern floras.[5] Brown's only comment was to suggest better ways of preserving plants, such as using brown paper rather than the more absorbent blotting paper, which fermented in the tropics.[6] Brown was another of Joseph Hooker's heroes and Hooker had studied with him before the *Erebus* voyage, so the older man's criticisms were embarrassing for the ambitious young traveler.[7] Later in the voyage, Hooker wrote to his father: "The more I think of my former collections, the more I fear they must disappoint you, though at the time I knew not how to improve them."[8] Their deficiencies were still on Hooker's mind two years later when he wrote to his friend the botanist George Bentham: "My first collections did not I know give any satisfaction at all, so the less I say about them the better."[9]

The shortcomings of Hooker's early collections illustrate his lack of localized knowledge; what worked in the Scottish highlands, where he had learned to collect alongside his father, clearly did not work in the tropics.[10] Although Hooker could draw on the experience of those, like Brown, who had traveled before him, it is clear that neither being told what to do nor reading about it were adequate substitutes for field experience. Many decades later, James Britten, keeper of the British Museum's botany department and a popular writer on botany, was still complaining that "in spite of all that has been written, it cannot be said that anything like uniform excellence has been attained, either in the collecting or drying of specimens: on the contrary, much carelessness is still exhibited in both particulars." Yet, even as Britten added another set of carefully written instructions to the pile, he realized that it was only through constant practice that the necessary skills could be acquired.[11]

As Hooker traveled, he gradually improved his collections with the help of some of his father's correspondents, the most important of whom were, as we have seen, Ronald Gunn and William Colenso. When Hooker met Gunn and Colenso (both of whom were older than he), they knew more about their local plants than he did. He made friends with them, gave them gifts, acquired specimens from them, and, with their help and local knowledge, improved his collections; in return, he had books, news, and formal botanical training to share with them (I will return to these exchanges and their significance). A key part of a naturalist's education took place during field trips, and of course, we have no way of knowing exactly what was done or said, which is frustrating since this kind of informal education (which remains crucial to most scientific training) was especially important for nineteenth-century naturalists. Fortunately, a careful study of naturalists' equipment and letters allows us to reconstruct some of this unwritten knowledge.[12]

Hooker's relationship with men like Gunn and Colenso continued long after the *Erebus* returned to England in 1843. Two years earlier, his father had been appointed first director of the Kew gardens, which had just been brought under direct government control. However, while the appointment brought William Hooker to the centers of scientific life in London, it reduced his income and he was unable to give his son much financial support. Fortunately, William's influential friends helped secure an Admiralty grant of £1,000 to cover the cost of the plates for Joseph's projected *Botany of the Antarctic Voyage*. However, its publication never made any money. In 1845, after failing to win the Edinburgh professorship, Hooker used his father's contacts to secure work at the Geological Survey, where he would meet another rising young naturalist, Edward Forbes, and form a lasting friendship. Hooker's position in London also brought him in regular contact with such scientific luminaries as Lyell, but he still had no permanent position.[13]

While he searched for work, Hooker built his reputation using the collections he had made aboard the *Erebus*, but his father's contacts and herbarium were even more important. Returning home to make use of these resources came at the cost of leaving the actual plants behind, but fortunately, he had tools with which to overcome his isolation. The plants came to him in Wardian cases (see below) or as sheets of dried specimens, but equipment such as microscopes or books could also aid in the acquisition of plants: such items were readily available in London but were hard to obtain in the colonies and thus became valuable trade goods that could be exchanged for specimens or used as gifts.[14] This barter reveals the first steps in a process by which the collectors themselves became tools that Hooker could use.

The Tools of the Collector's Trade

Naturalists' tools were mostly cheap and easily acquired, which contributed to natural history's popularity. Nevertheless, Hooker had to provide his own equipment when he left Britain; he recorded that "as botanist my outfit from Government consisted of about twenty-five reams of paper, of three kinds—blotting, cartridge, and brown; also two Botanising vascula and two of Mr. Ward's invaluable cases for bringing home plants alive, through latitudes of different temperatures. I was further, through the kindness of my friends, equipped with Botanical books, microscopes, etc., to the value of about £50, besides a few volumes of Natural History and general literature."[15] Similar lists survive for other collectors and include many of the same items: William Purdie, who collected for Kew in Jamaica during the mid-1840s, took specialized botanical equipment with him, which included three pairs of leather straps (used to secure a portfolio or plant press), six glazed lids (to construct Wardian cases), two India rubber seed bags, various types of paper and board, 1 large and 1 small vasculum, and a pocket lens. He also had various gardening tools, a few books, a map of Jamaica, and a pocket compass. William Grant Milne, who collected for Kew while aboard HMS *Herald* in the Pacific in 1852–56, also took six glazed lids for Wardian cases, plus "1 tin canister filled with putty" and "20 pieces glass." He had gardening tools like Purdie's as well as leather straps, one "large square vasculum," and two smaller ones. In order to perform basic meteorological observations he was given a thermometer, a dozen pencils, a journal book, and "one bottle ink." For writing notes he had fifteen quires of foolscap paper and "A grat [*sic*] quantity of loose paper for sending home specimens" and some books.[16] And finally, perhaps in recognition that the South Seas were potentially more dangerous than Jamaica, Milne also took "2 pistols and powder flask."[17] Analyzing the various items on these lists (where they were used, how hard they were to obtain or master, and the roles they played in bringing plants to Britain) tells us a lot about the practices of field collecting: not least how difficult it could be to make specimens that would be acceptable back in Britain.

The equipment lists contain five main categories of equipment: Wardian cases, vascula, paper, microscopes, and books. I will begin with the Wardian case because, at first sight, it seems the most obvious technology with which to move plants from the colonies to the metropolis.

Wardian Cases

Wardian cases (named after their promoter, Nathaniel Bagshaw Ward) were small, portable greenhouses used to transport living plants through potentially inhospitable climates (fig. 2.2).[18] Seeds were regularly collected, dried, and shipped, but unfamiliar species invariably proved difficult to germinate and cultivate. So, in 1833, Hooker's father had collaborated with George Loddiges, a commercial nurseryman from Hackney, in an experiment in which Wardian cases were used to send living plants to Australia (the nursery trade was vital to the development of the cases and used them to meet a demand for exotic plants).[19] They returned six months later with their contents intact, and Loddiges reported to William Hooker that "whereas, I used, formerly to lose nineteen out of twenty of the plants I imported, during the voyage, nineteen out of twenty is now the average of those that survive."[20] The Victorian craze for ferns and exotic flowers made Wardian cases famous, and when the *Athenaeum* welcomed a new edition of Ward's book *On the Growth of Plants in Closely-Glazed Cases,* it noted that their use "is so general that scarcely any one would think of bringing plants from abroad otherwise than in a closed glass case."[21] It might seem that the Wardian case should have ended Hooker's isolation from his plants, but "bringing plants from abroad" proved more difficult than the *Athenaeum* presumed.

One might assume that the advent of steamships made the colonies and their plants more accessible, and to some extent it did. In 1843, just a few years after the first transatlantic steam crossings, William Hooker

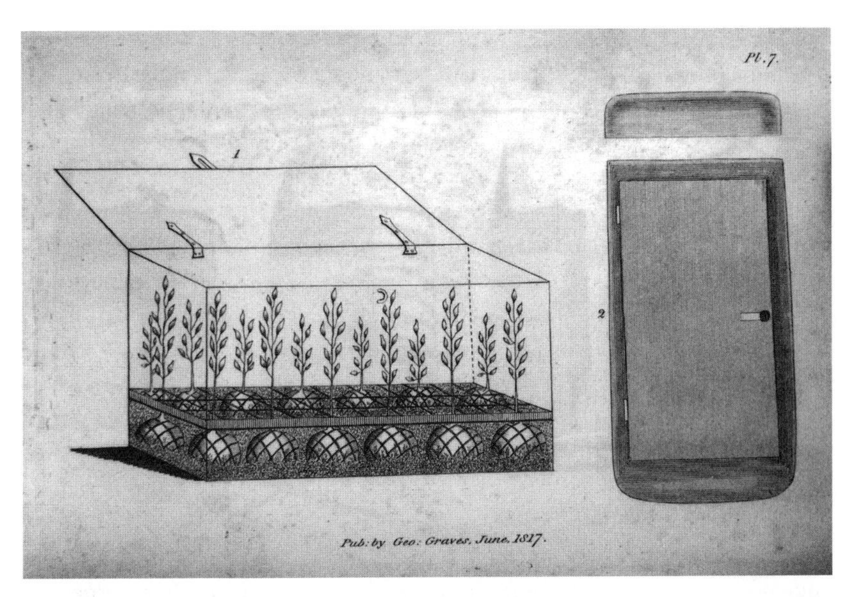

2.2 A case for transporting living plants (left) and a vasculum. Plate 7 from George Graves, *The Naturalist's Pocket-book* (1818). By permission of the Whipple Library, Department of the History and Philosophy of Science, Cambridge University.

told one of his collectors in the West Indies that his Wardian cases were "to come home as quickly as possible after being closed by the Royal Mail steamer."[22] However, it was not until the 1860s that steam began to compete successfully with sail on the Australasian routes, and another twenty years before the steamships were carrying the bulk of the world's cargo.[23] And even when cases were brought back rapidly by steamship, their record seldom matched Loddiges's optimistic claim of "nineteen out of twenty" plants surviving. Especially on longer voyages, considerable care was needed to ship plants successfully, which the ship's crew seldom provided. In the later 1860s, Joseph Hooker had to inform Sir George Grey, the governor of New Zealand, of the "disastrous condition in which the 6 cases of plants arrived:—there are not 6 living plants in them," which Hooker argued was evidence that "Wards cases are no easy things to pack *properly* & to transport safely." The losses in transit were so high that he confessed: "I often am minded to drop the whole system" but for the fact that from time to time "a case arrives as fresh as when it left." The reasons for the regular failures were unclear, but Hooker noted that "much depends on the condition of the plants not only when taken up & put in, but when sent off." Ideally, the collector should also be a good gardener, as Hooker's notes to Grey make clear: "The right amount of moisture to be left in the case, of earth, in which to plant the things, & of drainage, are all essential elements of success, & are rather matters of gardening experience than to be defined by notes."[24] However, the New Zealand governor seems to have lacked the right kind of experience, since Hooker had to write again a few months later to say that the latest case of plants had been received but "with, I grieve to say, the plants all dead, but one Conifer. We are quite disheartened with these 'Ward's cases,' on long voyages, except some *active able* person accompanies them."[25] An "active able" gardener would clearly have been the best person, as Hooker made clear to another New Zealand correspondent: "Pray never send Wards cases without a *trusty* hand who knows something of Gardening."[26]

Despite their problems, Kew continued using Wardian cases, not least because they could be sent to the colonies full of plants, accompanied by a request that the case be returned filled with local ones. Given Kew's limited resources, such nonmonetary exchanges were vital to establishing its network.[27] Even with high losses, the system regularly provided interesting and ornamental plants to enhance Kew's living collections—an important consideration when many still thought of the gardens as a royal park rather than a scientific institution. The government's support of Kew partly depended on its popularity—and the growing enthusiasm for gardening brought many visitors to botanic gardens.

Those who had no space for a garden could also indulge their taste by creating one in a Wardian case, which became a popular item of domestic furniture during the middle decades of the century. Ward's book explained their construction and use to the general public.[28] In 1853, the *Athenaeum*'s review of Ward's book commented: "In almost every city of Europe, among both the rich and the poor, where there is a love of plants, these cases have been introduced. Now that we have no glass duties and no window taxes, this mode of cultivating plants will, in all probability, become much more general."[29]

The popularity of Wardian cases exemplified the indistinct boundaries between the different groups with an interest in plants at this time: gardening manuals often stressed the need for apprentice gardeners to learn classification and the Latin names of flowers. Meanwhile, serious botanists were usually equipped with everyday gardening tools; Purdie took a selection with him to Jamaica, including one long green-wood saw, one small one, one hatchet, one chopper, one strong trowel, two hammers, two gimlets, four pocketknives, and one long cutting hook. And Milne had similar tools.[30] The widely read quarterlies contributed to this blurring by review-

2.3 A vasculum that Hooker took on the *Erebus* and presented to William Colenso upon leaving New Zealand. Courtesy of the Hawke's Bay Cultural Trust—Hawke's Bay Museum.

ing botanical and gardening books together, knowing that their readerships overlapped. This overlap suited men like William Hooker and John Lindley, who needed gardening enthusiasts to buy their books, but for Joseph Hooker's generation, the intersection between botanical and gardening skills compounded the difficulty of defining botany as a distinct science.

The fact that Wardian cases were associated in the public's mind with gardening may have fed Hooker's hostility to them, but their real drawback was their failure rate: he told James Hector that "we have hitherto been most unfortunate & I now call Wards cases 'Wards coffins'!"[31] Such problems reinforced Hooker's isolation from the plants he wanted to work on. Even when the cases were successful and the plants lived, many Australasian plants would not have grown in Britain except in greenhouses, and Kew's grounds could not have contained the 8,000 species Hooker examined while writing the *Flora Tasmaniae*.[32] Such ambitious works could only be written using dried specimens, but before they could become specimens, plants had first to be collected in the field, where the main tool was the vasculum, the metal box in which collectors carried their freshly picked plants (figs. 2.2 and 2.3).

The "Well-Filled Vasculum"

Around 1800, Patrick Neill, a Scottish printer and botanist, met George Don, superintendent of the Edinburgh Botanic Gardens, for the first time. Neill apologized for not having a letter of introduction, but Don pointed to Neill's vasculum and said "that is introduction enough to me."[33] The vasculum was not merely a practical tool; it identified its bearer as a botanist and provided a chance to strike up conversation. Fifty years after Neill's meeting with Don, the *Westminster Review* celebrated the hospitable nature of botanists: "The wanderer among them who shares in their tastes needs no introduction, not even the mention of his name, to ensure a warm welcome." Whoever the traveler was, "let him only exhibit his vasculum and folio, and he will be joyously received, nay, in many instances, find bed and board freely offered." A vasculum or a portfolio (an alternative means of preserving freshly picked plants) identified someone as one of the brotherhood of botanists, the "*vasculiferi*," a tool not just for collecting plants but for making friends; the reviewer claimed that, even when "in countries where we could ill express our wants in intelligible language," conversations might begin when the locals saw the traveler's "well-filled vasculum."[34] After Hooker attended the BAAS meeting in Newcastle (see chapter 1), he told his grandfather that "anyone who scorned the idea of being a naturalist before, was proud during the Association week to be seen with a hammer & vasculum."[35]

The popularity of botany meant that by the 1820s, vascula were mass-produced in Britain and were cheap enough for medical students, like William Hooker at Glasgow, to be able to buy one for their field trips.[36] Hence, they served as the ideal gift with which to mark a new botanical friendship, especially in the colonies, where they were harder to find (fig. 2.3). The pioneering Australian botanist Allan Cunningham met Colenso during a visit to New Zealand in 1837.[37] Back in Sydney, Cunningham had to apologize that "the larger vasculum must stand over until the next opp[ortunit]y., for I have been hurried as it is to send anything to you." However, he sent a "smaller vasculum" as a gift, which "I supply you for your pocket, begging your acceptance of it. I can well spare it."[38] Two months later, Cunningham was able, despite "the horrid depravity and drunken habits of the mass of the lower orders of artisans in Sydney," to send the larger vasculum he had had made for Colenso:

> and although it cannot boast of neatness in its construction, it has been executed as strong as can well be effected in time to meet the rough

> usage of the N[ew]. Zealander. It is an Improvement on mine, having the door or lid on the upperside, that can be opened to put anything within, without taking it off the natives' shoulders thus; [a similar design is shown in fig. 2.4]
>
> The straps will be found more convenient than those of mine where the native being enabled to adjust them instantly to the length he wishes them for the ease in convenience of the carriage of the box.[39]

Cunningham's description makes it clear that the vasculum was to be carried by a Maori servant and must have been a substantial affair, as large as a rucksack. And, as David Allen has commented, the longer the botanizing trip and the more serious the botanist, the bigger the vasculum.[40] More importantly, a cheap, mass-produced item in Britain had become something that had to be made to order in Sydney (Cunningham paid the substantial sum of £1.10 for "vasculum and straps" on Colenso's behalf), while it could not be obtained at all in New Zealand. As one moved farther from the center, the value of everyday items increased, and by helping Colenso get a tool he could not otherwise find, Cunningham cemented their friendship.

Nevertheless, even in the colonial context, the relative ease with which basic botanical tools could be obtained facilitated the creation of colonial collecting networks; as Cunningham and Colenso's correspondence shows, if a vasculum could not be bought, it could be made (even by the "drunken and depraved").[41] However, as with the Wardian cases, popularity was a mixed blessing for the professed botanist; once vascula became common, mass-produced objects, the serious *vasculiferi* could no longer identify one another among the numerous part-time enthusiasts.

The vasculum was not the only way to preserve fresh-picked plants, and a traveling botanist had to learn what worked best in a particular climate. Some botanists also carried an artist's portfolio, which served as a portable plant press by keeping specimens flat. Hooker learned its usefulness on his voyage, telling his father that at the end of the day's collecting, "I emptied my pockets into my travelling portfolio, which I may mention here is the only good way of preserving plants in the tropics, and were it not for the weight, ought to be looked upon as an indispensable addition to the vasculum. The poor withered herbs that I gathered on my previous excursions used on my return to be more crumpled still from the fiery heat of the sun beating on the vasculum, and sorry specimens they have made."[42] Knowing when to use a folio is an example of localized botanical knowledge—and perhaps scorning the increasingly commonplace vasculum in favor of the more cumbersome but less common folio helped the serious botanist

2.4 A larger vasculum (inset), designed to be worn on the back. Frontispiece and figure 5 from W. W. Bailey, *Botanizing: A Guide to Field Collecting and Herbarium Work* (1899). By kind permission of the Trustees of the Royal Botanic Gardens, Kew.

stand out in the crowd. In the 1860s, Bentham recommended "a light portfolio of pasteboard, covered with calico or leather," as being "better than the old-fashioned tin box (except, perhaps, for stiff, prickly plants and a few others)." [43]

Collectors also had to master different collecting techniques for various types of plants; seaweeds, for example, would rust a vasculum and make a folio soggy. Some books suggested the use of an oiled-silk bag, such as those used by fashionable women to carry their bathing costumes; an "oil-skin" bag was specifically recommended as suitable for gathering seaweeds by the Scottish naturalist the Reverend David Landsborough in his *Popular history of British Sea-Weeds* (1849).[44]

Once they had been collected, plants had to be dried and preserved, a surprisingly complex process that required several different kinds of paper, any of which could be hard to acquire.

Paper

One of the first problems a botanist faced was simply getting enough paper. Hooker was able to take twenty-five reams (12,500 sheets) with him on the *Erebus*, of three kinds: "blotting" (a soft, usually handmade, absorbent paper used in initial drying), "cartridge" (a high-quality white paper, also generally handmade, used for mounting specimens and drawing), and "brown" (a cheap, machine-made paper used for the later stages of drying).[45] In some cases, collectors were even provided with Bentall's Botanical Paper, a brown paper produced purely for drying botanical specimens, clear evidence of botany's popularity.[46] Bentham recommended that "plants with very delicate corollas [petals] may be placed between single leaves of very thick unglazed tissue paper," whereas most drying paper should be "coarse, stout, and unsized. Common blotting-paper is much too tender." [47]

Finding supplies of all these various specialized papers was no problem in London, which supported a range of shops dedicated to the various natural history crazes, but they were almost impossible to obtain in most of the colonies.[48] In 1845, Joseph Hooker and his friend the Irish botanist William Henry Harvey discussed the shortcomings of a West Australian collector, James Drummond. Harvey observed: "Drummond's plants are splendid—lovely things—but put up most piggishly. I would subscribe to buy the man a screw press, if one could induce him to use it. It is a thousand pities that he has not such a contrivance—& a better stock of paper. Fortunately he lives in an oven or his plants would all be rotten as well as crumpled." [49]

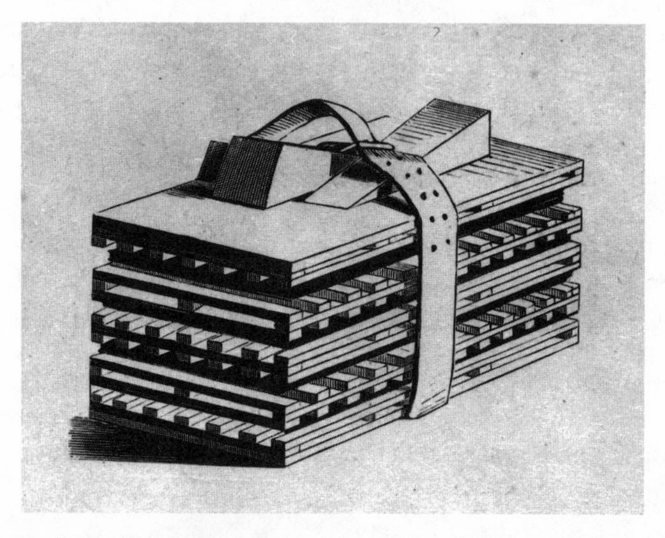

2.5 John Lindley's plant press, showing the use of leather straps and wedges to put pressure on the drying specimens. Drawing by Sarah Crease (née Lindley), March 1852. British Columbia Archives (PDP05046).

Drummond's specimens were unattractive, "crumpled," and "piggishly" arranged, as if made by an incompetent collector (fig. 3.3). Yet Drummond had been curator of the Cork Botanic Garden, Ireland, and a member of the Linnean Society of London before emigrating to Australia in 1829; he must have been experienced, yet the difficulty of obtaining tools (such as a press, which applied even pressure to drying plants; fig. 2.5) clearly prevented him from making good specimens. Obtaining a reasonable supply of paper was even harder, as Drummond acknowledged when he sent plants to William Hooker: "unfortunately when I shipped the specimens I could not get paper here to make them up in different sets, and there is none as yet for sale in the Colony."[50] This was a particular problem in the tropics, since—as Bentham noted—it was essential that "the paper used is well dried" and that, if it was not, it "will require changing very soon, to prevent [the plants] from turning black, for hot damp without ventilation produces fermentation, and spoils the specimens."[51] Even Colenso, despite being a printer by trade, found adequate supplies of paper hard to come by; his surviving specimens are still wrapped in proof papers from his press or pages torn from religious journals.[52] The lack of paper must have frustrated Drummond particularly, since he was a commercial collector and would have earned little or nothing for badly made specimens. However, as we shall see in the next chapter, even those collectors who refused to be paid wanted something in exchange for their specimens; every collector had to learn what their correspondents' expectations were—and then live up to them.

Even a collector as careful and experienced as Gunn could have problems. When he tried to collect seaweeds, for example, he found that, if he dried them in the usual way, "the coarser & larger kinds become brittle & break" with the result that they looked "so villainously ugly" that Gunn had decided to throw them back into the sea. As he told Hooker, "I am sure you would begrudge freight for such stuff"; even an unpaid collector like Gunn was anxious to produce good-looking specimens worth the cost of shipping. He asked Hooker "if there is any process by which they could be decently preserved," wondering whether "preserving them in *brine*" might help to ensure they arrived "more fit for examination."[53] Hooker had suffered difficulties himself when collecting seaweeds during the *Erebus* voyage and told his father that "their collecting was very tiresome as the water was very cold & the heavy swell on the rocks cost me many a hearty ducking; when fresh they appeared very beautiful, but the motion of the ship in that stormy harbor prevented me from laying them out so well as you would like."[54]

The problems of preserving such unfamiliar plants evidently worried Gunn, since he also mentioned these "ugly" seaweeds to William Hooker, telling him that they were "so perfectly horrible to the eye" that he had been "ashamed to send you any." He had tried again but was still concerned that "the present lot will do me no credit." Gunn was proud of his ability to produce attractive specimens, explaining, for example, that he used a portfolio instead of a vasculum "so as in most cases to retain the true colour of the flowers . . . they are pressed as naturally as I believe it is possible. I am sure no one can take greater pains with the specimens than I do."[55] Such comments emphasize the importance he attached to the aesthetic quality of specimens, as well as his pride in his skills and his desire to improve them. He asked for "any hints by which their appearance may be improved."[56] The appearance of specimens was so important that Gunn told Joseph he would "as soon think of biting my own nose off as offend your worthy Father."[57]

Gunn's concern that his specimens might do him "no credit" was well founded; after they arrived, Joseph Hooker told him that "there are good things amongst them but they must be laid out first," that is, separated and attached to individual sheets, work that was done by Kew's paid herbarium assistants.[58] Hooker sent Gunn's seaweeds to Harvey, explaining that the box was "all a lump, about 3ft x 2½—& 2 deep" and asking, "What is to be done with them?" As he could not afford to have them prepared, Hooker proposed that Harvey lay them out and keep a set for himself in exchange.[59] Harvey agreed but commented that they were "so dreadfully massed together that it is impossible to make much of them. . . . I must write Gunn a few hints on drying 'en masse.'" He suggested that Hooker also write to Gunn: "In your letter to him please write as strong as may be to collect the more delicate algae, & to have the goodness to dry *these* on paper, however he puts up the coarse ones."[60]

Hooker did as Harvey suggested, telling Gunn, "You must above all things send the large seaweeds," explaining that "you cannot tell what the lamentation we have made over those you told us you had thrown away because they looked so black & nasty." He explained that "one of those Laminarias is worth a dozen smaller more beautiful things" because "many are excessively curious & they are the least known of the *Algae*."[61] Under such promptings, a collector gradually learned not just how to preserve plants but also to identify those of particular interest.

The difficulty of making good specimens, especially of less familiar plants like the algae, is evident from the detailed instructions Hooker gave Gunn for drying them: "You must dry as characteristic a specimen as possible," and "also send one or two dried whole by hanging up in the air." Since Gunn was not paid for his specimens, Hooker had to phrase his instructions carefully, so as not to appear to be giving orders to an underling; he tactfully added, "you will I know do your best with them."[62]

Hooker did not specify what he meant by a "characteristic" specimen, but more details are given in Bentham's *Handbook of the British Flora*—its section "Outlines of Botany" was used by Hooker in several of his own works.[63] Bentham advised: "The specimens should be taken from healthy uninjured plants of a medium size. Or if a specimen be gathered because it looks a little different from the majority of those around it, apparently belonging to the same species, a specimen of the more prevalent form should be taken from the same locality for comparison."[64] From such instructions, and from Hooker's letters to his collectors, it becomes clear that a characteristic specimen was one that was intact, in good condition, well preserved, and attractively presented and labeled, but most of all, it was one that represented the most typical form of the plant; it could be neither too young, old, large, small, or in any way unusual. Yet even this list does not fully capture the elusive "characteristic specimen": as we shall see, that materialized gradually during the practice of collecting in the field. It would evidently take patience and experience to follow Bentham's instructions, for example, to recognize the healthy plants and distinguish atypical and "more prevalent" forms. And, as later chapters show, "characteristic" specimens were also shaped by such things as the conventions of botanical art and the classification system collectors used.

Learning to observe closely and preserve correctly was a laborious business, as is clear from William Hooker's "Short directions for collecting & preserving Seaweeds." These were written for Charles Wilford, who collected for Kew in China and Japan from 1857 to 1859, and (despite their title) were very detailed. Yet even after wading through many pages of comprehensive instructions, the collector would still have needed considerable expertise to put them into practice.[65] For example, Hooker noted that many species were washed ashore after storms, making them easy to gather, but added: "Faded or *bleached,* & decaying specimens should not be collected." How was an inexperienced collector to know whether or not the color had been bleached from a seaweed he had never seen before? Similarly, how should he interpret Hooker's comment that while coarse kinds could be brought home in a bag, delicate ones should go in a bottle of seawater and "an Indian rubber bag, a basket or a vasculum will do for the rest." Which were "coarse," which "delicate," and how were "the rest" to be defined? Joseph Hooker gave Wilford similarly lengthy instructions.[66] But once again, despite their detail, these instructions could be followed only by someone with the experience to know which "unknown seaweeds" were to be dried as common plants and which should or should not be put into freshwater. Apart from anything else, collectors needed an extensive knowledge of familiar plants in order to identify those that were unknown.

Both Joseph and William Hooker told collectors to wash seaweeds in a bath of water before drying. Once rinsed, the specimens were laid on white paper, placed on sheets of drying paper, and covered with calico or cotton and then with more drying paper. Successive layers of specimens, white paper, drying paper, and/or cloth were to be built up in this way. Once "a bundle is formed" it could be pressed "as other plants." The drying paper should be changed regularly (which would not be easy when paper was in short supply) and William explained that the cloth was to prevent the plant from sticking to the drying paper, but that if it did stick it should be left until quite dry, "when it can be readily removed." He noted: "It is intended that they should adhere to the white paper on which they have been displayed." Ensuring that the specimens stuck to the white paper but not to the drying paper or the cloth must have required practice, especially if the collector was thousands of miles from Kew, lacked specialized materials, and had to improvise solutions to deal with unfamiliar plants.

While many plants could be pressed and dried easily, some fleshy ones, like orchids, were harder to conserve. Bentham told collectors that "if the specimen be succulent or tenacious of life, such as a *Sedum* or an *Orchis*, it may be dipped in boiling water *all but the flowers*. This will kill the plant at once, and enable it to be dried rapidly, losing less of its colour or foliage than would otherwise be the case. Dipping in boiling water is also useful in the case of Heaths and others plants, which are apt to shed their leaves during the process of drying."[67] These may have been reasonable instructions for a well-off gentleman like Bentham, whose travels were confined to Britain and continental Europe, but they were perhaps a little harder to follow in the Australian outback. Collectors had to learn to place both orchids and some of the nonflowering plants, such as certain fungi (which were known as cryptogams), in spirits or pyroligneous acid, but that entailed carrying volatile liquids in the field, usually in heavy glass jars that were themselves expensive or hard to find (fig. 10.1).[68] Hooker complained that, on the *Erebus*, "not a single glass bottle was supplied for collecting purposes, empty pickle bottles were all we had, and rum as a preservative from the ship's stores."[69] Colenso could not obtain suitable wide-mouthed bottles in New Zealand and asked Cunningham to buy him some in Sydney, but the latter replied that they were hard to find: "they are not sent out, as there is no demand for them saving by chemists from whom I have purchased such as I have forwarded to you at I think a dear cost (3/6 per lb!)."[70] Even the most commonplace object could become a rare commodity in the colonies.

When bottles or jars could not be found, or the rum had all been drunk, it became impossible to ship certain plants. Some were too fragile or moist to dry successfully; others were too large, and most lost their shape and/or color when they were dried.[71] The common solution to all these problems was drawing, which is discussed in a later chapter.

The tools considered so far—Wardian cases, gardening tools, vascula, and paper—were used to bring plants to Hooker, but materials such as magnifying glasses, microscopes, and books could play the same role, since they could serve as both gifts, trade goods, and markers of status.

Microscopes

There was much more to acquiring a microscope than simply being able to afford one; both manufacturers and models proliferated enormously in the early nineteenth century, presenting aspiring botanists with a dizzying range of choices.[72] Most botanists used a simple microscope (i.e., one with a single lens) for basic examination and dissection of plants (hence it was also called a dissecting microscope), while those interested in matters such as physiology would use a more expensive compound microscope (with two or more achromatic lenses), which provided higher magnification but required specimens to be prepared and mounted prior to examination. However, an enormous range of accessories could also be bought, from micrometers for measuring objects to a camera lucida for drawing specimens.[73] Bentham recommended "lenses of ¼, ½, 1 and 1½ inches focus" as well as "a pair of dissectors, one of which should be narrow and pointed, or a mere point, like a thick needle, in a handle; the other should have a pointed blade with a sharp edge, to make clean sections across the ovary" (fig. 2.6).[74] Even without these accessories, 1840s prices for microscopes ranged from £0.7.6-£1.5.0 (for a "Botanic microscope, for flowers") to £2.0.0-£4.14.6 for a compound microscope.[75] By the 1860s, simple microscopes were rarer, and prices had shot up to as much as £22 for a large compound microscope (fig. 2.7) and £1.10 for a dissecting microscope; even the cheapest botanical microscopes cost £0.15.6.[76]

Given the diversity of microscopes, it was useful to have an experienced friend to give you advice, as Joseph Hooker discovered in 1845, when he set

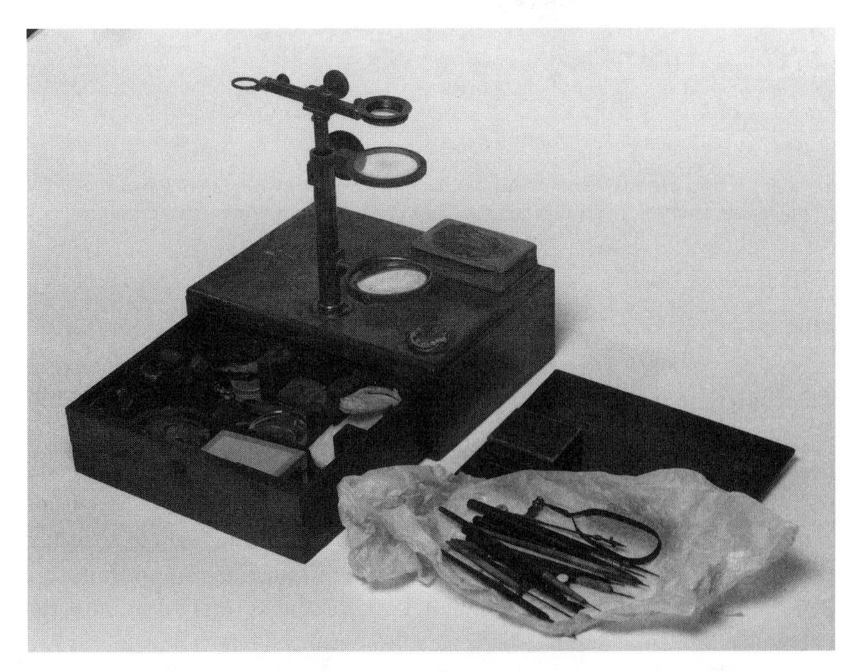

2.6 Joseph Hooker's dissecting microscope and some of the instruments used for dissection. By kind permission of the Trustees of the Royal Botanic Gardens, Kew.

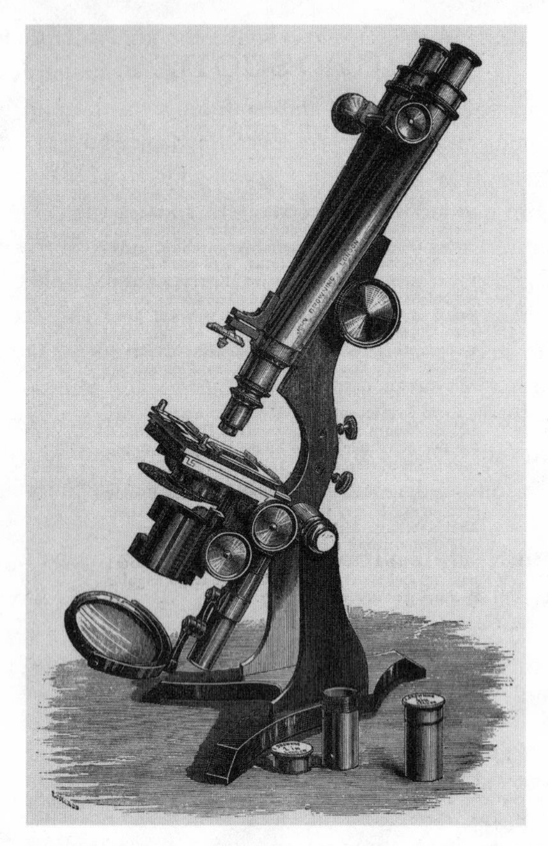

2.7 A large and expensive compound microscope: Browning's Extra Large Model Microscope (£42 in 1881). Figure 1a from John Browning, *Priced List of Microscopes* (1881). The Science Museum, London.

off to Paris for the first time. He had just received some money and Harvey suggested: "The £8 was very agreeable. It will just buy a Paris microscope. There is a man there, with an un-pronounceable German name which I forget, who makes most admirable cheap microscopes, better than Chevalier's by a chalk. Decaisne probably knows all about him; if not, & that you care to know, I can easily find out from Bergin, who has had two or three from him."[77] A microscope was essential for a botanist who wanted to study the smaller, more obscure members of the plant kingdom, the cryptogams. Bentham told his readers: "A compound microscope is rarely necessary, except in cryptogamic botany and vegetable anatomy."[78] As Anne Secord has shown, cryptogams were often studied precisely because of the difficulties involved; mastering such arcane knowledge conferred status.[79] And given that anatomy was one of the more philosophical branches of botany, the more complex and expensive compound microscope was very much the badge of the would-be philosophical naturalist.

In 1852, Hooker was delighted to receive some drawings of New Zealand mosses (passed on by his friend Grey) that had been made by Charles Knight.[80] Knight's drawings were received at Kew with delight, not least because it was often hard to persuade colonial naturalists to take an interest in such small and unglamorous plants. However, Knight said his studies had been hampered by "my microscope being imperfect, and useless for dissecting purposes"; he told Grey that he was awaiting "the receipt of one of Smith and Becks dissecting Microscope [*sic*] before proceeding in the investigation."[81]

Although Knight's geographical isolation meant a long delay while he waited for his new instrument, he had the advantages of being relatively wealthy (he was New Zealand's auditor general at this time) and having friends in Britain (where he had studied medicine); he had the money and contacts to acquire the latest-model microscope. Others were less lucky. Microscopes and even magnifying glasses were hard to obtain in the colonies, so colonial botanists had to rely on friends elsewhere to supply them. As Colenso had done when he needed a vasculum, he turned to Cunningham for assistance. However, after the latter returned to Sydney he was unable to help, explaining that "I shall not be able to obtain you a botanical microscopic glass to carry with you," since even such small, portable magnifiers "are not in request in the Colony and therefore they are not sent for" (fig. 2.8).[82] However, Cunningham found a solution to this problem; the following year, Colenso thanked him for "your always-to-be-borne in mind friendly present of a Bot. Glass, doubly enhanced to me in value from having seen it around yr neck and used by You."[83]

A hard-to-find item like a microscope made an especially good present. They were difficult to replace and too valuable to be given lightly; to earn one, a colonial botanist had to collect diligently and ensure his collections were of a high standard. One of Gunn's earliest letters to William Hooker suggests that obtaining a microscope may have been a motive for entering into correspondence. Gunn had been introduced to botany by his friend Robert William Lawrence, who already corresponded with Hooker.[84] When Gunn decided to approach Hooker to ask if he wanted a second Tasmanian correspondent, he mentioned: "The want of a microscope or strong magnifying glass is severely felt by me, as neither can be procured here." He added: "Should you be kind enough to select me as an *assistant Correspondent* (to Mr. Lawrence) perhaps you could procure me a small one,

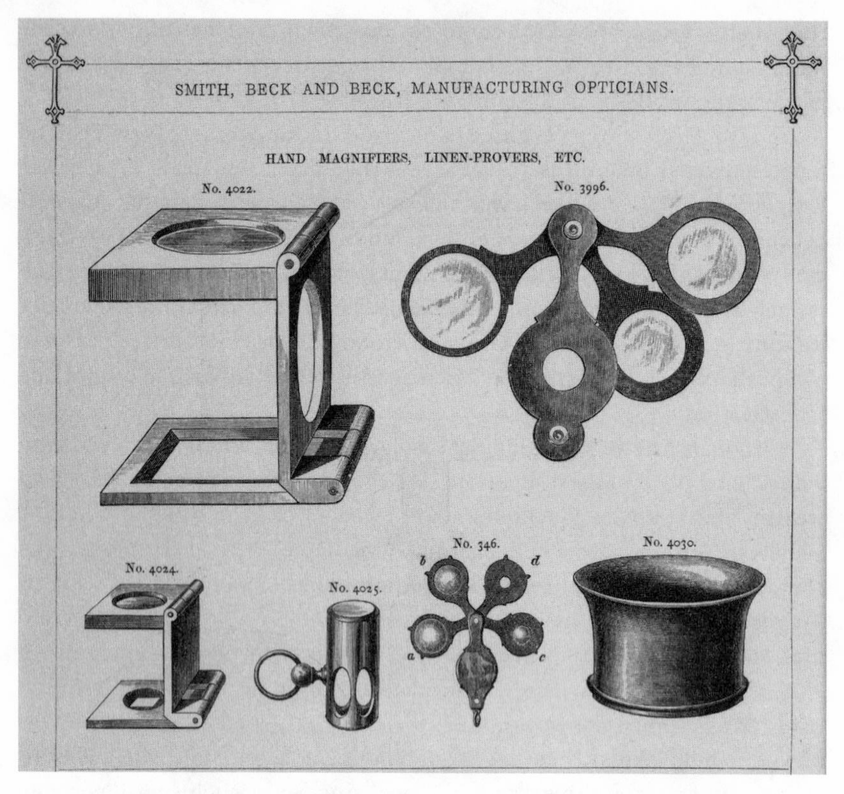

2.8 Hand-held magnifying glasses by Smith, Beck and Beck. From *An Illustrated Catalogue of Scientific Instruments Manufactured by Smith, Beck and Beck* (1866). The Science Museum, London.

and for which I shall willingly be your debtor,—as it will enable me to distinguish the various Mosses with greater facility."[85] Gunn's request seems to have been ignored at first; a year later, he was still waiting.[86] Indeed, he was still waiting ten years later, by which time he had apparently given up all hope of receiving such a valuable gift, and instead suggested that Kew might help him "to invest in a good Microscope" in exchange for the plants he sent.[87]

The diversity of microscopes made choosing a suitable one complex, so even a colonial botanist with money usually had to rely on a European correspondent for advice. Colenso wrote to Joseph Hooker in some detail about his "great want—a long-standing one, viz. a good Microscope, with which I can investigate our Mosses & Hep[aticae]." He mentioned having "2 *small* compound [microscopes], (one being an early present from a lady you well knew—*Lady Franklin*!) and one of these has done me great service."[88] However, he clearly felt the need for a larger, more powerful instrument yet had been hesitating "partly owing to cost." But he had decided to wait no longer: "Enclosed please find a D[ra]ft. for £40. for that purpose: £20.–£25 *may* suffice: if so, all right: I look to you to get me a *suitable* one, (which I dare say you can both well & *Easily* manage through Your *skilled* assistants, or friends). I have received lots of Catalogues, from Browning, Collins, and various opticians,—indeed too many, they bother me, both as to the great variety of instruments, and then prices, and qualities."[89] Colenso explained that he wanted "*such a Mic[roscope]. as you & your Ever honored Father, Mitten & Wilson have so long and successfully, used, and nothing more.*"[90] Colenso realized that such an instrument distinguished the committed botanist from the hoi polloi—asking for a microscope like those used in the metropolis was making a claim to metropolitan status.[91] He was not alone in seeing the instrument as an indicator of its user's status. The *Athenaeum* review of *Curtis' Botanical Magazine* commented that "we should like to see more dissection and a more liberal use of the microscope in the representation of the plants," on the grounds that "the botanist as well as the amateur cultivator of plants should be regarded in such works as these."[92] Like the other items Colenso sought from Cunningham and Hooker (a vasculum, magnifying glass, and books), the microscope illustrates his ambition to be a man of science, but its role as a badge of botanical status might make the metropolitan expert reluctant to distribute them too freely. Hooker complained to Huxley about a mutual acquaintance who "is a good soul, but is cursed with a Microscope," and thus inclined to see himself as a serious naturalist. To limit such claims, Hooker joked that he had "proposed a tax on microscopes some years ago, exempting Professors only."[93]

Microscopes made potent gifts; as markers of a botanist's status they were something the colonial botanist could aspire to. Since Hooker could not pay collectors, he was seldom in a position to tell them what to collect, but when they asked for help in acquiring items like microscopes, he was able to exert a little more influence over their collecting. He told Gunn that "you have collected so ably & well that there cannot be a large amount of Phaenogamic plants yet to be discovered, & we have as many duplicates of most as we know what to do with, I would therefore beg particularly to call your attention to the smaller things & lower orders." Mixed in with the praise ("you have collected so ably & well") and the politeness ("beg particularly to call your attention") was a certain amount of criticism: Hooker noted that "your specimens are all too imperfect for determination," and "in all your last collection of Cryptog. there was only one I had not gathered." "Imperfect," in this context, referred to specimens lacking the characters needed to identify and name them, such as their minute fruiting bodies.[94] Collecting such plants required not only sufficient anatomical knowledge to know whether specimens were complete but specialist equipment; as Hooker told Gunn, "you will require the pocket lens to detect species, especially of Jungermanniae & even the table microscope." From Hooker's perspective, vast collections of poor-quality specimens were worse than useless, as it took time and money to inspect and investigate them; he told Gunn, "I would earnestly request you to attend more to the quality than the quantity of the specimens of Cryptogamiae."[95]

As previously noted, part of what made cryptogamic botany so attractive to its devotees was precisely the fact that it was so difficult, thus lifting them above those they regarded as flower-picking part-timers. Yet, despite his enthusiasm for cryptogams, Colenso's specimens were often poor. In a letter to William Hooker, describing some new specimens, Colenso commented that he had just had a letter from Joseph and that "had it come earlier . . . I should, I think, have hesitated as to the propriety of troubling you with any more; and to this conclusion I should have been mainly drawn through D^r^. Hooker's own statement:—'I assure you that miscellaneous scraps such as you sent and all so carefully numbered are not worth the time and trouble of looking over.'—And . . . 'do spare yourself the trouble & expense of collecting such things.'"[96] Naturally, Joseph Hooker's own "time and trouble" shaped his desire to improve Colenso's work. He complained to Hector that Colenso's "masses of imperfect Algae & Lichens are simply not worth gathering. I cannot persuade him that he is wasting his time and mine."[97]

To judge from the phrases Colenso quoted, Hooker's letter must have been sharply worded and could have caused offense, especially as Hooker seems to have complained frequently about the quality of Colenso's cryptogamic collections. A year later, Colenso complained that "you *again* speak of the unsatisfactory state of many of my smaller Crypts" and explained that "I had vowed to collect and send you everything I could lay hands on! What an amount of Sisyphean-like, yet pleasing labor this vow of mine entailed upon me, I *alone* know."[98] Colenso's hard work and hard-won skill could make him touchy about his status, yet there was no breach between him and Joseph. On the contrary, Colenso continued the first letter quoted above, not by defending his efforts, but by apologizing for "having troubled you with such trash," which must be put down to "my ignorance" combined with "my over zealous efforts not to leave anything—even the minutest—unknown to you." Colenso closed deferentially, by promising

"I shall not willingly offend again in this matter. . . . please, my dear Sir William tell Dr. Hooker that I thank him greatly for his Letter."[99] Colenso's letters to William Hooker were always more formal and respectful than those to Joseph; nevertheless, the tone of this one gives a sense of just how badly he wanted to continue the correspondence, which offered a chance of acquiring scientific tools and the status that went with them. And, as we shall see, Colenso was lonely and (in the early 1850s particularly) in need of friends.

Collectors were especially hungry for botanical books, not least because a specimen's value (scientific or commercial) was increased if the collector could identify it and describe it with correct technical terms. Books were another commodity that were easier to acquire in the metropolis than in the colonies, and from Hooker's perspective, supplying collectors with books could save him time and money. Nevertheless, of all the botanists' tools, books most obviously disrupted the boundary between being a collector and being a professed botanist; like microscopes, books destabilized the hierarchy Hooker was trying to ascend.

Books

All collectors needed some formal book learning to complement their craft knowledge; the more ambitious the collector, the more he needed to know. Hooker's instructions to Gunn on seaweed collecting make this clear: he told the Tasmanian that "when you cannot preserve you must describe" and launched into a series of technical terms that he obviously hoped Gunn would either know or learn: for example, he told Gunn to specify whether each seaweed's root was "fibrous with the fibres conglomerate, or scutate i.e. of one shield like lump," and wanted to know if the stems were "round or compressed, simple or branched, & the form of the frond, color & texture."[100] Collectors were regularly reminded to use the correct botanical terms in the notes that accompanied their specimens; as is discussed below, doing so added precision and discouraged idiosyncrasy. Brevity and clarity saved time when sorting and classifying the specimens, so Hooker was happy to provide collectors with the books they needed.

A collector like Gunn, anxious about his "credit," realized that books would improve the value of his specimens; in his early letters he admitted being "ignorant of Botany," because he had "*no Books on the subject*—and none can be obtained here."[101] In 1844, he sent Hooker notes on his specimens and apologized that they had been written without reference books, "and in many cases you will probably find wrong words used to convey certain meanings." He also admitted that, being unable to devote himself to botany full-time, he had often forgotten those terms he had learned by the time he next returned to his plants.[102]

With books, as with microscopes, entering into a correspondence was a way of getting items that were hard to find in the colonies. Colenso shared Gunn's appetite for books and accumulated them rapidly; when fire destroyed his home in 1853, he reported that he had lost more than 100 books, despite the fact that his study and the bulk of his library were in a separate building that had been untouched.[103] Two years later, when his house had been shaken by an earthquake, Colenso told Hooker that all his books—he claimed there were now 4,500 volumes!—had fallen off their shelves.[104]

The collector's desire for books made them ideal gifts. Hooker had set off on the *Erebus* well "equipped with Botanical books" and recorded: "Of books also I have a good store and some for general reading, all Constable's 'Miscellany,' for instance. The rest are chiefly Botanical with a few on Zoology and Geology." [105] This floating library was a useful way of beginning friendships with the colonial botanists; when the *Erebus* was in New Zealand, Colenso acknowledged "your kindly offering to lend me Sprengel, Vaucher & Brown." [106] Like a missionary setting foot on a strange shore trying to win the natives' amity with beads or mirrors, Hooker had a supply of books to give those he met; when the expedition reached Tasmania, he told Gunn: "On board I have some books for you, which I have stowed in the Gunner's storeroom, the driest place in the ship." [107]

Collectors needed books to learn the language of botany, a project with which both William and Joseph Hooker were happy to assist. As well as botanical lexicons, they also provided complex works on classification and the more esoteric branches of botany. Consider the kinds of books that Purdie and Milne took with them: they included both practical volumes on gardening, such as *Millar's Garden Dictionary* and more complex systematic ones: Christiaan Henrik Persoon's *Synopsis methodica fungorum,* MacFayden's *Flora Jamaica,* J. C. Loudon's *Cyclopaedia of Plants,* and Lindley's *The Vegetable Kingdom.*[108] Once again, this mixture of practical gardening works and systematic ones illustrates the hybrid nature of the collector's skill. Not surprisingly, many of the books Purdie and Milne carried were the ones explicitly recommended by William Hooker in the botanical section of the Admiralty's *Manual of Scientific Enquiry.*[109]

Many of these books were used to identify plants in the field; without them collectors would not know if they were bringing home interesting new plants or old, familiar ones. The first step in identifying a plant using a work such as Lindley's *Elements of Botany* was to use the "analytical tables" (also known as a "key"), which required considerable botanical expertise. Bentham described their use in his *Handbook of the British Flora:* "to assist the student in *determining* or ascertaining the name of a plant . . . analytical tables are in this work prefixed to the Orders, Genera, and Species." "The student having a plant to determine, will first take the general table of Natural Orders, and examining his plant at each step to see which alternative agrees with it, will be led on to the Order to which it belongs, he will then compare it with the detailed character of the Order given in the text. If it agrees, he will follow the same course with the table of the genera of that Order to find the genus, and again with the key of the species of that genus to find the species."[110]

The key relied on discovering a plant's natural order and then its genus and species; in other words, the collector was required to make at least a provisional classification of it. Even labeling specimens could not be separated from classifying them, as Joseph Hooker's instructions to Wilford make clear; he was told to "ticket the specimens well with date & place where gathered, color of flowers, odor, and other remarks, as whether wet or dry ground, wood, meadow or cultivated ground &c. Seeds and specimens to bear the same number, name of Genus or Nat[ural] Ord[er] to be put on the seed packet."[111] However, for reasons that will become clear in later chapters, specimens were not to be identified with their species. Hooker argued that the final identification of specimens and, most importantly, the naming of new species could be done only in a metropolitan herbarium. Nevertheless, while I have separated collecting and classifying for analytic purposes, the complex relationship between them is symbolized by the fact that works like Lindley's and Bentham's, which contained a summary of the natural system of classification, were an essential part of the field naturalist's tools. Books illustrate the permeability of the boundaries between collecting and classifying, and that permeability would have important consequences for Hooker's philosophical project.

"Impracticable Specimens"

Joseph Hooker could not have written the books that made his name without his networks of colonial collectors, but despite the popularity of botany and its relatively inexpensive equipment, good collectors were hard to find. They needed a combination of craft skills, local knowledge, and book knowledge. Even dedicated and enthusiastic collectors like Colenso and Gunn often supplied unsatisfactory material; as Hooker told Gunn, "you cannot conceive how our home is filled with *impracticable* specimens, hardly a room is free from them."[112] Good specimens, properly prepared and labeled, saved time and effort—and only experienced, well-trained collectors could supply them.

The equipment that field collectors carried illuminates the craft of botanical collecting. Wardian cases, for example, make it clear why collections of living plants could not be used to study global vegetation, and—as with the gardening tools that were regularly carried—they reveal the sometimes uncomfortably close relationship between the botanist's and the gardener's skills. The vasculum shared some of the Wardian case's popular associations; what had once been the exclusive badge of the *vasculiferi* became a mass-produced and standardized item. Its cheapness allowed colonial botanists to join the natural history networks, thus becoming an invaluable resource for Hooker, but the popularization and democratization of botany was a mixed blessing to a status-conscious and ambitious naturalist. There was little about the professed botanist's tools that distinguished him from his lesser brethren.

Teaching colonial botanists what and how to collect involved training them in a uniform process of transforming plants into specimens, in which—at each stage of selecting, sorting, drying, pressing, pickling, drawing, and shipping—aspects of the original plant were destroyed.[113] Every effort to eliminate the idiosyncrasies of collectors also eliminated the individuality of the plants. Selecting only the "most characteristic" plant destroyed information about natural variability. Preservation usually destroyed some of the plant's three-dimensional structure. Dried specimens were cheap to make and were small and flat, taking little space to ship or store. Unlike the plants they represented, they required no space to grow in and neither water nor gardening skills to keep them in good condition. They had many qualities that made them useful to Hooker's project, but they were clearly not unmediated samples of "nature." The subtle complexities of making specimens, the difficulties of preserving cryptogams and orchids, and the meticulous instructions for collecting seaweeds, all illustrate once again that botanical specimens were artifacts constructed by botanical artisans.

The artificial nature of specimens is a reminder that whenever plants were turned into specimens, much was lost in the process, and so some botanists argued that specimens were an inadequate surrogate for living plants.[114] The delicate relationship between the craft practices of specimen making and philosophical matters such as classification is something I will discuss in later chapters, but specimen making reveals that there was no stable boundary between classifying and collecting. The exchange of specimens for such items as books and microscopes provided one of the means through which the right to classify was negotiated. If terms like center and periphery are taken as fixed and unproblematic, we overlook the real importance of field collecting's material practices—it was through them that the identity and locations of the center and periphery were negotiated and defined.

CHAPTER THREE

Corresponding

> Our acquaintance, my dear friend, has been, indeed, Brief! and that, too, under great disadvantages—but, believe me, I shall *ever remember* you; and though it is not probable that we shall ever *meet on Earth* again, Yet I endeavour to console myself with the hope of hearing from and corresponding with you.
>
> —William Colenso to Joseph Hooker, 1841

As the British empire grew rapidly in the 1830s, so too did the need for young men to administer it and protect it, cure its ailments, convert its heathens, and profit from its raw materials. Hooker began his career taking advantage of these new imperial opportunities, as did thousands of administrators, soldiers, sailors, missionaries, traders, and entrepreneurs. Luckily for an ambitious young naturalist, the Victorian enthusiasm for science extended into and beyond the colonies, creating a vast diaspora of scientific devotees. A determined metropolitan specialist—provided he had the talents and patience—could persuade such people to assist him. Gradually these diverse groups could be transformed into a correspondence network—possibly the single most important tool of the imperial scientific endeavour.

As we have seen, many of Europe's natural history museums and botanic gardens relied on gifts of specimens from travelers, explorers, and those who settled in the colonies. And both merchant and naval ships often carried rare or intriguing specimens home, without seeking payment.[1] Hooker's reliance on such informal, voluntary efforts was typical of the times—Darwin had a similar network—and without their collectors, men like Hooker and Darwin would have been cut off from the plants and animals they needed. Yet Hooker regularly criticized his collectors' efforts and exhorted them to try harder. What motivated them to assist him, given that they were invariably unpaid and must sometimes have felt unappreciated? Part of the answer, as we have seen, is that metropolitan correspondents bartered access to items that were rare in the colonies in exchange for specimens and information. However, such a barter economy required constant negotiations over the value of each participant's assets to ensure that everyone felt they were being treated fairly; some collectors realized that their resources (craft skills and access to plants) were difficult to replace and as a result developed expectations of returns that their metropolitan contacts could not meet.[2]

These negotiations were complicated by the issue of appropriate reimbursement for specimens. Those colonists who tried to earn money from selling specimens were the exception: colonial botanists were more often concerned with contributing to the advancement of science and of the empire or with raising their scientific or social standing. In many cases, refusing payment was central to their sense that they were gentlemen who botanized "con amore" (for love); even if Hooker had had the money to buy specimens (which he generally did not), many of his correspondents would have been offended by the offer. Once again, the ideal of scientific disinterestedness proved an obstacle in Hooker's path: in this case, the ideal made it hard for him to assert his authority over the colonials, not least because he *did* need to earn money from science, a fact that could potentially make him appear socially inferior to his collectors.

Personal friendships played a vital part in establishing, maintaining, and renegotiating the relationships between the network's members. As we have seen, Hooker's most important collectors, William Colenso (fig. 3.1) and Ronald Gunn (fig. 3.2), were those he met during his voyage and with whom he formed long-lasting friendships.[3] For both men, their friendship with Hooker was a reminder of the botanical friends who had first stimulated their interest in the science. Friendship begat botanists, and botany begat friendships. As we tease apart the often-conflicting interests—scientific, social, and economic—of the participants in these networks, analyzing each member's wants and strategies, it is important not to lose sight of this crucial affective dimension.

"To Procure Plants by Correspondence"

Many of the collectors mentioned in the previous chapter, notably William Milne, William Purdie, and Charles Wilford, were paid by Kew to collect. The equipment they took belonged to Kew, as did the collections they

3.1 William Colenso (1862). Courtesy of the Hawke's Bay Cultural Trust—Hawke's Bay Museum.

procured.[4] Paying collectors should have given Kew considerable control over them; but in practice such collectors often proved unsatisfactory. For example, William Hooker complained that Milne was idle yet had cost Kew over £400.[5] However, he was cheap compared with Purdie: £600 had originally been estimated to pay him, but he eventually cost £1,145.7.8½.[6] By contrast, as we have seen, Joseph Hooker's time on the *Erebus* cost the government very little and Kew almost nothing.[7] During the 1840s and 1850s, the gardens' budget was tightly controlled and under constant threat of retrenchment, so it is not surprising that Kew gradually stopped employing collectors during this period.

In the course of explaining to his government paymasters why Purdie's mission had cost almost twice the original estimate, William Hooker noted: "In order to render the collecting more effectual, as well as more economical, we were allowed to have a Partner in two of these Missions, who would pay a sum equal to ourselves & share with us in the proceeds of the Collections. In one case we were joined by His Grace the Duke of Northumberland, & in another by the Earl of Derby: hence their names will be found mixed up with this account." As botany's imperial and eco-

3.2 Ronald Campbell Gunn, looking every inch the gentleman he aspired to be. By kind permission of the Trustees of the Royal Botanic Gardens, Kew.

nomic role became more important, such noble partners seem to have become harder to find, a consequence perhaps of the declining influence of the aristocracy within the shifting social landscape of Victorian Britain. Meanwhile the rising cost of maintaining collectors in the field was becoming harder to justify, and Hooker concluded his letter by suggesting: "In conclusion I may observe that in consequence of the great number of intelligent & scientific men now resident abroad, we shall find it, as a general principle, more economical to procure plants by correspondence with such persons than to send out special Collectors."[8] Colenso and Gunn, like James Drummond and Charles Knight, were among the "intelligent & scientific men" William Hooker had in mind, and all were "now resident abroad" because of the rapid growth of Britain's empire during the middle decades of the century.

William Hooker's positions at Glasgow and then at Kew, together with his widely circulated publications, attracted correspondence from the colonies. Among those who wrote was Colenso, who introduced himself as "an entire Stranger, wishing to advance the Science of Botany," who "takes on himself the liberty of addressing you without an Introduction and also to send you a few Specimens of Plants."[9] Most botanical correspondences began with such a gift: when Gunn joined the network in 1832, he also sent William Hooker a parcel of specimens. Gunn's interest in botany had first been aroused by his friend Robert Lawrence, who collected for and corresponded with Hooker. Gunn felt that Lawrence's "attainments in Botany and indefatigability and perseverance in Collecting" were such that he doubted Hooker would need another Tasmanian correspondent: "therefore in sending you the present package I do it with a view that should you *not* desire a second correspondent in this Colony, to recommend my humble services to some Botanist Friend who will in return forward me a few good works to advance me in the Science, (of which I am as yet totally ignorant,) and also—seeds of any Plants, useful, remarkable, ornamental or which have not yet been introduced into this Colony and of which also a partial list is sent."[10] As we saw in the previous chapter, Gunn hoped to begin a mutually beneficial relationship of exchange, and books were his immediate want.[11] They were expensive in the colonies, but as with microscopes, having money to buy them did not solve the colonial botanist's problems; not knowing which to buy or where to obtain them often left them relying on their metropolitan contact's goodwill and expertise.[12]

Joseph Hooker spent considerable time buying books for his collectors, and from his perspective it was time well spent, since books helped improve collectors and thus collections.[13] However, as collectors acquired books, they became experts in their own eyes and those of their fellow countrymen. Gunn and Colenso developed their own local networks of collectors, which ran on barter like the global ones; Gunn asked for "any other stray Books you think useful," adding that "an extra copy would be useful as enabling me to make a profitable exchange with Collectors."[14] As Colenso's and Gunn's networks, libraries, and herbaria grew, so did their confidence, and (in Colenso's case particularly, as we shall see) their developing expertise created problems for Hooker.

Local Knowledge

Whatever they wanted in exchange, collectors' access to plants was their main asset. As Hooker told Gunn: "If you have any Magellan plants & would not object to lend them pray send them by the first opportunity, & they shall be returned quite safe, you have little idea of the immense rarity of these things, I would give a guinea for a single carpel of the umbelliferous plant."[15] In fact, Gunn had a clear sense of the "immense rarity" of his collections, given the difficulties Hooker would otherwise face in obtaining them. He noted of one consignment: "You will perceive by the above list that the far greater Number are *very rare plants*." "Against *these* I shall draw *handsomely* because I am sure that it would cost very many Pounds to get any one in V.D.L. to go so far for them." But as this and other letters make clear, by "drawing" against the plants Gunn meant acquiring books, not cash.[16]

Colenso was equally conscious of his unique access to New Zealand's plants; in 1854, he reminded Hooker that when he had first contacted Kew, he had been "your *only* collector in a new field."[17] However, such isolation was a fragile asset; when Hooker criticized the quality of some of Colenso's specimens, the latter responded that "without doubt, had N.Z. not become colonized, and the writer of this been your only N.Z. collector, his specimens, whether old or young—mouldy or imperfect—would have been more highly valued."[18] As colonization gathered pace—and clipper and steamships improved communications with Britain—colonial collectors needed to increase the value of their assets if they were to continue to barter successfully. Learning to make better specimens was one way to do this, while another was getting to know local plants better than any newly arrived or visiting botanist could hope to do. The latter had to rely on local collectors for assistance, and the locals could be quite scornful of the sojourners' ignorance.[19] When Johan Karl Ernest Dieffenbach, naturalist to the New Zealand Company, published a successful book about his travels in New Zealand, Colenso was irritated not just by Dieffenbach's errors but by the latter having become well known on the strength of information he had gleaned from Colenso and others.[20] Among other things, Colenso had shown Dieffenbach "the only locality" of a particular plant, which the latter would never have found on his own. Only a botanist permanently resident in the colony could acquire such detailed knowledge, and Colenso was upset when Hooker did not record it in print, especially as Colenso explained that he had "very particularly mentioned their present, only known localities; in such a way, too,—by *correct* spelling of the native names,—as would be of great service to any future Collector."[21]

Local knowledge had three distinct facets: detailed knowledge of plants' locations and habitats and of the colony's geography (*topographic knowledge*); familiarity with living plants, which allowed colonists to identify their country's unique species (*endemic knowledge*); and contact with groups like the Maori or Aborigines, which allowed some colonial naturalists to learn their plant lore (*indigenous knowledge*).

Despite his hunger for books, Gunn realized that his first need was for endemic knowledge. He wrote proudly to tell William Hooker that "I am every day more satisfied with myself at my progress in Botany—and am still continuing the formation of my Botanic Garden." He was convinced that "as my knowledge increases my ability to discover new plants will be much increased."[22] Growing native plants allowed numerous, good-quality specimens to be made easily, thus creating a stock of trade goods for future barter. Ready access to living plants was vital because, as George Bentham explained to budding botanists: "A botanical **Specimen,** to be perfect, should have *root, stem, leaves, flowers* (both open and in bud) and *fruit* (both young and mature). It is not, however, always possible to gather such complete specimens, but the collector should aim at completeness. Fragments, such as leaves without flowers, or flowers without leaves, are of little use."[23] Since few plants possess buds, flowers, and immature and ripe fruit simultaneously, supplying "such complete specimens" would require growing them or making numerous visits to the same sites over several months. Knowledge of flowering seasons was therefore another aspect of endemic knowledge.[24] James Drummond noted that once he had despatched a consignment of plants, "if I were to wait to hear the fate of the specimens sent home, I should lose a whole seasons collecting, and I have determined on drying the same plants over again, and to take my chances of disposing of them," adding that "the season for collecting is not yet over, but I did not like to let the opportunity pass, without sending what I had got by me" (fig. 3.3).[25]

A colony's indigenous people were another potential source of local knowledge; an Aborigine named Calgood accompanied the pioneering Western Australian collector Georgina Molloy on some of her collecting trips, and perhaps as a result, she got help from other Aborigines. However, her husband objected, expressing the widespread view that the Aborigines were treacherous savages.[26] His all-too-common attitude may explain why so few efforts were made to learn from the Australian Aborigines.[27] Things were somewhat different in New Zealand, where Maori were an object of greater curiosity, not least because their military resistance to the white intruders had earned the latter's respect.[28] Colenso's first letter to William Hooker emphasized: "From my situation in this Land, I have very many opportunities of collecting Specimens, when travelling among the Natives, which a Botanist merely visiting N. Zealand, for a short period cannot possibly have."[29] Through his missionary work, Colenso also acquired a knowledge of Maori language and customs and, unlike many settlers, some degree of respect for them. As he knew Maori and possessed the only printing press in the country, Colenso printed the Maori text of the Treaty of Waitangi (17 February 1840). However, his protests that many Maori did

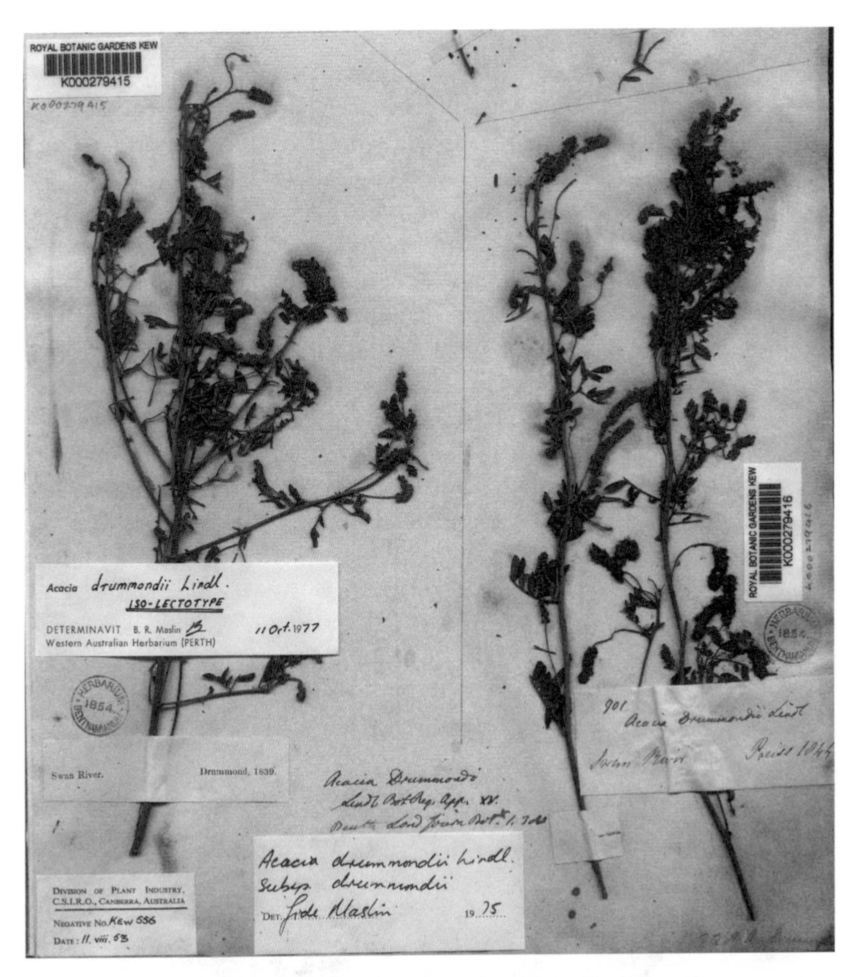

3.3 Detail of a specimen of *Acacia drummondii* collected by James Drummond, 1839. Note the numerous notes added by later taxonomists. By kind permission of the Trustees of the Royal Botanic Gardens, Kew.

not understand its implications were ignored.[30] His letters to both Hookers made frequent references to Maori issues and he exhibited a rare degree of sympathy for those he regarded as an "ancient and ill-used people."[31] Unlike other missionaries and settlers, he never enriched himself by buying Maori land, which—combined with his command of their language—made it possible for him to travel to regions considered dangerous and to learn from the Maori.[32]

Colenso's pride in his knowledge of Maori is apparent in his pamphlet *On Nomenclature* (1883), in which he renewed his attack on Dieffenbach by ridiculing the German's attempts to include a Maori vocabulary in his book.[33] He also criticized the Reverend Richard Taylor (fig. 3.4), one of his fellow missionaries in the Church Missionary Society, who had written a brief description of New Zealand's flora, *A Leaf from the Natural History of New Zealand* (1848), which included some Maori plant names.[34] In discussing the country's many pine trees, he mentioned their traditional uses and gave the native (but not the scientific) names of about a dozen species.[35] Colenso was distinctly lacking in Christian forbearance in his criticisms of his fellow missionary's efforts, noting that he had given "the (*pseudo*) Maori name of a small plant" as *Te-pua-o-te-reinga,* which Taylor translated as "the flower of Hades (or hell)." According to Colenso, no Maori had ever connected the plant with Hades, not least because their language contained no such concept as hell; the mistaken name was a "strange jumble of ideas wholly foreign to the little plant" that had been "evolved from Taylor's mind." Colenso believed that, as this plant was small and hard to find, the Maori with Taylor had called it *Pua reinga,* which simply means flower "sought after, or desired," just as the Maori term *Wahine reinga* means the "woman eagerly followed, sought, &c." [36]

3.4 Richard Taylor. By permission of the Alexander Turnbull Library, Wellington, NZ.

Colonists with firsthand local knowledge were understandably critical of those whose learning was superficial or secondhand, but Colenso's persecution of Dieffenbach and Taylor (both of whom were dead by the time he published his pamphlet) seems somewhat excessive. It gives a sense of how passionately he felt about the Maori and their language; as we shall see, he protested vehemently against those he felt were destroying their language. Colenso offered his indigenous knowledge as justification of his attempt to name an additional species of *Phormium* (the New Zealand flax); he told Hooker, "I have also taken the *universal* distinctive uses of the plants into consideration; and no New Zealander [i.e., Maori] would (or could) ever use one sp. for the other."[37] However, Hooker refused to concede the usefulness of indigenous knowledge (or, as we shall see, any other form of local knowledge) and he ignored Colenso's proposed names.

Local knowledge—whether of the colony's climate, plant localities, indigenous people, or its other collectors—was a precious asset to a colonial botanist. Like craft skill, local knowledge could be acquired only through firsthand experience, which often entailed considerable effort.[38] Collecting was also a physically demanding activity, part of the way a collector proved his seriousness. From the colonial botanist's perspective, local knowledge was perhaps the best kind to have, since Hooker matched them in craft skills and excelled them in his formal botanical education—hence Hooker's repeated denials of the relevance or usefulness of such knowledge; he overruled it because acknowledging its validity would have tilted the trading terms too far in the colonist's favor.

"Enthusiasts Like Ourselves"

As we have seen, Hooker sometimes had to beg his correspondents for specimens; he told Gunn: "You see how wretchedly I am off for some of the Auckland Isld plants I would have given a guinea a specimen (if I could have afforded it) for those you have."[39] But he could not afford a guinea—even a shilling a specimen would have been impossible, given that, as Hooker explained, the Admiralty "have me on Asst. Surgeon's single pay of £130, on which I grow uncommon fat as you may suppose."[40] Gunn responded that Hooker's "account of the Rewards bestowed upon Science & learning in England is not encouraging," but that "it hardly required your letters to satisfy me that Natural History must be followed for its own sake alone by enthusiasts like ourselves."[41] Gunn saw himself as a gentleman who scorned payment because science must be "followed for its own sake alone"; his phrase "enthusiasts like ourselves" exemplifies the way bartering—as opposed to selling—specimens allowed him to think of himself as Hooker's friend and social equal.[42]

Some years earlier, Gunn had initially received no reply when he sent specimens to John Lindley and admitted to William Hooker that he had a "slight feeling of annoyance," especially "when I compare your Conduct to his." Gunn explained that his motive for collecting "was purely taste, and a mind bent upon some pursuit, and not necessity or for a livelihood & I was afraid Mr. Lindley whom I only knew from his public name, might forget those points." Gunn's stress on his "taste" for botany and his description of it as a "pursuit" again suggest he was anxious to be seen as following it in a gentlemanly way. Gunn did not want to be regarded as a commercial collector: "If my collections are worth the freight & a few seeds in return," he explained, "it was all I looked for."[43]

Refusing payment was essential to Gunn's idea of himself as a gentleman of science; in a letter to Joseph Hooker, in which he complained that John Edward Gray of the British Museum had not reciprocated his gifts of specimens, Gunn wrote: "I do not desire to profit by Natural History, but I cannot afford to follow it extensively at my own sole expense merely 'for the glory of the thing'—without even books to guide me in my researches—and which a public institution like the British Museum could surely afford to give."[44] The following year he claimed: "My excessive fondness for Collecting requires no stimulating from money—but I cannot afford to buy Books and expend money on collecting too."[45] A "fondness" for collecting was of course the mark of a vocation, not a profession, and Gunn urged Hooker to "tell Mr. Gray that I do not & did not complain from any mercenary motives."[46]

Gunn asked Hooker to name and return the duplicate specimens he sent, adding that "if you will do that I shall promise to collect," but threatening that "I shall certainly *stop* the supplies unless you give me the names of ***everything.***"[47] This tongue-in-cheek threat was Gunn's way of making it clear that he expected his gifts to be reciprocated. By refusing both unreciprocated gifts (patronage) and payment, Gunn was claiming equality with his correspondent.[48] He concluded his complaints about Gray by observing: "I wish I was independent—I would then work away con amore."[49] Being financially "independent" (as Hooker himself wished to be) would enable him to be a disinterested botanical enthusiast, a gentlemanly "amateur" in the traditional sense of one who pursues a subject for love, but both Gunn and Hooker knew that a gentleman needed an income before he could afford to refuse payment.

Anne Secord has analyzed the role of gift exchanges in overcoming the social distance between naturalists of different social classes. The artisans she has studied could join scientific correspondence networks by adhering to gentlemanly standards of conduct—such as giving gifts and declining payment. Gifts were reciprocated in kind and cash payments were often refused by artisans, who knew that money played no part in exchanges between gentlemen and were too proud of their botanical knowledge to be placed in the role of employees.[50] Unlike Secord's artisans (who were generally not trying to change their social status), Gunn's refusal to accept payment probably reflected his aspiration to emulate his lamented friend Lawrence, the son of a wealthy landowner. When they met, Gunn held a poorly paid government post as superintendent of convicts, and his professed desire to follow botany "con amore," as a "pursuit," and not out of "necessity or for a livelihood" may well have been motivated by a desire to emulate his late friend and act the gentleman. William Archer, another of Gunn's friends, was also a landowner who pursued botany in a disinterested, gentlemanly way.[51] Archer became a specialist in orchids, but—like Knight—it seems to have been his artistic talents that initially allowed him to contact the Hookers by sending them some drawings as a gift (fig. 3.5). Archer, who described botany as his "darling pursuit," refused payment for his illustrations; he and Gunn shared an ideal of botany as something to be pursued for its own sake rather than for gain.[52]

Colonial collectors who held or aspired to genteel status highlighted the ambiguous standing of those in the metropolis who were paid men of science. Had it been possible, someone like Hooker would undoubtedly have been more comfortable relying on paid collectors, which perhaps explains why he sometimes blurred the details of his relationships with his collectors in letters to metropolitan contacts. Within a few years of returning to Britain, he told James Ross that "Gunn and Colenso are still employed in making collections in all parts of these islands and are paid by my father and self for doing so, from our private pockets."[53] The colonial botanists would have been dismayed by the implication that they were employees, especially as they were clearly being reimbursed for their expenses rather than being paid. Nonetheless, Hooker's references to Gunn suggest that at times he thought of the Tasmanian more as an employee than as a fellow enthusiast. In 1844 Hooker told William Harvey that he had just sent Gunn "an immense envoi of books, partly presents & partly the recovered £50," and noted that "he ought to send us lots of things."[54] His use of the word "ought," together with his additional remark that "I have written

3.5 Watercolor illustration of a *Macdonaldia* orchid by William Archer. Gunn attempted to name a new genus, *Macdonaldia,* in honor of his friend Charlotte Smith (née Macdonald); *Macdonaldia* was only recognized as a "distinct section" of the genus *Thelymitra* by Hooker (1859: 3-4). See fig. 4.3. Courtesy of the Linnean Society of London.

him most fully about the [specimens] we want so badly," suggests an order being given to Gunn, so that he could meet the obligations being imposed by the books and "presents"; a gift given in expectation of reciprocation is not really a gift, and a gift that is perceived as more valuable than any that might be offered in return is a way of putting someone in your debt.

Not only would Hooker's correspondents not accept the subordinate status that went with direct payment, but in several cases they even became Hooker's patrons by using their positions in the colonial governments to obtain financial support for his publications.[55] In 1862, Colenso and Knight helped secure a grant of £500 toward the proposed *Handbook of the New Zealand Flora* (1864), and Colenso told Hooker that he, Knight, and others had privately agreed to find an extra £100 if required.[56] Almost a decade earlier, in 1854, Gunn and Archer had helped encourage the Tasmanian government to grant Hooker £350 toward the cost of publishing the *Flora Tasmaniae.* Hooker told Gunn he had been "both gratified and surprized" by the news: "Mr Archer tells me now that it is through your representations that this has been effected. Indeed my dear friend I do feel most deeply obliged, & I am sure I speak truly in adding, that I hardly know how to express my feelings. In the first place I keenly feel your personal friendship and kindness, secondly the great honour done me & the testimony it procures to the value of my labors from total strangers."[57] Hooker's sense of being, quite literally, indebted to the colonial collectors (which contrasts strikingly with his description of them as employed or obliged to collect) further complicated his relationship with them. By refusing payment, men like Gunn and Archer claimed a degree of social equality, but by obtaining money for his work, they had almost made Hooker into their client.

Although Colenso lacked Gunn's social ambitions, he shared the ideal of practicing science for love not money. Cunningham had probably instilled this ideal in Colenso, urging the missionary to follow botany in the same spirit he did: "what you do, do well, with all your heart. Cherish a feeling for investigations of these kinds that will urge you to go about them con amore."[58] Colenso followed his advice and, like Gunn, never asked for payment. Colenso wrote to Hooker after his ordination and commented that "I think you know my mind—to be devoted to the welfare of the poor Natives. *As a recreation,* however, Botany is, and will be, my darling pursuit."[59] Colenso made one attempt to gain a paid appointment through the Hookers' patronage, but this was during a difficult period in his life, when his religious vocation was threatened. In 1843, he had married Elizabeth Fairburn, but the marriage—which had been insisted on by Bishop Selwyn as a necessary prelude to Colenso's ordination—was not a success. In 1848, Colenso began an affair with a young Maori woman, Ripeka Meretene, and when Elizabeth found out in 1851, she left with their children; Colenso never saw her or them again. The church suspended his licence to preach in 1852, and he lived as a virtual recluse for about four years.[60] It was during this period of suspension that he tried to find a paid position; in 1854, Colenso wrote to Joseph Hooker: "If you, or Sir William . . . could procure me a suitable appointment . . . or pay my expenses,—or, pay for any plants (Dried or living,) I might send you . . . I am at your Service. I would willingly accept an appointment of (say) £200. a year."[61] However, once he had been reinstated as a vicar he returned to practicing science for love and made no further efforts to pursue a paid scientific career.[62]

Some naturalists (such as Colenso and Cunningham) had scientific aspirations, whereas for others (such as Gunn, Archer, and Lawrence), social advancement seems to have been more important. Yet it is impossible to divide the collectors into "scientific" and "genteel" categories; all refused direct payment for their collections, yet all needed an income before they could pursue botany. Their situations mirrored Hooker's, torn between the necessity of accepting "honourable compensation" for his work and the desire to be "independent."

Botany was an established route to social advancement in the early nineteenth century. Joseph Paxton was perhaps the most successful, rising from gardener for the Duke of Devonshire to Sir Joseph Paxton, MP.[63] Such opportunities were even greater amid the fluid social relations of the colonies; it was largely through natural history that Gunn became secretary to Tasmania's governor, and through his friendship (originally founded on botany) with the Lawrence family that he moved from being a minor government official to managing their estates and eventually becoming a landowner himself.[64] Gunn needed an income before he could afford to work con amore; however, his growing prosperity seems to have coincided with a gradual loss of interest in botany. Perhaps managing his estates and those of the Lawrence family became too time-consuming, but he may well have felt that he had finally become a gentleman and consequently had less need to engage in aspirational, gentlemanly pursuits.[65]

"The Purchase of a Set of Plants"

Although the gentlemanly ideal was widespread, it was often hard to pursue. When James Drummond first arrived in the Swan River colony, he was the honorary government naturalist and founded the colony's first botanic garden. However, funding for the garden ran out and he turned to commercial plant collecting to support himself and his family; his children collected for him and he claimed that his youngest daughter, Euphemia, knew the local orchids as well as he did.[66]

Drummond's few extant letters to William Hooker show how difficult it was for a commercial plant collector to make a living, especially given the slow and uncertain communications between Australia and Britain. In response to a letter from Hooker, which had presumably requested plants, Drummond explained that he would have to collect and dry duplicates of a large collection he had made in 1837 and shipped to James Mangles, a British nurseryman:

> Mr Mangles had promised to dispose of these specimens for me at the price per hundred sp. which you mention in you letter, but about the time I expected they would reach London I received a letter from Mr Mangles saying that his health was so bad that he could not manage to dispose of the specimens, and recommending me to address them to Mes^rs Loddiges, at Hackney but it was then too late to give them any other direction, under these circumstances I do not know how the specimens have been disposed of.[67]

Mangles (like the Hookers) could not afford to send a plant collector to Western Australia, any more than Kew could at this period, and so he had to rely on whoever happened to be in the colony; when he found—as Harvey and Hooker did—that Drummond's specimens were of poor quality, it was purely by chance that he was put in contact with Georgina Molloy.[68]

Drummond was not the only colonial botanist whose circumstances forced him to turn to commercial collecting; Augustus Oldfield's career seems to have followed a similar trajectory. In 1854, William Hooker told Ferdinand von Mueller, government botanist in the colony of Victoria, that he had "just received a most rich set of Cryptogamae from a Mr Oldfield in Van Diemen's Land."[69] Given that it was rare to find a collector interested in these plants, Hooker was understandably delighted with Oldfield's collections, which appear to have been a gift, presumably intended to begin a mutually beneficial correspondence.

A year after his first collections had arrived, Oldfield wrote again to say that he had sent another collection: "I offer it (as you no doubt will receive it) as the contribution of a poor student who chiefly laments his poverty as preventing him acting as he would fain do in the cause of science."[70] Despite his poverty, Oldfield initially appears to have pursued botany in the same disinterested way as Gunn. Joseph Hooker would later say of Oldfield that "he gave specimens and information of the most valuable description . . . without return of any kind, and placed his knowledge at the disposal of naturalists in the most enlightened manner."[71] But during his last years in Australia (he returned to Britain in 1862), he gradually gave up offering specimens "without return of any kind" and tried to turn his local knowledge into cash. In 1861, he told Hooker he was about to go collecting but was "forced to depart from my general rule, and seek, by the sale of specimens, some return for the great expense I shall be under in this undertaking" and was therefore writing "to ask whether you can further my views by the purchase of a set of the plants I may there collect, and collections of seeds."[72]

Oldfield, apparently never a wealthy man, was "forced to depart from my general rule" by poverty and illness.[73] He went blind in the 1860s and by the time he died in London, in 1889, was again short of money.[74] His sad death is a reminder that for every Gunn, successfully bartering his specimens for genteel status, there were probably several Oldfields, who failed to advance either science or themselves.

"I Have No Botanical Friend"

Perhaps the commonest reason to join a collecting network was a desire for a like-minded companion; the importance of loneliness and the resultant need for friendship are essential to understanding how colonial connections were established and maintained. Hooker's relative isolation aboard the *Erebus* doubtless heightened his pleasure at making new friends when he met Gunn and Colenso. And, as noted, Gunn joined the Kew network through his friend Lawrence, who had, as Gunn told William Hooker, "exited [*sic*] in me a taste for Botany and Collecting."[75] Gunn met Lawrence in the early 1830s and they soon became friends, but in October 1833, Lawrence's wife died—less than a year after they had married and soon after giving birth to their daughter. According to Gunn, Lawrence was so distressed by the loss that he "was carried off in a fit of apoplexy" soon afterward, husband and wife being buried within a fortnight of each other. This tragedy affected Gunn deeply. He told William Hooker that he had been to Lawrence practically

> his only friend on earth, and we were almost brothers to each other,—Our pursuits and feelings alike, and it will be long ere I shall be able to fill the blank his death has made. I owe much to his memory as he led me to commence the study of Botany, in which I have spent many happy hours, and yet look forward to years of pleasure in the same pursuit. His loss to you will also be most severe, as he was years ahead of me in experience . . . of Botany . . . I can only however promise to do *all I can*, and trust time will improve me.[76]

The idea that a death left "a blank" in one's life was a conventional one; the same image was used by Elizabeth Gaskell in *Mary Barton* (1848): "the children, when they die, are hardly noted by the world; and yet to many hearts, their deaths make a blank which long years will never fill up."[77] Nevertheless, Gunn's use of the metaphor shows how deeply he felt his friend's death. One reason he was initially drawn to Joseph Hooker, was the hope that he might have found a botanical companion to fill-up the blank Lawrence had left, and of course Hooker had just learned of the death of his brother, William, so may also have been eager for a friend.

Lawrence's death moved Gunn to continue his work; he seems to have thought of his botanical collections as, in part, a memorial to his comrade. Similarly, Colenso's interest in plants was also aroused by a botanical friend, Allan Cunningham. As we have seen, after his return to Sydney, Cunningham wrote to Colenso, answered his botanical questions, bought him collecting equipment, offered homely advice on diet and health, and sent him gifts.[78] Cunningham identified the plants that Colenso had gathered in the months after they had met, noting one that "is perfectly new, and is a very remarkable species." He promised to cite Colenso as its discoverer when he returned to England and published his collections.[79]

Colenso was overwhelmed by his new friend's generosity: "Really, I have no little weight of obligation on me now! How shall I make a shadow of return?"[80] Cunningham replied:

> If, as you enquire, you wish to know in what way you can make me (not a "shadow") a solid substance of return for the little civility and attention, I may have shown you since we last saw each other at Paihia, I'll just tell you—by writing to me at your leisure as long as I am in this roasted colony; by sending me . . . any little vegetable novelty, and by bearing in mind this humble request of mine, viz. Not to lose sight of the vegetation of the Land you live in, and not to scatter to the winds, that little you gather'd regarding the peculiarities of those vegetables, when I was with you. Let these investigations be your recreation after the more important miss[ionar]y. duties of the day are done.[81]

These words must surely have struck a chord with the evangelical missionary, not least because by the time Colenso's reply arrived in Sydney, Cunningham was dead and Colenso's final letter was "returned unopened." It was the one in which he had thanked Cunningham for the magnifying glass, and he signed himself "your most sincere well-wisher and disciple and friend, William Colenso."[82] Cunningham's "humble request" thus became his last testament to his "disciple." Just as Gunn was partly motivated by Lawrence and inspired by his death to carry on his work, so Colenso's pursuit of botany resulted from his determination to carry out Cunningham's last wishes and prove himself worthy of the botanical friend he had so unexpectedly lost.

Joseph Hooker's correspondence with both Gunn and Colenso began with the friendships they formed during the *Erebus* voyage. Soon after the ships had arrived in Hobart, Joseph wrote to tell his father: "Except to the Governor's I have been nowhere save to my constant companion's Mr. Gunn, with whom I spend almost every night, & we have had several excursions together. . . . He is a most excellent fellow, full of enthusiasm & cares for nothing but his plants. On the first Sunday after our arrival, Mr. Gunn came on board in the morning & . . . dined with us. Since that, I go constantly to his house every evening about 6, & we examine plants together."[83] From the outset, Gunn's letters to Hooker were characterized by a relaxed, friendly tone; he told his new friend—whom he was soon addressing as "My dear Hooker"—that "you cannot please me better than by coming in your plainest garb—as my only object in wishing the pleasure of your company is to enjoy a quick yarn upon Botanical matters."[84]

The friendships that grew from these brief meetings were to be long-lived and intense, perhaps—as Colenso's words at the head of this chapter show—because each man knew he would probably never see the other again. Although Colenso and Hooker corresponded for sixty years, until Colenso's death in 1899, they never did "meet on Earth again." Gunn's farewell was similarly intense: "God knows when—if ever—we two may meet again. Never mind my dear friend—you will at least always carry with you my warmest & most sincere wishes for your welfare & happiness—& I hope when you have leisure by & bye—you will not forget to spin an occasional yarn to *One* in V.D. Land who will always remember you with pleasure."[85] It is hard to imagine what such partings felt like. For Hooker, he was heading home to family, friends, and botanical colleagues; but the colonial botanists were isolated from the centers of European science and had few friends in the colony who shared their interests. Not surprisingly, they were often nostalgic about the visit of the *Erebus*. Within a year of Hooker's departure, Gunn wrote reminiscing about their collecting trips together and the plants they had seen and asked whether Hooker remembered "drawing Corks out of *Bottled ale*."[86] Hooker had enjoyed the trips and the beer as much as Gunn: "What would I give to have you here, old friend, for one little day? I often wonder how your game leg is; & whether bottled ale is good yet."[87] However, such memories were also a reminder of the distance that now separated them. Gunn replied: "Bottled ale agrees uncommon-well—but alas I have no Botanical friend to crack a bottle with.—Verily I would walk a round number of miles to see your face again but I see little chance of it just now."[88]

Hooker's relationships with his collectors need to be considered against the background of Victorian ideals of male friendship and masculinity.[89] Friendships between men form the centerpiece of many Victorian novels: Mortimer Lightwood and Eugene Wrayburn in Dickens's *Our Mutual Friend;* the narrator, Phineas Fletcher, and John Halifax, the eponymous hero of Dinah Craik's 1856 novel. Charles Kingsley used intense friendship between men to exemplify his ideal of muscular Christianity in his novel *Yeast* (1848); when the hero, Lancelot Smith, has lost both the woman he loves and his fortune, he and his former gamekeeper, Paul Tregarva, swear friendship. Smith says, "The only man I utterly love, and trust, and respect on the fact of God's earth, is you." The two then set off on a pilgrimage together (such shared and lofty goals being a reminder to the reader that there was nothing inappropriate or unwholesome about their relationship).[90] As John Tosh argues, for some Victorian men, the friendship between Alfred Tennyson and Arthur Henry Hallam, celebrated so poignantly by *In Memoriam* (1850), epitomized this ideal of male companionship: the climactic moment of the poem coming when Tennyson is able once again to touch his dead friend.[91] The circulation of such ideals in contemporary literature may have shaped the way such friendships were expressed, since they were largely conducted by correspondence and would have been governed by conventions of writing rather than those of conversation.[92] Tennyson and Hallam shared, among other things, an interest in geology, and the natural history field trip was an opportunity to combine healthy exercise with masculine companionship at a time when it was common for men to take walking or other holidays together.[93]

Camaraderie in the pursuit of higher goals was not merely a convenience but an ideal to be pursued. Such ideals help us understand why Hooker and Colenso's friendship continued long after Hooker had stopped working on New Zealand's plants and Colenso was too old to collect (fig. 7.2).[94] A similar warmth is evident in Gunn's letters, of which dozens survive. Even though each man knew what he wanted from the relationship, these friendships cannot be reduced to self-interested attempts to maximize the value of one's botanical assets; on the contrary, I think the friendships help us understand how the networks survived.

To be almost the only person in a vast new colony who cared about its plants created a distinctive loneliness that made men like Gunn and Colenso almost desperate for friends with whom they could share their interests. In the mid-1830s, Thomas Keir Short arrived in Tasmania with a letter of introduction from William Hooker that allowed him to make Gunn's acquaintance and gain his trust.[95] Gunn—whose wife had recently returned to her family in Dublin, in the vain hope that they might cure her alcoholism—was delighted to meet a new friend and told Hooker, "He now lives with me & I find him a companion suited to join in my Natural History pursuits."[96] However, his delight did not last; Short left the colony after running up debts in Gunn's name. Gunn was angry about the money, especially as he had lent Short the vast sum of £200 to get home.[97] But he also felt that his friendship had been betrayed; after telling Hooker a great deal about Short's failings as a gentleman and a botanist, Gunn added, "I must now say something on the other side":

> We never had a difference and many acts of kindness passed between us—I believe him to be much & sincerely attached to me—and indeed he ought to be—He has a goodness of heart which compensated for much, and would have made him a pleasing companion had not his vanity led him astray—He never willingly said or did anything to annoy me—and therefore I feel every wish to benefit him—and sincerely wish him well—but as I do not believe you knew him so well *as now do I* consider it only an act of Justice to you to inform you on the many points in the early part of this letter.[98]

But soon after this letter, Gunn discovered the extent of the debts Short had incurred and became more hostile toward him. Soon after Short had arrived, Gunn discovered "that he was living at a ruinous rate at an Hotel—that is about £500 a year—I invited him to join me in my humble fair [*sic*] & I would make him up a bed." From then on, "my house was his Home and my friends were his friends." He recounted once more the tale of his own kindness, Short's deceptions and profligacy, and the debts that he ran up, all of which were so painful because, as Gunn explained, "he had experienced my kindness for months—and in all things was more a brother than otherwise."[99] As we have seen, Gunn had earlier described himself and Lawrence as "almost brothers to each other," and he had probably hoped that Short would "fill the blank" that Lawrence's death had made. Four years after Short's betrayal, Gunn tried to persuade Hooker to leave the *Erebus* and stay in the colony for a few years: "all I can say is that in my House I offer you a Home, & every assistance in my power, & that I for one shall be rejoiced if you can remain with me for a year or two."[100] Perhaps he hoped Joseph would finally provide the friendship he had lost when Lawrence died. Loneliness could leave a man open to exploitation, and

Short's ungentlemanly behavior wounded Gunn and made him mistrustful for some years afterward; when Harvey was planning to visit Australia in 1854, Hooker felt it necessary to tell Gunn that "I am sure you will be delighted with him, he is no Short! I assure you."[101]

Colonial women also found botany a useful way to make friends. When Georgiana Molloy immigrated to Western Australia in 1830, she took a *hortus siccus*—a book of pressed flowers, which provided opportunities to strike up conversations when her companions asked to borrow it.[102] Correspondence was another way to break the isolation; Molloy was lonely and relished the chance of a friend, even a distant one, who shared her interests.[103] Yet these were strange friendships, which sometimes seemed to heighten rather than alleviate the sense of loneliness. As Molloy observed to her British correspondent James Mangles, "Our Acquaintance is both singular and tantalizing, and somewhat melancholy to me, my dear sir, to reflect on. We shall never meet in this life."[104] In a letter to William Hooker, Gunn mentioned Charlotte Smith (née Macdonald), the wife of a local shopkeeper (fig. 3.5), who had collected some of the algae he was sending and changed the drying paper for his specimens while he was away from home.[105] Gunn enjoyed her company, and when she and her husband left the neighborhood permanently he wrote sadly that "I am again alone."[106]

Smith's contributions were acknowledged by Joseph Hooker, but many others were not. Hooker may not have even known their names; such collectors were an asset to a local botanist, who relied on them to extend the geographical range and scope of his collections, so their identities may have been deliberately concealed.[107] Archer implied that there was some competition for collectors when he told Hooker that "Müller seems to have gotten hold of Oldfield. I think you told me he had."[108] In other cases, their contributions may have been considered too minor to acknowledge, or their status too low. One of Joseph Hooker's closest friends during the *Erebus* voyage was David Lyall, his counterpart on the *Terror*, whose collections were acknowledged in both the *Flora Antarctica* and the *Flora Novae-Zelandiae*.[109] Lyall continued to send plants to Kew long after the Ross expedition, including North American species, collected while he was surgeon to the British Columbia Boundary Commission (1858–61). Although Lyall received credit for them in print, his letters make it clear that many were collected by a Sapper Buttle, a private in the Royal Engineers.[110] Lyall complained that Buttle was "by no means of a pliant disposition," "self willed and conceited," and "sometimes acting in direct opposition to orders," but beyond these few comments Buttle remains an enigmatic figure (even his first name is unknown); he is doubtless representative of a larger

TO

RONALD CAMPBELL GUNN, F.R.S., F.L.S.,

AND

WILLIAM ARCHER, F.L.S.,

This Flora of Tasmania,

WHICH OWES SO MUCH TO THEIR INDEFATIGABLE EXERTIONS,

IS DEDICATED

BY THEIR VERY SINCERE FRIEND,

J. D. HOOKER.

ROYAL GARDENS, KEW,
January, 1860.

3.6 The dedication page from Hooker's *Flora Tasmaniae*, acknowledging the help of Gunn and Archer. A similar page, acknowledging Colenso, David Lyall, and Andrew Sinclair, appeared in the *Flora Novae-Zelandiae*. Reproduced by kind permission of the Syndics of Cambridge University Library.

number of collectors whose contributions were rarely recognized.[111] After receiving some New Zealand specimens from Dr. David Monro, in Nelson, William Hooker wrote to him to say that "you and the shepherd have quite astonished and delighted Dr. Hooker by your discoveries." The shepherd was Roderick McDonald, who Monro acknowledged as having gathered alpine species while tending his flocks in New Zealand's highlands. However, Hooker always referred to him as "the shepherd" ("the shepherd too has done uncommonly well and should be encouraged to continue by all means"), and MacDonald was not mentioned by either of the Hookers in any of their publications.[112] By contrast, the Edinburgh-born and educated Monro was of a rather different status to McDonald (his eldest daughter, Maria Georgiana, married Joseph Hooker's friend James Hector), which may help explain why he was prominently thanked in the *Flora Novae-Zelandiae*.[113]

Collectors were generally not acknowledged if they had no interest in the scientific world, unless they were of high social status.[114] Most of those I have discussed—Archer, Colenso, Gunn, Knight, Lyall, Monro, Smith, and Taylor—were acknowledged by name in Joseph Hooker's publications (fig. 3.6), but the names "Buttle" and "MacDonald" did not appear; nor did those of Gunn's children, Colenso's native informants, or Gunn's "good old horse *'Ball'*"—the companion of many dozens of tough journeys," whom he described as "a capital botanical horse as he follows me like a dog."[115] A good botanical horse (presumably one who did not eat rare species before they could be collected) was as essential to the collector as diligent shepherds and sappers, but no more worthy of acknowledgment.[116] Drummond was not thanked, nor did his name appear in Hooker's survey of those who had contributed to the exploration of Australia's botany. His collections may have been too small to warrant recognition, but it is more likely that his status as a primarily commercial collector cost him an acknowledgment.[117] Where there was payment, there was no debt of gratitude.

"The Sight of Your Well-Known Hand-Writing"

Friendship was often the glue that held informal networks together and needs to be considered alongside more conventional analyses of the scientific, social, and economic interests of participants in the natural history sciences. The need for a friend can help explain why (and occasionally even how) they collected; lots of them felt desperately lonely and botany provided a route to alleviate that loneliness. Colonial collectors were a long way from Britain, which they frequently still regarded and referred to as "home," and their interests in the sciences often compounded this sense of isolation, since they seldom found other colonists with whom to share their enthusiasms (fig. 3.7). This loneliness expressed itself as a longing for a botanical friend, or nostalgia for a bottle of ale; sending regular parcels of specimens provided an excuse for writing that would in return bring letters that assuaged such longings. What might be called the loneliness of the long-distance botanist was a resource someone in Hooker's position could utilize, but only if he had the skills—empathetic and diplomatic—to do so. His businesslike requests for more specimens were regularly accompanied by gifts (which were sometimes needed to sweeten some criticism he had made of their earlier specimens). Some were botanical (such as books, magazines, or collecting equipment), but others were more personal (such as photographs)—knowing what gift a particular person might appreciate was useful—but for many of the colonists a long letter from their absent friend was the best gift Hooker could send them.[118] In his old age, Colenso wrote to Hooker that "the sight of your well-known hand-writing on a newly-arrived letter from you acted like a semi-electric shock! and both surprised & delighted me."[119] Part of Hooker's success, despite his sometimes abrasive manner, was that these were reciprocal relationships; he had been

3.7 The loneliness of the long-distance botanist? This illustration, from a book by one of Gunn's friends, captures the colonial naturalist's isolation. Fern Tree Valley, Van Diemen's Land. From James Backhouse, *A Narrative of a Visit to the Australian Colonies* (1843). By kind permission of the Trustees of the Royal Botanic Gardens, Kew.

lonely while on the *Erebus,* and in later years he often recalled the kindness he had been shown on his travels. He clearly empathized with his friends in their solitude.

The affective aspects of collecting also demonstrate that, while craft skill and botanical learning were hard to replace, enthusiasm and dedication were even rarer. Both Gunn and Colenso were distinguished by a loyalty that prompted them to make intense efforts for their distant friend. Hooker and Harvey were most impressed when Gunn hired a steamship for one of his algae-collecting trips; as Harvey said, "he is a capital fellow, and delightfully zealous to hire a Steamer to go a seaweed groping—a magnificence in marine botany hitherto unheard of." Unfortunately however, Gunn's "magnificence" had not paid off in specimens, and Harvey passed on detailed instructions as to where he should look in future.[120] Nonetheless, Gunn's enthusiasm contrasted strikingly with the laziness of some paid collectors. Even if Hooker had been able to pay Gunn (and even if Gunn would have accepted payment), no amount of money could have bought such energetic loyalty (fig. 3.8).

As we have seen, when Hooker first met Gunn and Colenso, his collecting abilities were still developing and they knew their colony's plants better than he did. And Hooker initially shared their lack of a paid position and hence their financial worries. The fact that both Gunn and Colenso refused payment for their specimens emphasizes the relationship between gentlemanly and scientific status in this period: neither Gunn nor Hooker sought to improve their scientific status in order to be better off, nor was

3.8 A photograph of Hooker that he sent Gunn as a gift, the envelope signed "from his very old friend, JD Hooker." By permission of the Alexander Turnbull Library, Wellington, NZ.

a higher income merely a route to higher scientific status—scientific and social status were inseparable aspects of being a gentleman. Gunn's phrase "enthusiasts like ourselves" was an apt one.

Hooker raised his status by giving gifts, thus creating a distance between himself and the collectors, a process that they sometimes resisted. As we shall see in the next two chapters, as the herbarium grew, so did Hooker's authority. Yet when Hooker met Gunn and Colenso, he was all but unknown; his warmth and generosity, combined with their aspirations and loneliness, began friendships that were essential to sustaining a long-term connection. The affective quality of the networks is especially important in understanding the early part of Hooker's career, since it helps us understand how he accumulated the authority that is so apparent in his later life.

Focusing on friendship and botanical barter forces us to reexamine the idea that imperial men of science, like Hooker, were simply exploitative of colonial botanists.[121] This analysis results from an outdated view that colonies like Australia and New Zealand were passive recipients of Britain's scientific knowledge. Attending to their material practices shows that Hooker had to barter with skilled colonial collectors in order to get specimens. The result was not a one-way flow of plants or authority from periphery to center but a complex negotiation in which each side bartered its assets according to its interests and in the process defined who was central or peripheral and why. Such exchanges did not, of course, take place between equals, but that did not make the colonial actors powerless.[122]

The collectors were a disparate group: Drummond and Oldfield needed to earn a living, while Georgiana Molloy, like so many of the collectors, mainly wanted a friend, but it is important to note that in her case, isolation probably made her intense relationship with Mangles possible; her husband would no doubt have been concerned by it had there been any chance they would ever "meet in this life." By contrast, Colenso aspired to scientific standing, yet he shared with Gunn a sense of botany as something one did for love, not money. Gunn seems to have been more preoccupied with his voluntary status than Colenso—the latter, being an ordained minister, perhaps had no need to demonstrate that he possessed a vocation, whereas Gunn aspired to being a gentleman and felt the need to stress that he pursued his "taste" for botany as an inclination and not from "necessity or for a livelihood." Gunn's and Colenso's aspirations and anxieties were another bond with Hooker, who shared many of their goals; a reminder that in the early part of his career, Hooker had much more in common with his collectors than just a love of plants. Finally, friendship kept the collecting network running smoothly while Hooker was encouraging colonial botanists to comply with his classifying standards, which are the subject of chapter 5, but before we examine them, we need to consider a further aspect of collecting: learning to see like a botanist.

CHAPTER FOUR

Seeing

> Apparatus having been provided, the student should select for examination as perfect a specimen as he can obtain; and should carefully study every part in the order hereafter explained. In doing this he *must on no account guess*, but be *certain* that he sees correctly what is before him.
> —John Lindley, *Descriptive Botany* (1858)

Becoming a naturalist meant learning to see like one: learning to draw and to use a microscope were two closely related practices that trained the budding naturalist to see the world in a new way. Drawings could record details that were lost when drying specimens, but the activity of drawing was also a way of investigating nature: as botanists learned to draw they learned to look closely at the structure of a plant and to understand the shape and function of its parts and how they worked together. As one would expect, the nineteenth-century rhetoric of scientific illustration emphasized drawing only what one saw, creating a dispassionate record of the facts; one way artists demonstrated they had done this was to erase their individuality, getting rid of any quirks or idiosyncrasies that might be seen to compromise their objectivity. Yet, precisely by learning the conventions of depiction in order to identify their pictures as *scientific* illustrations, naturalists were internalizing a set of rules that helped establish what they could—and could not—see. Drawing is another form of tacit knowledge and a reminder that such knowledge is liberating: as scientific workers master an "instinctive" sense of how to manipulate equipment, take measurements, and record their results appropriately, they are freed from the necessity of constantly thinking about these mundane tasks—but they may also be deflected from imagining a completely new experiment. Treating the making of botanical drawings as an experimental practice shows how drawing shaped the ways Victorian naturalists saw—and thus collected, classified, and understood—the plant kingdom.

Drawing was not just a matter of looking but also of producing; it was a step in creating everything from quick sketches and watercolors to wood engravings and lithographs, from simple images in cheap newspapers to lavishly hand-colored plates in magazines or expensive books. Innovations in printing and papermaking technology, together with the abolition of newspaper stamp duty and massively increasing markets for print, all contributed to an immense expansion in printed materials, many of which were illustrated.[1] Professed naturalists were only one, rather small, group involved in the production of this dizzying array of images, and they formed only one of the audiences who consumed them. So this proliferation of images provided another important reason for conforming to the conventions of scientific art: learning the rules allowed serious botanical practitioners to differentiate their works from those of the flower painters and Sunday-afternoon watercolorists, just as using the correct scientific terms separated them from writers on gardening or flower arranging (see chapter 7). The questions of *who* made and consumed botanical images, as well as *how* they were made and consumed, have complex implications for the status of botanists and their discipline.[2]

Illustrations made books and journals attractive, boosting sales and audiences. Printed images were more dependable, portable, and widely available than either living or dried plants could be, especially as new technologies made pictures increasingly affordable. Moreover, Joseph Hooker had access to one of the period's most prolific and highly respected botanical artists, Walter Hood Fitch. One would therefore expect Hooker's books to be lavishly illustrated—and many of them were—yet he complained regularly about the necessity to illustrate and deliberately avoided illustrations in some instances. Amid the proliferation of printed images, their absence is telling and provides another insight into the ways plants were studied philosophically.

"Copy the Botanical Illustrations"

Given the widespread Victorian rhetoric of fidelity to nature, one would expect trainee scientific illustrators to begin their lessons out in the field, drawing living plants without preconceptions, but in practice it was invariably pictures of plants, not the plants themselves, that provided beginners with their first exemplars.

When Hooker and his older brother attended their father's botanical lectures at Glasgow, they were regularly accompanied by Walter Hood Fitch, who worked for Hooker's father. Fitch had been apprenticed to a landscape painter, but when William Hooker met him, he was working in the drawing office of a cotton mill, designing patterns for fabric-printing. The mill's owner, noting the boy's talent and knowing that his friend Hooker needed some help, recommended him and Fitch began working in Hooker's herbarium during the evenings. Hooker was so impressed by the boy's draughtsmanship that he bought him out of his apprenticeship and employed him as a botanical artist for over thirty years.[3]

Like William Hooker's other students, Joseph and Walter Fitch shared "an oblong folio of twenty-one lithographed plates, with descriptions of the organs, &c. of upwards of three hundred plants." The plates were done from William Hooker's own drawings and he had had them printed specially for his students. Joseph Hooker recalled: "A copy of this work was placed before every two students in the class during that portion of each day's lecture which was devoted to the analysis of plants obtained from the garden and placed in the students' hands for this purpose."[4] The combination of pictures and plants helped impress the important points on the mind's eye; the printed images serving to clarify, even to correct, nature if the specimens were imperfect or unusual in any way.

In 1837, William Hooker produced an expanded version of the Glasgow volume, and Fitch—who was now employed full-time by Hooker—drew and lithographed more than a thousand pictures for this volume, the *Botanical Illustrations* (fig. 4.1).[5] Hooker described the book's usefulness when he was to give some lectures in Newcastle, telling his host, the geologist William Hutton, that he would "feel the want of a Botanic Garden" to provide specimens for his classes: "But this deficiency will be in some degree remedied by a very splendid set of drawings which I have by me & to which I am continually making additions, on a larger [?Artist's]-folio size. I have moreover just completed a new Edition of my 'Botanical Illustrations' with above a 1000 figures, prepared solely for the use of my Classes." Not only could pictures take the place of specimens if necessary, they were in some ways superior to actual plants, as Hooker explained when he told Hutton that he had found the illustrations "of infinite service in giving rapidly a knowledge of the different parts of a plant."[6] The clear, bold style of the drawings enhanced the speed with which the essential features of a plant could be grasped—which was a crucial part of the drawings' appeal. And the fact that every student would have been looking at exactly the same picture, rather than at necessarily variable specimens of recently collected

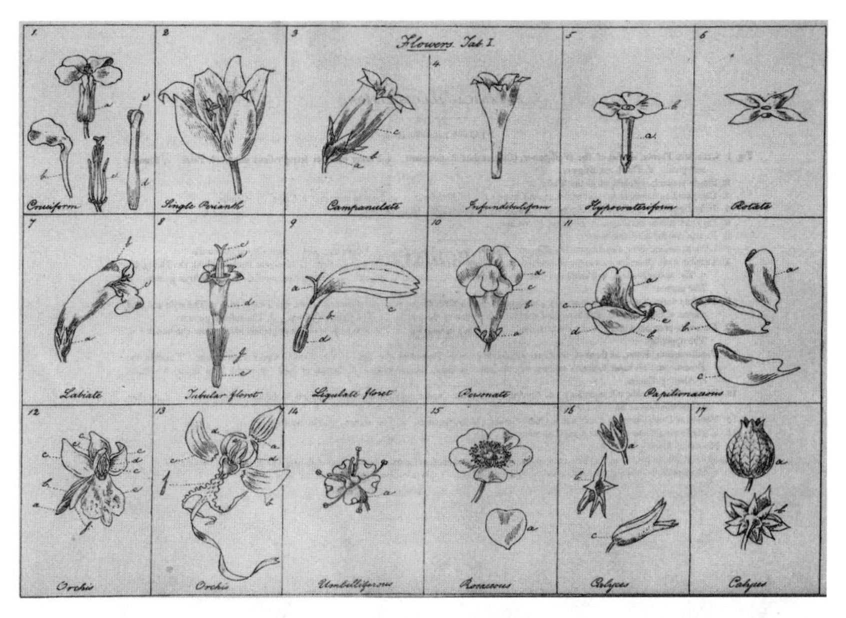

4.1 A plate from William Hooker's *Botanical Illustrations,* intended as a model for students of all kinds. Reproduced by kind permission of the Syndics of Cambridge University Library.

plants, would have helped ensure that the students retained identical ideas of each species and its characteristics.

In addition to the printed images, Joseph Hooker described how his father "made large coloured drawings" of the "principal medical and economic plants," which "were hung in the class-room when the Natural Orders to which they belonged were being demonstrated." As with the printed books, these wall charts were used in conjunction with "dried native specimens . . . taken from his herbarium, or living ones from the garden when they were to be had."[7] The size of the charts helped give them considerable impact, which must have helped impress the relevant details on the students' minds. Similar charts were used by many botanical teachers to make clear which visual features of the plants the students should memorize in order to be able to distinguish the natural orders. Robert Kay Greville, an Edinburgh botanist, delivered popular lectures on botany and used "100 Elephant folio" sized drawings to illustrate them.[8] When John Stevens Henslow began lecturing at Cambridge University in 1827, he took advice from Greville and William Hooker and prepared his own large, colored wall charts.[9] Charts became integral to botanical lecturing

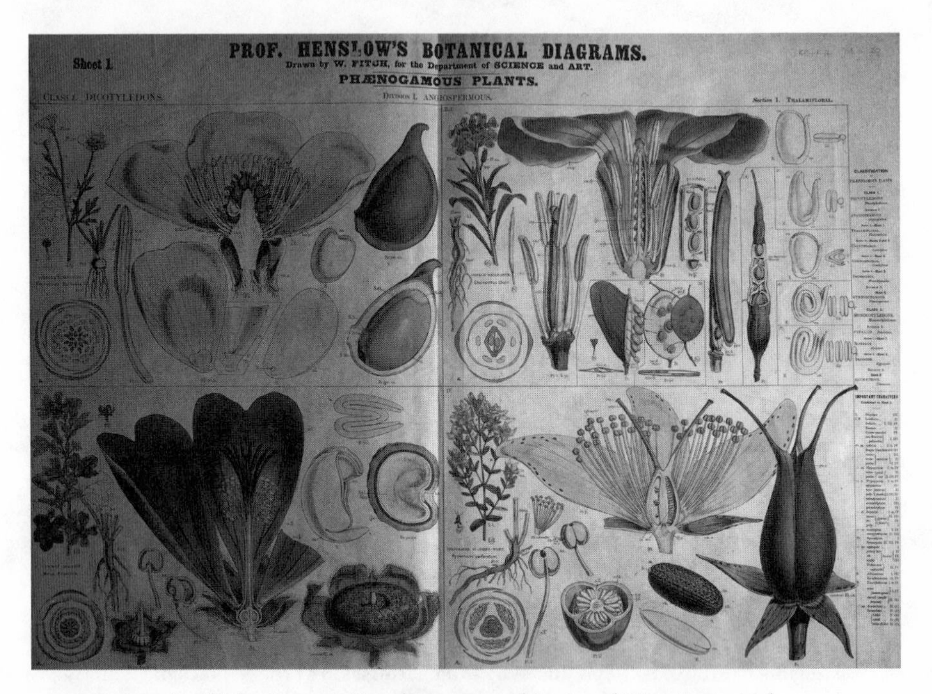

4.2 An example of John Stevens Henslow's printed *Series of Nine Botanical Diagrams* (1857), lithographed by Walter Hood Fitch. Reproduced by kind permission of the Syndics of Cambridge University Library.

because they helped overcome the problems of either not being able to obtain enough (or any) specimens of a plant or of not being able to obtain it in flower (a common problem, since the British academic year traditionally begins in the autumn). The demand for charts soon led to them becoming printed and sold; for example, Fitch helped Henslow produce a printed *Series of Nine Botanical Diagrams* (1857) for use in schools (fig. 4.2).[10] Wall charts, like printed illustrations, were permanent, always available, and could be shared by numerous students; they helped create a standardized mental picture of what the plant "ought" to look like.

Those attending lectures were encouraged to copy the wall charts and printed illustrations, helping them memorize relevant details and teaching them the standard to which botanical drawings should conform and which they should emulate. Twenty years after the appearance of his father's *Botanical Illustrations*, Joseph Hooker recommended the book to one of Kew's collectors, Charles Wilford, telling him that, whenever no specimens were at hand, his drawings could be improved if he would "copy the Botanical Illustrations."[11] Hooker had probably been given similar advice himself, since in 1845, while he was in Edinburgh, he wrote to Fitch asking to borrow some finished drawings to copy.[12]

It is unsurprising to find that copying was a major way in which botanists learned to draw; it was the standard method for anyone who wanted to learn drawing at the time, used by everyone from the most serious professional artist to the humblest part-time enthusiast. Students thronged the nation's art galleries, copying the landscapes and portraits of the great masters in order to master their techniques. And, of course, as they copied, they also absorbed the conventions of the genre, just as a botanical artist would. As with botanical collecting, the popularity of drawing as a pastime was increased by easy access to materials, which were cheap, portable, and—especially in the case of paper—mass-produced.[13] Numerous manuals were produced specifically to teach flower painting and drawing. Most of these stressed the decorative aspects of flowers, but James Sowerby's *The Botanical Drawing Book* (1788, 1807) was an exception: it applied the progressive technique of the books that taught landscape drawing (building up from single strokes, to individual objects, from outline to shading to color, eventually to finished landscapes) to botanical art. Sowerby's book also used botanical terminology to describe the parts of the flower, and Sowerby made it clear that he saw drawing as a way to learn botany.[14]

The Victorian revolution in printing helped feed this fashion for copying. The combination of stereotyping (which duplicated wood engravings so more copies could be produced), steam-powered presses (which produced many more copies per hour), and cheap, machine-made paper, all helped make illustrations readily available to almost anyone who wanted them. In advising parents on how to encourage their children to draw, John Ruskin commented that "a limited number of good and amusing prints should always be within a boy's reach: in these days of cheap illustration he can hardly possess a volume of nursery tales without good woodcuts in it, and should be encouraged to copy what he likes best of this kind." Ruskin advised parents both to encourage and to correct, "pointing out where a line is too short or too long, or too crooked, when compared with the copy; *accuracy* being the first and last thing they look for."[15] He was far from unique in giving such advice, as Ann Bermingham has shown: toward the end of the eighteenth century, a new style of landscape painting came into fashion, which urged artists to pay closer attention to the specific details of individual trees, rocks, and places. Yet, despite its claim to produce drawings that depicted nature more accurately, the chief method advocated for learning this new style was to copy pictures rather than copying from nature. Drawing manuals regularly argued that nature was too complex for novices to grasp and that they should therefore begin by copying illustrations and practicing techniques indoors, including drawing dried specimens of leaves or branches, before venturing out to tackle whole trees.[16]

This widespread practice of copying drawings and prints would have been particularly familiar to the Hooker family, since they had close connections with the fine arts, especially landscape painting. Joseph Hooker's grandfather Dawson Turner was a book collector (especially of illustrated volumes) and patron of the arts—of the watercolorist John Sell Cotman in particular. Cotman accompanied Turner when the latter moved from Norwich to Yarmouth in 1811 and was a frequent visitor at Turner's home, where he gave lessons to William Hooker's mother, Lydia Vincent (whose cousin George Vincent was also a landscape painter who studied with Cotman).[17] Joseph Hooker later recalled visiting his grandfather's house, "the walls of the room being covered with pictures of which my grandfather had a small but very choice collection," which included a Titian "and one or more Cotmans."[18] He would probably have met Cotman on these visits and would certainly have known his work. Like most of his contemporaries, Cotman had learned his trade copying drawings, and in an effort to support his growing family, he founded a circulating library of drawings which were rented out in order to be copied.[19] So, within the Hooker family, drawings and prints were routinely treated as exemplars for copying, but were also commodities whose sale and circulation could earn one a living.

The ubiquity of copying has two consequences. First, using existing, and often mass-produced, images as models helped create and strengthen a genre of recognizably scientific images, which helped to distinguish the works of professed botanists from the many other kinds of images of flowers that were in circulation. Second, adherence to the conventions of the genre also served to create and strengthen a particular concept of the nature of species themselves, a conception that served Hooker's imperial scientific project.

"The Female Pencil"

By the mid-nineteenth century, a distinctive genre of technical botanical illustration had emerged (fig. 4.3). Its most obvious characteristic was the isolation of a single species: the pictures in technical works, such as Hooker's, invariably showed a single specimen on a plain white background; Fitch commented that "in strictly botanical drawings a background is seldom given."[20] This convention produced images that looked like herbarium sheets (which are discussed in more detail in the next chapter), and I will

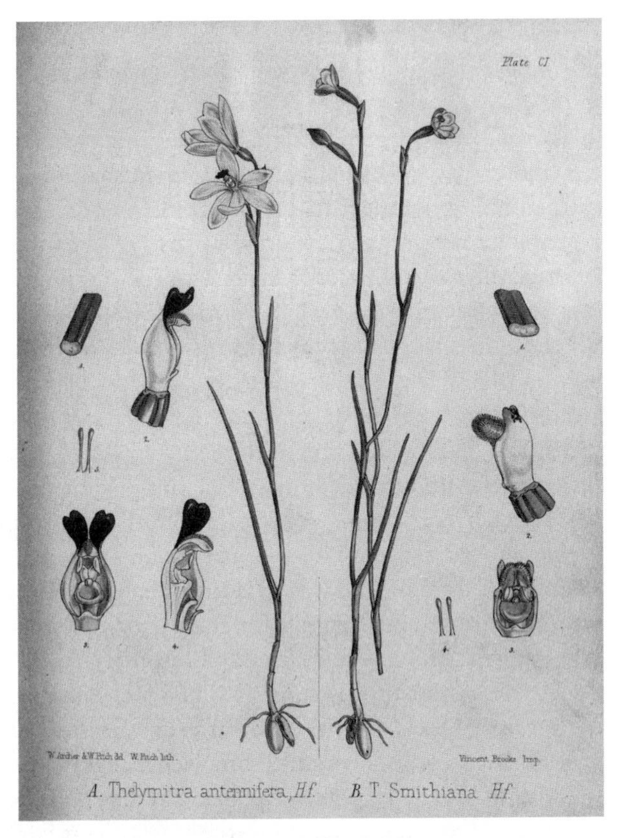

4.3 A "strictly botanical" drawing: a single species depicted with dissections but without a background. *Thelymitra antennifera* (from Hooker's *Flora Tasmaniae*) lithographed by Walter Hood Fitch from William Archer's drawing (fig. 3.5). By kind permission of the Trustees of the Royal Botanic Gardens, Kew.

refer to them as herbarium-style images. By contrast, flower paintings usually showed a number of species, grouped for aesthetic reasons and often accompanied by a naturalistic background or similar details. These visual details helped to distinguish specialized taxonomic works from the flower paintings found in florilegia, pattern books (which depicted schematic flowers for use in embroidery and other decorative arts), horticultural and gardening periodicals, or advertisements.[21]

One of the ways in which the markets for works dealing with flowers were divided was by gender: men were assumed to be the natural audience for serious systematic works, while women were thought more likely to enjoy the lavishly illustrated magazines and books. The latter works were connected to traditional feminine accomplishments such as flower painting and embroidering floral designs.[22] Meanwhile, the strenuous nature of fieldwork was considered to disbar women from some aspects of natural history.[23]

As we have seen, the *Edinburgh Review* had described a typical botanist's day in the early 1830s (chapter 1) and noted that, once the male botanist returned from his collecting to display his treasures, "the female pencil is called forth to copy the new or rare plant."[24] Over the course of the nineteenth century, drawing became one of the few ways in which women could participate in the male-dominated world of the full-time botanist. Many women worked as botanical illustrators, including Miss S. A. Drake, who illustrated Lindley's *Sertum Orchidaceae* (1837–42) and his *Ladies Botany* (1834) and produced plates for the *Transactions of the Royal Horticultural Society* and the *Botanical Register*.[25] Joseph Hooker's daughter Harriet Ann produced many illustrations for the *Botanical Magazine* after Fitch's resignation, and Matilda Smith took over from her, with Fitch's nephew John Nugent Fitch doing the engraving.[26]

However, for each woman who was paid to produce botanical art there were many more who sketched purely for their own enjoyment or who were simply consumers of the various kinds of flower images. The English nineteenth-century flower-painting genre, with its focus on numerous, artistically arranged species (as opposed to illustrating the important details of just one), symbolized a type of botanical knowledge that was particularly attractive to those women who regarded a broad but superficial knowledge of plants as more appropriately feminine than a deep, specialized one. For many women, a general knowledge and love of flowers was a superior accomplishment, since it allowed botany to be a part-time activity. In Elizabeth Gaskell's *Wives and Daughters*, Mrs. Gibson notes that her daughter Cynthia's talents are "not for science and the severer studies," whereas her stepdaughter, Molly, who takes an interest in natural history, "reads such deep books—all about facts and figures: she'll be quite a blue-stocking by and by." This conversation occurs as the two girls are competing for the attention of the same young man, and it is clear that their mother considers Cynthia's ability to recite poetry more likely to attract a mate than Molly's knowledge of natural history: as Mrs. Gibson says, "it is all very well not to be a blue-stocking, because gentle-people do not like that kind of woman."[27] Women, like Cynthia, who simply wished to know the names of the plants they painted might have preferred to use the simple Linnaean system of classification and would probably have thought it unfeminine to be like Molly and devote oneself to mastering Latin, plant anatomy, or the complexities of the natural system of classification (these rival systems are discussed in more detail in the next chapter). Finally, botany's associations with healthy exercise, the beauty of flowers, and its pious associations with natural theology made it an improving recreation, unlikely to lead young ladies into unsuitable thoughts or activities.[28] Small wonder that, as the *Monthly Review* observed, women were "often much more attached than the stronger sex to botanical pursuits."[29]

The association between botany and women undoubtedly contributed to its relatively low standing in the male-dominated councils of scientific societies like the BAAS; it is no coincidence that almost the only scientific societies that admitted women for most of the century were the Botanical Society of London and the Horticultural Society.[30] Ann Shteir has argued that the process of professionalizing botany in the nineteenth century tended to exclude women from science by relegating them to its amateur or popular forms.[31] Yet, as we have seen, during the early decades of the century, neither the amateur nor the professional existed as stable categories; they emerged from the activities of men like Joseph Hooker as they struggled to raise their status and that of their discipline. As with other aspects of the accepted professionalization story, we should be cautious about assuming such changes were intended, given that none of the supposed would-be professionals had a clear sense of what kinds of botanical careers were going to be possible.[32]

As noted in the introduction, Joseph Hooker's publications were mostly intended for a specialized scientific readership; as a result, their illustrations conform to the strict scientific style and there are few images of plants in their habitats. One exception was *Rhododendrons of Sikkim Himalaya* (1849), which was largely produced by his father. By contrast with Hooker's other works, *Rhododendrons* included a few plates that showed these exotic new species growing in their natural habitats; the book created a sensation and helped start a widespread fashion for growing rhododendrons in British gardens (fig. 4.4).

Given Joseph Hooker's difficulties in finding full-time work, it might seem surprising that he did not produce other commercial works like *Rhododendrons*, especially in light of his family's association with a renowned artist like Fitch. Yet when the Admiralty granted Hooker money to pay for the plates for his *Botany of the Antarctic Voyage*, Hooker complained to Ronald Gunn: "After a desperate deal of work with the Admiralty & Treasury, my father at last procured a grant of £1000 to go towards defraying the expense of a work on the Antarctic & other Southern plants of our voyage. . . . The govt stipulated that for the money we provide 500 plates, done on stone, which we have engaged to do, & as they are valued

4.4 Illustrations for a broad audience typically included background and habitat details instead of dissections. *Rhododendron Dalhousiae* lithographed by Walter Hood Fitch from one of Joseph Hooker's sketches. From J. D. Hooker, *Rhododendrons of Sikkim Himalaya* (1849). Courtesy of the Cory Library, Cambridge University Botanic Garden.

at £2 a plate by the artist the whole of the money will be thus expended."[33] The need to spend his entire government grant on illustrations clearly rankled a little. A few months later, Hooker told Gunn that he was sending him "a plain copy, but shall get you a set of colored plates: thus you will have it useful & ornamental too" (fig. 5.2). His comment implies that although illustrations were useful, coloring was mere ornament—an opinion that would have been strengthened by the fact that, as he explained, "I have to purchase all the coloring of the work: & have to give colored copies to the *nobs* who do not care a straw for me or my book." Meanwhile, his publisher, Lovell Reeve (later Reeve & Co.), was paying him nothing for the work.[34]

As Hooker told his colleague the botanist William Wilson, he would have preferred to issue the book with "outline lithograph octavo uncolored plates. But Govt. does not like that, they must have shew for their money."[35] Although colored plates helped sell his book, they mainly attracted "private gentlemen" rather than professed botanists, or helped impress "nobs" such as his father's titled patrons and influential friends.

Given the choice, Hooker would have ignored such audiences and kept his costs down by using cheaper line drawings, which were all that were needed for plant identification—one of the major uses for botanical images. In a letter to William Hooker, Gunn admitted that he was "not so clever a Botanist as to be able to read off the description of a plant & at once recognise it." Were he able to "devote much of my time to my plants no doubt I could do [so] pretty well," he added, but the need to earn a living left him too busy to develop such a skill. Instead, he found "that illustrated books are so useful to me as, having a capital eye, I am enabled to know plants by sight without the toil of poring over the descriptions."[36] Just as in Hooker's botanical classes, the pictures were intended to supplement a knowledge of living plants; in an earlier letter, Gunn had requested seeds for his garden, "so that I may acquire a knowledge of the different Genera by sight, as also of the various natural orders, as until that is acquired I find I shall be labouring much in the dark."[37] Gunn's "capital eye" was an asset both to him and to the Hookers, but it was most useful when the larger, flowering plants were to be identified. A dozen years later, despite much study and many hours of botanizing, Gunn complained to Joseph Hooker that he found studying the smaller, flowerless plants—such as mosses—much more difficult and told Hooker that "I have not got such a confounded microscopic eye as you have to detect differences."[38] Given that cryptogams were sometimes collected precisely because they were hard to identify, possessing a "microscopic eye" was a badge of honor for those who had mastered the skill. By contrast, while the ability to recognize plants by sight, without the aid of a microscope, was useful, it required less proficiency than being able to "read off the description of a plant & at once recognise it," which helps explains why some botanical books deliberately avoided the use of illustrations.

As Anne Secord has shown, there was some mistrust of illustrations among serious botanists. In a letter to William Hooker, William Wilson complained that colored plates encouraged superficiality: "these often do harm instead of good & are more likely to make 'knowers of species' than sound botanists."[39] Some nineteenth-century theories of perception and education assumed that color appealed to the senses, but that "sound botanists" needed to be governed by their minds, hence Wilson's appeal for outline drawings (woodcuts or lithographs) that showed dissections of the plants rather than colored figures of the whole, growing plant.[40] Fitch made a similar point in the articles on "Botanical Drawing" that he wrote for the *Gardeners' Chronicle:* "Facility in colouring is easily acquired, but a correct eye for drawing is only to be rendered by constant observation." He dismissed those who were anxious to begin coloring before they had mastered drawing as engaged in mere "paper staining."[41]

Depending on the intended audience for a book, even uncolored illustrations could be considered a little suspect, to judge by Joseph Hooker's decision to omit all illustrations from his *Student's Flora* (1870). The lack of pictures is another contrast with the book's rivals, two of which were illustrated: Lindley's *School Botany* was profusely illustrated and George Bentham's *Handbook of the British Flora* contained wood engravings by Fitch illustrating every species listed.[42] Omitting illustrations was one of the ways in which Hooker tried to achieve his aim of ensuring that his book would be "more complete & philosophical" than its rivals. Hooker described Bentham's volume as "not scientific enough for a classbook,"[43] and Bentham himself acknowledged in the *Handbook*'s preface that it was "specially destined to assist the unscientific botanist in the determination of British plants." "In the present edition, in order still further to facilitate that object, the publishers have called in aid the experienced talent of Mr. W. Fitch, who has supplied original drawings of every species included in the Flora."[44] Pictures here are explicitly linked to the needs of the *unscientific* botanist; so their absence from Hooker's *Student's Flora* was another strategy that helped exclude the unscientific and thus defined the book's genre and audience. By the time the book appeared, there was clearly a new audience of more serious botanical students, thanks in part to the time Hooker and others had devoted to reforming scientific education and raising the status of their studies. The absence of pictures forced these students to actually read Hooker's words and develop a more analytical approach to plant identification.[45]

"Combining the Peculiarities"

I claimed earlier that illustrations helped to form and stabilize a particular species concept. To understand precisely how this worked, we first need to look a little more closely at how botanical illustrations were actually made.

Fitch produced at least 12,000 published drawings in his lifetime, as well as countless unpublished ones.[46] His images are widely admired and collected, not least because of the way in which the plants come to life so vividly on the page, their leaves catch the light, their petals glow, one can almost feel the Himalayan breezes blowing through some of his rhododendrons (fig. 4.4). Yet, in a great many cases, Fitch never saw the plants he drew. He worked from dried specimens and sketches that others had made in the field. Moreover, most of his finished images show plants that never grew at all—they are usually composite images, drawn from numerous different sketches and specimens, which Fitch combined using his unique mix of artistic and botanical expertise. His images depict the typical form of the plant, a form that might never be seen, since he shows us, not what the plant looked like, but what it *ought* to look like.

Fitch stressed that, far from simply drawing what one saw, it was vital to acquire some theoretical botanical knowledge *before* one began drawing:

> I may premise that a knowledge of botany, however slight, is of great use in enabling the artist to avoid the errors which are occasionally perpetrated in respectable drawings and publications, such as introducing an abnormal number of stamens in a flower; giving it an inferior ovary when it should have a superior one, and vice versa. I have frequently seen such negatively instructive illustrations of ignorance—quite inexcusable, for a little knowledge would enable them to be avoided.[47]

Such technical knowledge was especially important when novel plants from distant countries were being drawn. A knowledge of plant anatomy and structure was vital for an artist who had never seen the living plant and was trying to deduce its living appearance from dried specimens. Gunn wrote to William Hooker in the 1830s complimenting him on the recently published illustrations of plants which Gunn had collected, but he noted that they were "very correct except Corraea Backhousiana—the flowers of which are *pendulous* & *not erect*.—They have been rubbed upwards in drying by the Gent^{n}. who sent them to me from Woolnorth" (fig. 4.5). To avoid such problems in future, Gunn explained that "I now endeavour to dry my specimens as much as possible like nature—and prevent their shrivelling up as much as I can before submitting them to pressure."[48]

Gunn's comments are another reminder of the importance of skillful collecting, but even good specimens required interpretation and restoration before satisfactory drawings could be made from them. The first step was usually to soak the dried plants in water to restore some of their original shape. Sir John Hill described the technique used to produce the illustrations for his *Exotic Botany* (1759). First, the dried specimens "were brought

4.5 Illustration of *Corraea Backhousiana*, drawn from a poorly prepared specimen by a lithographer who had never seen the living plant: hence the upward-pointing flowers, which should be hanging down. From W. J. Hooker, Icones Plantarum, vol. 1 (1837). Reproduced by kind permission of the Syndics of Cambridge University Library.

to the state wherein they are represented in these designs, by maceration in warm water." He described in detail how to place the plants in water, which was then heated slowly, so that "the plant, however rumpled up in drying expands and takes the natural form it had when fresh," adding that "even the minutest parts appear distinctly."[49] The same technique was still in use a century later, as Bentham explained to his readers: "To examine or dissect flowers or fruit in dried specimens it is necessary to soften them. If the parts are very delicate, this is best done by gradually moistening them in cold water; in most cases, steeping them in boiling water or steam is much quicker."[50] However, Hill's confident claim that soaking ensured that the plant regained "the natural form it had when fresh" was somewhat misplaced; different plants responded differently to the process. The Irish naturalist William Harvey wrote to Hooker during their collaboration on the algae for the *Flora Antarctica* that "it is difficult, in things which do not *perfectly* recover their original form on moistening, to determine what allowance to make for *drying*. Two of your sketches which represent the tubes very sinuous appear to me to be distorted by drying & not perfectly restored."[51] Even an expert like Hooker could slip up when working with a relatively unfamiliar group of plants.

A further drawback to soaking was that, as Hill noted, the "specimen is destroyed by this operation," even though each specimen had briefly shown itself "in full perfection," so much so that he wished he could have saved "some of these but they were sacrificed to the work."[52] The destruction of the specimen meant several were needed for each illustration, but even then, the artist had to work rapidly before the specimen disintegrated and as a result mistakes could be made unless the artist had a clear sense of what the living form should be.

The difficulty of restoring dried plants made Fitch especially proud of his own skills in this area. He told budding artists to "proceed systematically in flower drawing" so that "by dint of zealous application" "he may become qualified even to draw a dried specimen from the herbarium—an effort which will test his judgment, and call forth all his knowledge of perspective and adjustment."[53] Such drawings required the artist to construct a single image from a sheet of dried specimens, to create a "typical" lifelike plant from a series of flattened dead ones. Joseph Hooker noted that the figure of *Rhododendron arboreum* was drawn from several specimens.[54] And in discussing a plant from the Azores to be included in William Hooker's *Icones Plantarum* (1836–64), the Yorkshire-born botanist and phrenologist Hewett Cottrell Watson (fig. 9.3) noted that a good drawing would necessitate "combining the peculiarities of the few specimens, rather than giving a portrait of one only," a job that Fitch would have to do since Watson himself lacked the drawing ability to produce the illustration.[55] Fitch is often considered to be the first botanical artist to have produced really satisfactory drawings from herbarium specimens, partly because of his ability to construct the image of a single plant from several different specimens.[56] A unique combination of botanical and artistic knowledge was required, which Fitch evidently understood the value of, since he claimed that "sketching living plants is merely a species of copying, but dried specimens test the artist's ability to the uttermost." He claimed that it was from drawings made from herbarium specimens that he would hope to "be judged a correct draughtsman."[57]

To make the test of the artist's "patience, and correctness of eye," even harder, the finished illustrations had to include dissections of the plant made from the same desiccated samples. Fitch claimed that "one of the finest exercises" of patience "with which I am acquainted, is the analysis of a dried flower, from an herbarium specimen, perhaps very small, worm-eaten and gluey, and having no apparent analogy to any known plant."[58] By "analysis," Fitch referred to dissections (fig. 4.3), usually made under a simple or dissecting microscope (fig. 2.6), which revealed how the parts of the plant were connected to one another. Given sufficient skill and practice, even a completely flattened and dried specimen, once it had been restored using water, could be cut apart to reveal the structure, providing valuable clues as to its original appearance and thus its classification. Dissection required both manual dexterity and botanical knowledge, qualities Fitch possessed in abundance; as Bentham noted, with regard to the drawings for the *Handbook*, "The dissections which accompany each figure have been generally taken from the artist's own specimens, the organs selected by him for illustration being those which appeared to him the most characteristic of the species or genus."[59]

The inclusion of dissections marked the seriousness of an illustration, revealing that the plant had been properly analyzed not merely sketched. In 1805, a critic had described the *Botanical Magazine*, which at that time did not include dissections, as merely a "drawing book for ladies." To avoid such sneers in future, William Hooker added plant dissections to the plates when he took over its publication in 1826.[60] The inclusion of dissections was also a product of a shift to new methods of classification (explored in more detail in the next chapter). While the earlier Linnaean classification relied exclusively on the parts of the flower, the natural system of Antoine-Laurent de Jussieu and Augustin-Pyramus de Candolle aimed to use all the parts of the plant, including details of its structure. One of Lindley's innovations was to use embryological characteristics, such as seed structure, for classification, and so he began adding details of dissections to the plates of his *Botanical Register* well before they appeared in the more popular *Botanical Magazine*.

Dissections are one example of how important an artist's botanical knowledge became when resurrecting dried specimens. Fitch noted, for example, that however irregular a flower, the relationship of the teeth of the calyx (the lower, usually green part of the flower) to the divisions of the corolla (the petals) was always fixed for any given species. The correct relationship must be strictly observed since a mistake would shake a botanist's faith in the accuracy of the artist because "however beautiful his

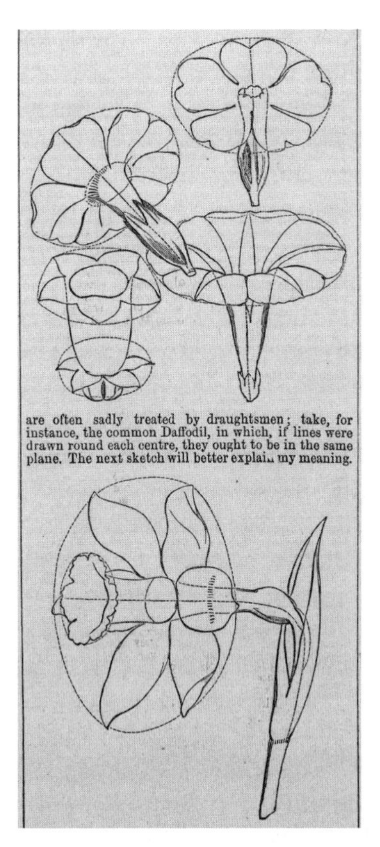
are often sadly treated by draughtsmen; take, for instance, the common Daffodil, in which, if lines were drawn round each centre, they ought to be in the same plane. The next sketch will better explain my meaning.

4.6 Drawings by Walter Hood Fitch to illustrate his introductory articles on botanical illustrations. The simplified forms allowed him to emphasize the underlying geometry. From the *Gardeners' Chronicle,* February 13, 1869, p. 165. Reproduced by kind permission of the Syndics of Cambridge University Library.

works might otherwise be . . . it betrays carelessness, which is worse than ignorance."[61] Fitch illustrated this and similar points with simple, outline drawings that were not intended to depict any particular plant. The generalized character of Fitch's drawings is inevitable in a brief article intended for beginners, but they are nevertheless a reminder of the different ways in which the artist needed to know the "typical" form of a plant before he or she could draw it (fig. 4.6).

In addition to revived and dissected specimens, Fitch used other people's field sketches to complete his images. In the case of *Rhododendrons of Sikkim Himalaya* (1849), these were mostly done by Joseph Hooker, who was delighted with the results and positively raved about Fitch's drawings when the first part reached him in Darjeeling.[62] For the later *Himalayan Plants* (1855), Fitch worked mostly from sketches by Indian artists, and Hooker explained in the book's introduction that Fitch had corrected "the stiffness and want of botanical knowledge displayed by the native artists who executed most of the originals" (fig. 10.3).[63] Indigenous artists understood neither the conventions of European botanical art nor the science that underpinned it, hence the need for such "corrections." The book's introduction also included a glowing tribute to Fitch's skills from Hooker, which referred to the earlier rhododendron plates as "a series of drawings that have been justly pronounced as unrivalled excellence in an artistic point of view. No pains have been spared by the same incomparable Botanical Artist to render the plates now published worthy of imitation."[64] Given the emphasis on copying as a pedagogical technique, "worthy of imitation" was almost certainly intended literally. The artistic conventions that Fitch had initially acquired as a boy, sitting alongside Hooker in the Glasgow lecture theater, copying wall charts and prints, were one of the many forms of expertise he brought to bear when creating his plant portraits, and their "unrivalled excellence" ensured that the next generation would copy them and reproduce the conventions they embodied.

"An Act of Wanton Impertinence"

Even Fitch's prodigious skills would have been of no use without specimens and field sketches to work from, and it was field collectors, not artists, who supplied them. Given that the various means of preserving plants did not retain all their features equally well, drawing helped capture the colors and the characteristic shapes of the living plant, as well as those aspects of the three-dimensional structure that were lost when the plant was flattened out and dried. Drawing was also an essential technique for plants that were either impossible to preserve or too large to be sent intact. No matter how limited their talents, every botanist needed to be able to draw, it was a crucial part of their scientific skill—unsurprisingly, Joseph Hooker told Wilford to "practice drawing incessantly."[65] Hooker himself had done the same aboard the *Erebus*, as we have seen; while the ship was at sea and there were no plants to draw, he had spent hours drawing marine invertebrates, telling his father that he had done so because "though I never intend to make anything but Botany a study, I do not think I can do better than I am doing; it gives me a facility in drawing which I feel comes much easier to me."[66] Lindley also told beginners to learn by taking common flowers, examining them carefully, and "making careful drawings of what you see."[67]

Botanical travelers were also encouraged to acquire a specific style. For example, Hooker told Wilford: "It is better to copy on rough pieces of paper and to sketch spoons tumblers &c in *perspective* than be idle."[68] Drawing in perspective helped capture the three-dimensional form of plants, but it was not something everyone could master. William Wilson told Hooker that "I am not surprised to find that my diagrams of the new species are objected to by Mr Fitch & yourself on account of their want of perspective." As Wilson acknowledged, this was "a department beyond my reach without a course of study that I have no leisure to pursue."[69] Wilson's comments are a reminder that no amount of botanical expertise could compensate for a lack of draughtsmanship. In describing how to draw flowers, Fitch stressed that knowing the underlying geometry was vital to getting the perspective right.[70]

Botanical drawings also needed to be more than simply accurate. Wilson complained of one botanist's work that "there is a lack of bold free outline, & general want of spirit."[71] Wilson was not alone in expecting to see both aesthetic and scientific criteria satisfied; in telling Wilford how to draw, Hooker urged him: "Be much more careful to make an accurate free bold outline than to shade. Do not hold the pencil too tight nor press too heavy or hard, but learn to make free strokes."[72] The bold outline, drawn with free-flowing strokes, was the most prominent characteristic of Fitch's work (which he derived from William Hooker's style of drawing), and Hooker's comments are further evidence that Fitch's style was taken as exemplary by Hooker and that existing drawings, not living plants, were the ultimate guide to what a good illustration should look like (fig. 10.3).[73]

As with the collecting techniques I have already discussed, what is being sought was compliance with a specific set of standards, standards that were defined in the metropolis and that sought to reimpose objectivity on idiosyncratic spectators of plants by turning them into orderly botanical observers. However, these standards involved much more than simple uniformity: the fact that botanical illustrations were expected to be "bold," "free," and full of "spirit"—as well as "accurate"—is a useful reminder that the ideal of objectivity which lies behind such pictures is not that of a photographic or mechanical objectivity.[74] Instruments such as the microscope were vital to improving one's objectivity, but it took skill to use and master them. Early in the nineteenth century, the English natural philosopher William Hyde Wollaston had patented a device he called the camera lucida, a mechanical aid to drawing which it was claimed eliminated the need for artistic ability. In the 1840s, Sir John Robison described a "Camera lucida for drawing flowers" that he had recently exhibited at the Society of Arts for Scotland. Since it would only work with small objects, he claimed that it was "particularly calculated for the accurate representation of botanical specimens, which it will render in their true proportions."[75] Robison's device was simply a sheet of plate glass, set upright in the middle of the drawing board, with the specimen on one side and the paper on the other. The user positioned his head on the same side as the specimen so that the image appeared superimposed on the paper, allowing him to simply trace the reflection. Similar devices were created for attachment to microscopes and remained in use until well into the twentieth century.

The mechanical accuracy of tracing might seem the ideal way to serve the Victorian naturalists' Gradgrind-ish desire for facts and to eliminate the idiosyncrasies of individual perception, but in practice tracing was considered inadequate. As Wilson noted, "With the help of a camera lucida anyone ought to draw correctly," which accounted for their popularity, but he went on to observe that "neither with it nor without it will an unskilful draughtsman draw beautifully."[76] In another letter to Hooker, Wilson observed that a drawing by one of their colleagues was "a poor one which I am not so much surprised at now I know that all his drawings are got up with the aid of a camera lucida applied to his microscope." Wilson added, rather haughtily, that "I suppose he is no draughtsman independent of such help" and observed that, while "the camera lucida is a very good instrument," it was no substitute for "a certain degree of skill in drawing."[77] Unsurprisingly, men like Wilson, Hooker, and Fitch were not willing to admit that the expertise they had spent years acquiring could be replaced by a simple, inexpensive gadget. No machine could confer artistic skill or scientific expertise, but more importantly, mechanization substituted an individual plant for the universal plant that was the real goal of botanical drawing.

For all his expertise, an artist like Fitch was regarded as little more than a superior kind of camera lucida by his botanical masters. In his old age, he wryly noted that "Botanical artists require to possess a certain amount of philosophical equanimity to enable them to endure criticism for as no two flowers are exactly alike, it is hardly to be expected that a drawing should keep pace with their variations in size and colour." The variability of form and color was especially obvious with orchids, which Fitch described as seeming "almost to have been created to puzzle botanists, or to test an artist's abilities, and consequently they are all the more worthy of a skilful pencil in endeavouring to do justice to them." Yet the wielder of the skillful pencil must not forget his place. An artist who struggled to "render correctly any specimen put into his hands" had better be sure that he is not looking at "any abnormal growth"; otherwise, "he is liable to have his veracity called into question." Woe betide the artist who took the rhetoric of fidelity to the object at face value and was "rash enough to represent what may be regarded as impossible by some authority who has made Orchids his speciality": "It might upset some favourite theory or possibly destroy a pet genus—an act of wanton impertinence which no artist endowed with a proper respect for the dicta of men of science would ever be wilfully guilty of!"[78] However skillful the artist and however keen his eye, only the botanist could decide what forms were or were not "possible" within a given genus. A drawing must be shaped by expert judgment on matters of plant anatomy and classification—and no machine could exercise such judgment. Yet, as we have seen, that expert judgment had been shaped as much by looking at drawings as by looking at plants.[79]

Any set of artistic conventions shape the way the world is seen.[80] Catherine Morland, the heroine of Jane Austen's *Northanger Abbey,* realizes that because she knows nothing of drawing she cannot really see the landscape; by contrast, her new friends, the educated Tilneys, "were viewing the country with the eyes of persons accustomed to drawing, and decided on its capability of being formed into pictures with all the eagerness of real taste."[81] The increasing popularity of drawing as a pastime in the early nineteenth century helped form new ways of seeing the landscape and so altered people's understanding of and relationship to the countryside around them; as Austen put it, people saw the landscape as beautiful only insofar as it formed appropriate raw material for art, if it was capable of "being formed into pictures." I would argue that botanical illustration had a parallel effect on the vision of botanists: its conventions helped the trainee learn one of the most elusive aspects of collecting—knowing what a "good" specimen was.

Hooker told Wilford that "good specimens (but not many duplicates) of all should be dried," just as he told Gunn to send "as characteristic a specimen as possible," but in each case he gave no further guidance as to what was needed.[82] In the absence of more explicit guidelines, drawing—and copying drawings, in particular—would have provided a de facto definition: a good specimen looked like a botanical drawing or was one that would make a good subject for drawing; it would need to be a mature, undamaged plant, complete with flowers and/or fruit, exhibiting those characteristics of the plant that were most useful in identifying and classifying it. What was being absorbed by botanists as they learned to draw plants was a particular way of seeing plants: rather than looking at flowers with the eyes of poets, artists, or gardeners, they learned to observe scientifically.

"To Picture to the Mental Eye"

Learning to draw plants was supposed to familiarize the novice with their structure and anatomy, yet—as numerous experts stressed—botanical drawing required a prior technical knowledge of botany. As Anne Secord has shown, although expert botanists such as John Lindley urged students to learn drawing, it was a study to be taken up *after* they had learned to recognize and classify plants: the drawings were intended to fix in the students' minds precisely those characters they had learned to use in their classifications. As a result, Secord argues that drawing was only a learning process for those who already knew what to look for.[83] Her argument is a reminder of the complex relationship between book learning and depiction, but perhaps underestimates the role drawing played in the education of botanists; men like Hooker certainly knew what they were looking for, but that was in large part because drawing had been a vital part of their training and had thus helped shape the ways in which they perceived the vegetable kingdom.

For the professed naturalist, the act of drawing was one point in a cycle of observation, mimesis, inscription, and memorization; a cycle that helped create an image, not of any one plant, but of the species itself in his mind. Whenever a naturalist named a new species, he would designate a single specimen as the "type specimen," which provided the ultimate reference point for those who wanted to know what the new species was. I argue that, before that type was designated, the naturalist would form a mental type specimen, an ideal image of the "typical" form of the species, as it would appear if it were possible to eliminate the idiosyncrasies of individual plants and specimens; this mental image was a better representative of the species than any actual physical specimen could be.[84] In forming this image, visual depiction was closely linked with the written descriptions that accompanied (or sometimes replaced) pictures: the botanist Cuthbert Collingwood, lecturer on botany to the Royal Infirmary medical school at Liverpool, argued that learning to describe plants with correct terminology allowed one "to picture to the mental eye a vivid and accurate image of an absent plant."[85] Both images and written descriptions were intended to make fragile plants permanent—transportable, fixed, and knowable—but they also reduced the variability of actual plants to the manageable confines of clearly defined species.[86] I have separated these two kinds of inscription—drawing and writing—into separate chapters for clarity's sake, but they were closely related practices. The botanists' "vivid and accurate image" of the plant, whether derived from a picture or a description, shaped both their sense of what they ought to be collecting and how they would eventually classify it—not least because naturalists selected which of a plant's characters needed to be depicted or described based on the characters, such as the flowers, that were used as the basis for classification. As a result, drawing helped shape each naturalist's concept of what a species was.

As Lorraine Daston and Peter Galison argue, the visual type represented in a botanical plate is a generalization, a characteristic form distilled from numerous distinct individuals, none of which it actually resembles. Yet, they argue that it was not an ideal, in that it does not represent some preexisting archetypal form, but is perhaps best thought of as a statistical artifact, like a graph. The typical form depicted in an image represents the mean value for the species.[87] However, we need to remember that these generalizations are shaped by where they are made and how they are made; Fitch's work required numerous specimens from all over the world, and in sufficient quantity that several could be boiled and discarded. Without Kew's resources, its imperial reach, a very different picture—both literally and metaphorically—of the world's plants would have emerged.

Learning to create and use botanical illustrations helped botanists to identify and name unfamiliar plants, so copying illustrations was training not merely in the conventions of depiction but also in memorizing the details that distinguished one species from another, the basis of classification. In the 1830s, the botanist Robert Wight (coauthor with the Hookers' friend George Arnott of the *Prodromus Florae Peninsulae Indiae Ortentalis*) commented that "the insufficiency of language alone, to convey just ideas of the forms of natural objects, has led naturalists, ever since the invention of engraving, to have recourse to pictorial delineation to assist the mind through the medium of the senses."[88] For beginners in particular, the pictures that accompanied written plant descriptions helped them to *see* the plants, to distinguish crucial details from distracting irrelevancies; the convention of depicting a single plant on a plain white background, for example, was an unspoken argument for eliminating the local effects of climate, soil, and competition from neighboring plants from one's understanding of the identity of that species.[89]

The conventions of herbarium-style botanical illustration embodied the elimination of various kinds of local knowledge, which—as we have seen—Hooker considered too idiosyncratic to be useful. The style of illustration did not merely educate collectors as to what a good specimen should look like: as we have already seen, in making specimens many of the features of the living plant were omitted or destroyed, either by choice or by necessity. The same was true of drawn plants; details deemed distracting or irrelevant were removed, and in the process, collectors were encouraged to ignore them too. The illustrations thus encapsulated not merely the ideal specimen but a particular conception of a species; they exemplified an imperial vision of the plant; what was depicted was not the indigenous product of Australia or India but a standardized, generalized, idealized plant, defined in the metropolis and exported back to the colonies to tell their inhabitants what "their" plants really looked like. The persuasive power of these beautiful pictures surely explains their presence in many of Hooker's books; even though they were expensive, potentially distracting, and encouraged superficial habits in trainee botanists, no written description could convey Hooker's conception of a species so vividly or immediately.[90]

CHAPTER FIVE

Classifying

Names have a curious potency: to know something's name—its *correct* name—is to know what it is, to have grasped something more than an arbitrary label. The power to confer names is a kind of magic. In the Judeo-Christian tradition, God created the beasts of the earth and the fowls of the air, but then he "brought them unto Adam to see what he would call them: and whatsoever Adam called every living creature, that was the name thereof." The phrase that King James's translators chose, "that *was* the name thereof," resonates with a subtle sense that Adam's names were not capricious, that he knew what the names should be. Almost as if he knew what God himself did not. Hence our sense that, despite William Shakespeare's oft-quoted opinion, that which we call a rose would almost certainly not smell as sweet if it were called the Lesser Stinkwort—at the very least, it would have been less likely to have found its way into the language of poets, philosophers, and lovers.[1] For as long as humans have been conferring names on plants and animals, they have been guided by this elusive sense that there are right and wrong names, and so naming is not to be undertaken lightly, nor is it a power to be conceded readily.

Given the imperial context of Joseph Hooker's career, it is unsurprising to find him arguing that species could be named only in the metropolis, by naturalists like himself who commanded global specimen collections. When William Colenso tried to name some supposedly new fern species, he received a sharply worded rebuke from Hooker: "From having no Herbarium you have described as new, some of the best known Ferns in the world."[2] In this and similar clashes Hooker routinely offered his vast libraries of books and specimens as an argument for keeping the right to name in the metropolis (fig. 6.1).

Accumulating specimens and expertise at the empire's center was, as numerous historians have noted, a strategy for building the power and authority of those at the center and extending their imperial reach.[3] Classifications clearly played a key role in this accumulation: their concepts were devised, implemented, and enforced by those at the center, and so they tended to reflect and advance the interests of their makers. However, by considering classification as a set of material practices, we realize how much work was required to build, organize, and maintain a large metropolitan collection. This was work that was also being done in the field, on the empire's periphery, but in that context it entailed a different set of practices. These two distinct sets of practices affected and constrained the concepts that were developed and used in each context; where you did your classification shaped both how and why you did it—ultimately, one's whole classificatory scheme.

Classification was a basic tool of Victorian natural history, the key to providing stable names that facilitated scientific communication and the efficient exploitation of the world's natural resources, but its importance ran even deeper. The apparently simple act of conferring (or refusing to confer) a name is, in effect, a pronouncement on one of the most fundamental questions in biology: what, if anything, is a species? Not surprisingly, Darwin was fascinated by classification and relied on Hooker for both information and criticism. It was a constant topic of debate between the two men, becoming a source of evidence for and against natural selection. Hooker helped shape Darwin's thinking on classification, and as we might expect, Darwin's published pronouncements on the topic drew considerable support from Hooker, not least because they provided him with a philosophical justification for his classificatory practices. Nevertheless, as we shall see, Hooker remained troubled by some of the practical implications his friend's theory might hold for the daily work of classification.

Ticketing

The first steps toward classification occurred in the field; whenever a specimen was collected, any details that would not be preserved in a dried specimen had to be sketched or noted. George Bentham told his readers that if there was no time to examine a specimen immediately, it was essential that "a note should be taken of the time, place and situation where it was gathered; of the stature, habit, and other particulars; of the kind of root it has; of the colour of the flower; or of any other particulars which the specimen itself cannot supply, or which may be lost in the process of drying."

The notes supplemented the information the specimen itself supplied, and so it was essential that they "should be written on a label attached to the specimen or preserved with it."[4] I want to argue that these notes were not a neutral record but were governed by formal conventions that, as with the conventions governing drawing, helped standardize the specimens being made, ultimately fitting them to imperial, rather than local, purposes.

In 1842, William Hooker's instructions to his collectors emphasized the importance of recording basic details for each specimen. For example, he told Dr. David Monro, who had arrived in New Zealand earlier in the year, that several specimens of each plant he collected should be dried and attached to a single sheet, and "under one of the specimens of each kind, put a ticket about the size of this [c. 20 × 50 mm, or 1 × 2 inches], with a number and the general habit on. If you have any remarks to offer, you can add it on the label. Then I preserve each label with the specimens, and send you the name of that number."[5] By "general habit," Hooker referred to the overall form of the living plant (tree or shrub, bushy or climbing, etc.) and it is evident from the size of the label that Monro was not expected to write much. There was no mention of naming plants, presumably because Monro was inexperienced and the vegetation of the district still relatively unknown; as Hooker explained, numbering specimens would allow him to educate his new collector by sending back the names.

Hooker gave similar instructions to William Purdie when he set sail for Jamaica to collect for Kew: "As much as possible & particularly if you do not know the name of a plant, take care to [?paste] a ticket with a number (1, 2 or 3 & so on) to living plants or seeds & the same number to the dried specimen that tallies with it. This is *very* necessary to be done."[6] Kew's ability to correlate the different kinds of specimens—dried or living plants and seeds—was dependent on accurate labeling, but living examples of most plants could neither be collected nor housed at Kew. In most cases a dried plant was all that was available, so it was essential that the label preserve details that might be lost in making a specimen. Joseph Hooker told Charles Wilford to "ticket the specimens well with date & place where gathered, color of flowers, odor, and other remarks, as whether wet or dry ground, wood, meadow or cultivated ground &c."[7] Such details aided both in the classification and in the cultivation of the plant.

Joseph Hooker also told Wilford that he should "practice writing full notes on plants and Botanical subjects." Hooker warned that, at first, these notes would be "too diffuse and full and too large," but experience "will soon teach how to make them full, brief, terse and yet quite intelligible."[8] The key to this economy was to use appropriate botanical terminology

Would you like some of these?

No. Campbell's Island. Antarct. Exp. 1839–1843. J.D.H.	No. Kerguelen's Land Antarct. Exp. 1839–1843. J.D.H.	No. New Zealand Antarct. Exp. 1839–1843. J.D.H.
No. Campbell's Island Antarct. Exp. 1839–1843. J.D.H.	No. Hermite Island, Cape Horn Antarct. Exp. 1839–1843. J.D.H.	No. New Zealand Antarct. Exp. 1839–1843. J.D.H.
No. Lord Auckland's Islands Antarct. Exp. 1839–1843. J.D.H.	No. Hermite Island, Cape Horn. Antarct. Exp. 1839–1843. J.D.H.	No. New Zealand Antarct. Exp. 1839–1843. J.D.H.
No. Lord Auckland's Islands Antarct. Exp. 1839–1843. J.D.H.	No. Hermite Island, Cape Horn Antarct. Exp. 1839–1843. J.D.H.	No. Tasmania Antarct. Exp. 1839–1843. J.D.H.

5.1 Hooker used printed sheets of plant labels to record brief details and encouraged others to do the same. Letter from J. D. Hooker to W. H. Harvey, [before 21 June] 1845, KEW (JDH/2/3/5). By kind permission of the Trustees of the Royal Botanic Gardens, Kew.

(see chapter 8); doing so helped make notes more precise, and standardized terms also discouraged idiosyncratic ones that took time to decipher. Conforming to appropriate standards of brevity and clarity would save Hooker a great deal of time when sorting and classifying the specimens, and so—as we have seen—he was happy to provide collectors with the books they needed to learn the terminology.

Using the correct terms and encouraging brevity were aspects of encouraging compliance; collectors were actively discouraged from giving too much detail. One way of doing this was to print standardized tickets (of an appropriately small size); a letter from Hooker to William Harvey was written on the back of a printed sheet of such labels, each of which was 30 × 75 mm (1¼ × 3 inches; see fig. 5.1). Hooker has written across the sheet's corner "Would you like some of these?"[9] John Lindley informed beginners: "In small local herbaria, printed forms of tickets are sometimes used, in which the name and all other particulars are included; such tickets should be pasted (not glued) upon the lower-right hand corner."[10] The printed tickets Hooker offered Harvey were also used when duplicate specimens were distributed to other herbaria, including those of colonial collectors, after they had been named and identified at Kew; by sending named specimens back to the colonies, Hooker was encouraging the use of standard practices and—more importantly—of the names he had decided were correct. In the preface to the *Flora Novae-Zelandiae*, Hooker had urged that "the knowledge obtained" (ultimately from metropolitan naturalists like himself) by colonial botanists should "be fixed, accumulated and distributed, by forming and naming collections of dried plants, and depositing them in private colonial schools and libraries."[11] And if standardized labels from Kew were used, so much the better.

As historians have noted, standardizing such things as methods of collecting, labeling, and preserving specimens helped exchange networks to develop and flourish.[12] Collectors were generally happy to follow such guidance, since accurate labeling—including identification of localities—was among the factors that added to the value of a specimen (including, in some cases, its commercial value).[13] In any case, by urging brevity on his collectors, Hooker was following his own practice; he generally did not make extensive field notes but relied on printed (later, mimeographed) labels featuring only the name, altitude, and locality, especially of materials distributed to other herbaria.[14]

Nevertheless, the deceptively simple and apparently generous act of returning sets of named duplicates was a way of exporting metropolitan classification schemes to the colonies, a process that ultimately devalued the colonists' local knowledge. Not surprisingly, when Hooker urged William Colenso to be briefer in his notes, the latter responded, "I quite agree with you, that *'it would have been far better to have noticed habitat, &c., on the ticket of ea. specimen,'* had I only given such information; but, in very many instances, I have made long remarks, much more than could have been written on a ticket."[15] Colenso hoped his information would convince Hooker that there were significant differences between the plants he was describing, but from Hooker's perspective, such detailed remarks, especially about habitats, were not merely irrelevant but made his global surveys more difficult—hence, his repeated instructions to use *small* specimen tickets, which left no space for "long remarks."

The concise form of the specimen label was mirrored in the published descriptions of the plants once they had been classified. As previously mentioned, Hooker eschewed the use of both analytic keys and illustrations for his *Student's Flora*, preferring to rely on "curt diagnoses."[16] These diagnoses formed the bulk of each of his floras: from the *Flora Antarctica* (1847) to the *Student's Flora* (1870), the most obvious similarity to the casual reader is that page after page is filled with brief, rather-cryptic descriptions of plants. For example, Hooker described the New Zealand fern *Lomaria* (fig. 5.2) as follows: "*Sori* frondibus distinctis, lineares, continui; capsulis demum superficiem totam pinnulae contractae operientibus. *Involucrum* marginale, scariosum, continuum, intus liberum v. dehiscens." Hooker had received a

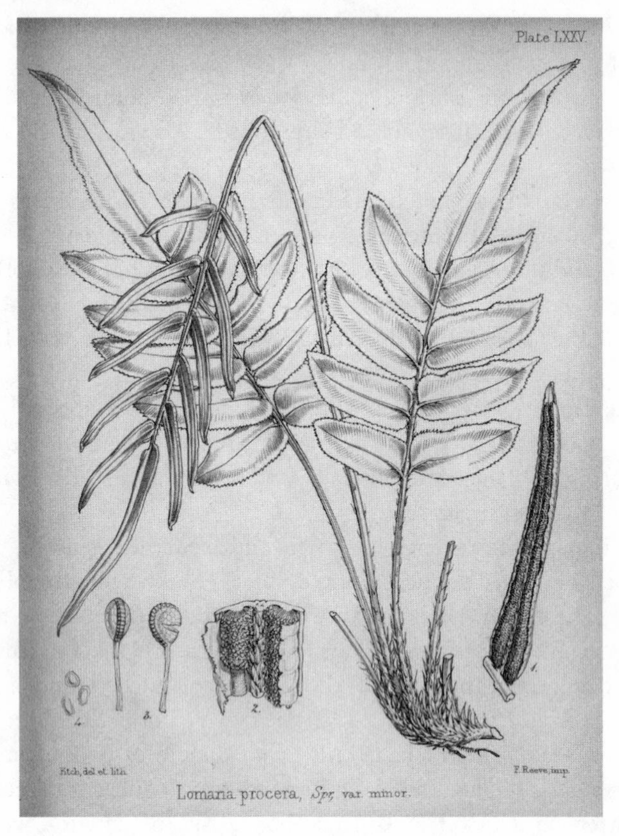

5.2 "That very Protean fern, *Lomaria procera*," an example of an uncolored plate from Hooker's *Flora Novae-Zelandiae.* By kind permission of the Trustees of the Royal Botanic Gardens, Kew.

classical education at Glasgow High School and read both Latin and Greek with ease, but the classical languages would have served as a considerable barrier to many who wished to be botanists.[17] Hooker and Bentham wrote the massive *Genera Plantarum* (1862–83) entirely in Latin, but Hooker realized that the "N.Z. or V.D.L. floras . . . must be more in English for our colonists benefit."[18] So, in the *Flora Novae-Zelandiae,* the Latin passage quoted above was followed by a broadly equivalent passage in English, although not an English that most readers would readily make out:

> A large tropical and south temperate genus of Ferns.—*Fronds* tufted, usually pinnatifid, or simply pinnate; the central ones in the tufts bearing fructification, the rest barren, with broader piling; sometimes one side

> only, or a few pinnae; of the frond only are fertile. *Sori* as in *Pteris,* but generally occupying the whole under surface of the pinnule, as in *Stenochlaena,* which however has no true involucre. *Involucre* marginal, scarious, continuous, often reaching the costa. (Name from λωμα, *a fringe;* in allusion to the scarious indusium.)[19]

This was what Hooker meant by "the simplest language that can be applied to Botany," but a high level of expertise was obviously needed simply to read such books.[20] As Hooker explained to his New Zealand readers: "it is impossible to write Botanical descriptions which a person ignorant of Botany can understand; although it is supposed by many unacquainted with science, that this can and should be done."[21]

The expertise displayed in such impenetrable prose would have impressed the colonial naturalist into accepting the judgment of the metropolitan expert, a subject I shall return to, but the diagnoses were also the pattern that beginners were expected to follow when learning to label their plants. Just as we have seen when learning to draw, learning to write like a botanist established one's credentials but also internalized the standards being set in the metropolis.

"Peculiarly Local": Colonial Classification

When we consider classification, our first thought is probably of books, great systematic treatises that are conceived in quiet dusty rooms within museums or botanic gardens. However, very few works of this kind are purely theoretical; Linnaeus's *Philosophia Botanica,* for example—which is one of the founding documents of all modern, scientific classification—contains detailed practical instructions on how to describe, measure, and preserve plants, as well as how to organize a botanical outing and where to build a greenhouse.[22] In the same vein, Hooker's introduction to the *Flora Novae-Zelandiae* began with "a few remarks to my readers in the Antipodes," in which he argued that it was only by studying botany "for himself, [that] the beginner rapidly acquires a knowledge of the structure and anatomy of Natural Orders, Genera and Species." He therefore recommended that beginners should start by acquiring "the scientific names of a few conspicuous New Zealand plants." They should then take an elementary work—he recommended Lindley's *School Botany* and *Vegetable Kingdom,* even giving their prices—from which they would learn to "refer these plants to their Natural Orders."[23] As these instructions make clear, classification was not some esoteric matter, only to be approached by the experienced adept, it was one of the first things a beginner had to tackle, because you could not do anything with plants until you could recognize them—and you could not recognize them until you could classify them.

In works like Hooker's, which used the natural system of classification (which is discussed in more detail in the next chapter), the detailed descriptions and/or illustrations of plants were organized according to the natural order or family the plant belonged to. In practice, this meant that even with the most elementary botanical textbooks, readers had to make a preliminary classification of a plant before they could discover any more about it. The first step for a beginner was to master the use of an analytic table or key. As figure 6.2 shows, keys were based on a series of questions, each of which led to another question, until the family or order of the plant was determined. Although they were designed to be easy to use, George Bentham had to acknowledge that, however carefully metropolitan experts like himself framed their analytical keys, "it cannot be hoped that they have been rendered so precise as to preclude doubt. The beginner especially will often be at a loss as to which alternative agrees the best with the plant he is examining."[24] It required considerable expertise to use one, both in mastering the terminology and in knowing precisely what it meant in practice. Just as we saw with drawing in the previous chapter, there was no possibility of simply going out into nature and looking around you; the beginner needed to take a book with him, so that he would know what he was looking at.

Metropolitan experts like Hooker had ample motivation for encouraging their correspondents to master the art of identification. Hooker complained regularly about the burden that "badly named" specimens imposed on him. As he told James Hector, director of New Zealand's colonial museum, when he was complaining about the quality of some of Colenso's collections: "as you know well naming these things is dreadful work." Such work was especially frustrating when collectors sent further examples of the same species; Hooker grumbled that "I am utterly sick of going over the same untitled species of Lecidae & Confervae etc. that come over and over again."[25] By "badly named" Hooker primarily meant that collectors had not used existing names, either leaving the specimen unnamed or (which was worse) giving it a new, and in Hooker's view, superfluous one. He therefore urged collectors to learn to identify their specimens and use existing names wherever possible.

However, the books Hooker sent sometimes disappointed their colonial readers; Ronald Gunn complained that "I have seen Loudon's Enc[yclopaedia]. of Plants but it does not meet my expectations, and is particularly deficient in the Plants of this Colony."[26] Like most of the available botanical works, Loudon's had been written in Europe by a botanist who had never visited Australasia, and the defects exhibited by European experts and their books suggested to some colonial botanists that they knew more about their own colony's plants—a suspicion that was to have serious consequences for Hooker.

As we have seen, one way to identify a plant was to match it to a picture, but the collector was more often required to match its characteristics to a concise description or diagnosis. If he was unable to do so, he could not give it a name: the collector might (through carelessness, ignorance, or lack of books) have failed to recognize a familiar species; he might have selected a different set of characters from those a book's author had emphasized; or perhaps the nameless plant was a species previously unknown to science.[27] How was a collector to know if he had merely failed to identify a well-known plant or was looking at a new species?

As their confidence and expertise grew, local collectors sometimes became aware of a potential defect in European books: the published descriptions were supposed to be used to identify living plants, but—like many of the illustrations—were often based on dead, herbarium specimens.[28] Not surprisingly, some botanists, particularly those in the colonies, argued that dried specimens were inadequate for identification. The Australian-based naturalist Ferdinand von Mueller told William Hooker: "A good chara[cter] for distinguishing Tetrathecae is offered also by the direction of the sepals in a fresh state; I adopted it in my own diagnosis of Tetrath. baueraefolia, but neither Steetz nor Scuchhardt could make use of it as they saw only dried specimens."[29] While metropolitan experts had more access to books, colonial ones had better access to the living plants. Despite having few books, Gunn soon had enough confidence in his firsthand knowledge to argue with Joseph Hooker about how many Tasmanian species of *Tetratheca* there were; he commented, "the Tetrathecas bother me a little I must confess, but I do not despair [of] proving by & bye that we have *at least* 4 species."[30] His confidence was based on the fact that the genus *Tetratheca* (a group of small flowering shrubs, some of which look a little like heathers; fig. 5.3) is found only in Australia; as both Mueller and Gunn realized, although a European botanist might have seen more pictures, books, or dried specimens, he could not have examined more living plants than they had.[31]

Despite never setting foot in Australia, George Bentham wrote an enormous *Flora Australiensis*; as one of its reviewers commented, Bentham's diagnoses, despite being based entirely on "dead materials," were written

5.3 *Tetratheca procumbens* from Hooker's *Flora Tasmaniae.* The genus *Tetratheca* is found only in Australia, which gave local naturalists confidence that they knew the plants, especially in their living state, better than any European naturalist. Reproduced by kind permission of the Syndics of Cambridge University Library.

so as to "prove useful both to beginners, and to proficients who have fresh specimens to work upon," because they were based on "characters that are both constant and discernable" in living and dried plants.[32] Yet, while it was obviously desirable to use such characters, it was not always possible, and Mueller (who had hoped to write the *Flora* himself) was typical of those who doubted that descriptions "drawn from dead materials" could ever be as useful as those based on living plants.[33]

Colenso had relied on the characteristics of living plants when he attempted to name what he felt were new species of New Zealand ferns. As we have seen, Hooker rejected these classifications, as did one of Hooker's colleagues at Kew, John Gilbert Baker, with another set of ferns. Colenso told Hooker that his knowledge of the living plants gave him an expertise Baker lacked: "I *know* my N.Z. ferns (sp. nov.) to be very distinct from those long known, & which he supposes them to be. At the same time, had I but *dried spns. only*, as you have had there, I am pretty sure I should have made the same, or greater, mistakes."[34]

Classifying living plants rather than dried ones was a major difference between colonial and imperial classificatory practices. Working with live plants helped local naturalists resolve the difficulties they often found in using European books; their inability to recognize a particular plant from the published descriptions or depictions was the result, not of a lack of expertise on their part, but of the writers' reliance on "dead materials." Many were also delighted to find that working from living plants allowed finer distinctions to be made; different forms of the plant, which the distant experts dismissed as mere varieties, seemed—when studied closely enough—distinct enough to be classified as species and given new names. This was an added attraction to men like Colenso, who hoped to see their names in print. It was pleasant to be acknowledged as a plant's discoverer or better still to have a species named after you, but actually naming a species conferred a little immortality on its namer: the abbreviation "Col." (for "Colenso") would always appear whenever the new species was referred to.

Each naturalist's degree of mobility—both geographical and social—also affected their classificatory practices. Men like Gunn and Colenso, who did not earn their living from botany, generally had little time for traveling and collecting. Botanizing had to be squeezed into a few spare hours, and so, inevitably, the part-time botanist came to know the plants that grew close to home better than any others. Of course, they supplemented their local collections through exchange and correspondence, but lack of time, money, and opportunity generally restricted their networks to other naturalists within the colony. As a result, colonial naturalists were usually forced to make the most of what was close at hand. Studying and collecting a relatively restricted set of plants over many years, however, allowed them to notice small differences: it might, for example, gradually become clear that the ferns that grew in Colenso's back garden were slightly different from the ones that grew along the nearby riverbank, even though European experts insisted they were the same species (fig. 5.4). Repeated collection and examination of fresh specimens were enough to persuade Colenso that they were separate species, hence his attempts to name them.

As will be discussed in the next chapter, there were no universally

5.4 *Paihia from the Islet Moturangi,* 1837: William Colenso's drawing of his home in Paihia, on the shores of New Zealand's Bay of Islands. The building under the two birds on the right was Colenso's printshop. By permission of the Alexander Turnbull Library, Wellington, NZ.

accepted rules for defining species at this time; each naturalist had to exercise his judgment and experience to decide how great a difference between two plants was enough to merit naming a new species. Many colonial botanists took great pride in their different kinds of local knowledge, and this shaped their de facto definition of a species. In 1852, Colenso complained to Hooker: "In your Letter, you speak of 'species, whose numbers,' you say, 'I greatly exaggerate in many instances'—take Cardamine for example: your firm belief is, that I 'have but *one* species, identical with your *corymbosa*, and that *that*, and all, are varieties of *hirsuta*.'" In response to this, Colenso claimed that, while some of these plants might be mere varieties, he was convinced "that we have more than one, and more than two, species of that Genus. *Three*, at least (if not 4) I feel pretty confident will be considered as species."[35]

One reason Colenso cared so much about *Cardamine* (or bittercress, a worldwide, mostly commercially valueless genus of rather small weeds; fig. 2.1) was that he hoped to prove that his adopted country's flora was not impoverished; he believed it contained not merely many species but many that were endemic (i.e., unique) to New Zealand. In the *Flora Novae-Zelandiae's* introduction, Hooker asserted that "compared with any other countries in the same latitude," New Zealand's was "a very scanty Flora indeed, especially as regards flowering plants."[36] He had clearly made some similar comment in an earlier letter to Colenso, since the latter complained that "I do *not* altogether think, with you, that our "Flora is a *scanty* one." Colenso acknowledged that the colony's flora was "anything but gorgeous" but was convinced that "it will be found to be peculiarly *local*" (i.e., rich in endemic species).[37] He substantiated this claim by arguing that he was better placed to judge than Hooker was: "I know that several plants are only to be found in one or two isolated spots throughout the whole island. Hence I am of the opinion, that when the immense & dense forests of N. Zealand shall have been opened, and her morasses & glens explored, her Botany will shew itself to be of a proper magnitude, and bid fair to compete with the Botany of Islands of a similar size and parallel."[38] Colenso's wish that New Zealand's flora could "compete" with those of other islands illustrates his pride in the plants of his adopted country and his desire to make them better known. There were still relatively few white settlers in the country at this time, so much exploration remained to be done, which bolstered the hope of finding many more new species. Gunn had similar hopes, telling Hooker, "I am sure that I shall yet add largely even to flowering Plants, and vastly to the Cryptogamiae."[39] He also wanted Tasmania's endemic species to be recognized by ensuring that their precise locations were published.[40]

Colonial botanists clung to the endemic species that validated their prized local knowledge. When Hooker criticized Colenso for supposedly misnaming ferns, the latter responded, "I am well aware that I know very little indeed (save from books) of the Botany of any Country except N.Z., still, I fancy, I know the specific differences of many N.Z. plants"; the less-than-subtle implication was that most of Hooker's knowledge of New Zealand's plants had been drawn from one brief visit, from books, or from collectors like Colenso. The latter's knowledge, on the other hand, had been acquired directly and laboriously, in the field. Colenso reminded Hooker that one of his ferns had been found "in *one* locality, & that in the densely wooded & untrodden centre of this Island, (& that, too, after 20 years incessant travelling with all its concomitant poking & peeking)."[41]

While Gunn made fewer attempts to name plants, he did occasionally dispute Hooker's classifications. For example, he commented: "My thing that you think a sp. of Ceratella or Trineuron you will more probably find to be Forstera. At all events it is not Compositae."[42] A month later, he wrote again on the subject of "*Ceratella versus Forstera!*" convinced that he would prove Hooker wrong once he had had the chance to "dissect some recent specimens & report the results to you"; by "recent specimens" he meant freshly picked ones, which might retain characters Hooker could not see in his dried specimens. Nevertheless, Gunn—in properly gentlemanly fashion—acknowledged "that my authority in these matters is worth nothing—in fact my opinions are not worth the ink in which they are written."[43]

"Carefully Formed, Perfectly Kept, and Correctly Named": Metropolitan Classification

For the colonial naturalist, the main point of classifying was to know your local plants, to recognize the familiar, and—with a little luck—discover some new species. Meanwhile, those at the empire's center had different goals, of which the most important was to give clear, unambiguous names that improved communication within the empire. Hooker told his colonial readers that at the Great Exhibition of 1851, where he had been a juror, "the vegetable productions of our Colonies" were "almost invariably so badly named, that the often valuable information given with them, and collected at great cost and trouble, is, in most cases of novelty, useless in England."[44] Among the "vegetable productions" Hooker complained of were some sent by one of his own collectors, Richard Taylor, which were listed in the exhibition's official catalog simply as "Box of sundries, list enclosed, Rev. R. Taylor."[45] Taylor's manuscript list of the items he had sent reveals that he had given details, including scientific names, of only some of the products; other descriptions were distinctly sketchy, such as "[item] XIX. specimen of the Toatoa bark. [Item] XX d[itt]o. of the Hinau bark both used as dyes."[46] At a time when much of Britain's wealth was generated by the trade in plant-based products, this lack of precise names meant that the presence of potentially valuable plants in the colonies might be overlooked.[47]

In addition to accurately cataloging the empire's natural resources, metropolitan naturalists were interested in the closely related project of comprehending the distribution of the world's plants, in part to facilitate their exploitation (a topic I shall return to). To further these vital imperial projects, Hooker devoted much of his life to naming and renaming plants from the colonies, removing duplicate names (synonyms), and abolishing what he thought were spurious species. This was work that could only be done in one place: the herbarium (fig. 6.1).

Many nineteenth-century botanical manuals told their readers how to create a herbarium; John Lindley's *Elements of Botany* (1841; fig. 6.2), which Hooker used aboard the *Erebus,* was just one of dozens of such works.[48] Lindley included a long list of instructions on how to collect, examine, draw, name, dissect, dry, and label plants, explaining that "a collection of dried plants is called a *Herbarium.*" Lindley went on to observe that, "if carefully formed, perfectly kept, and correctly named, [a herbarium] is as invaluable to a student as it is indispensable to a Botanist."[49] Forming and keeping a herbarium was hard work, and while naming and classifying were central to metropolitan herbarium building, as they were to colonial field collecting, a strikingly different set of skills and materials were required.

Lindley's detailed instructions for creating a herbarium were typical of those given in similar manuals and—as with collecting instructions—they reveal the craft involved. Various materials were required that might be hard to obtain in the colonies: several different kinds of paper and card; purpose-built wooden cabinets; and specific glues and poisons. The latter were used, as Lindley explained, to protect specimens from insects; other herbaria, such as that at the British Museum, used masses of camphor (mothballs) to the same end.[50]

The British Museum's James Britten explained that once protected, the specimens had to be glued onto sheets: "a mixture of gum tragacanth and gum-arabic (the former dissolved in the latter), in about equal parts, is used for this purpose," while "very coriaceous [leathery] specimens are secured with glue" and "the stems and ends of branches are usually also secured with straps," which were made of paper and gave additional support.[51] Lindley also suggested straps and explained how to attach envelopes for seeds and loose materials.[52] In addition to the right materials, a botanist would need considerable experience to know how each different kind of plant should be treated.

Herbarium building used a lot of paper, which, as we have already seen, could present considerable problems for colonial collectors. First, there was the paper onto which the specimens were actually fixed, which needed to be "rather stout, and large enough to admit the full representation of the species"—that is, to hold numerous specimens, showing every part of the plant; Britten bemoaned the "wretched scraps with which some collectors content themselves." For reasons that will become apparent in a moment, it was also essential that every species in the herbarium be mounted on the same size sheet: the Kew herbarium's were 16½ × 10½ inches (419 × 262 mm), while the British Museum's were slightly larger (17½ × 11¼ inches; 445 × 286 mm), but Britten acknowledged that "the former will be found amply sufficient for our purpose."[53]

For a large herbarium, an enormous number of such sheets would be needed, since as Lindley told his readers, they should "never put more than one species on the same half-sheet." Bentham gave identical instructions almost twenty years later, insisting that "no more than one species should ever be fastened on one sheet of paper."[54] Once the specimens were mounted, the sheet could be labeled, "in some convenient spot near the specimen itself," with the species name, the location where it was gathered, and other relevant details (figs. 2.1 and 5.1).[55] The sheets of separate species were then inserted into folders made from "sheets of stout brown paper" (yet another kind of hard-to-find paper). The brown folders represented the genera: one folder per genus, containing a number of specimen sheets. Britten concurred, suggesting: "Each genus will require a separate cover, which may well be of some stouter paper than that on which the plants are mounted; the name of the genus should be written at the left-hand corner."[56] Arthur Henfrey wrote virtually identical herbarium-building instructions: specimens should be glued to white paper, which "may be half sheets of white cartridge paper of folio size, and all the species of a genus should be placed in a sheet of the same, having the name of the genus written at the left-hand corner. These sheets may be tied up in bundles, all the genera of an order together, or simply be laid in the drawers of a cabinet."[57]

Finally, "wooden cabinets are constructed, with shelves, on which the covers can be placed according to their natural orders."[58] Britten illustrated the type of cabinet used in the botanical department of the British Museum, adding that "at Kew the cabinets are somewhat similar, but their height is greater and the shelves are fixed."[59] (Keep these seemingly trivial differences between the practices of Kew and the British Museum in mind; as we shall see in chapter 10, they were actually of considerable importance.)

Apart from emphasizing the dexterity and specialized materials required, the most interesting aspect of these instructions is that they show a classification being constructed, quite literally, from the specimens. Classificatory systems were hierarchical, with each group containing those below it: plants were united by common characters into species; species that shared common properties were united in a single genus; genera into orders; and orders into families.[60] The herbarium was a physical manifestation of this hierarchy: a species was both an abstract concept and a sheet of specimens; a genus was a broader concept and a brown paper container for species; a family was a group of genera that shared both affinities and a drawer in a polished wooden cabinet in Kew.

However, the most important benefit of arranging the herbarium as these instructions describe was that it allowed the classification to be changed easily. If a botanist combined two species, one sheet had its name crossed out and another written in; with a scratch of the pen, what some botanists thought of as a natural kind simply ceased to exist. With another quick scribble, a new species was born. Plants were reclassified by simply moving them from one folder or drawer to another (hence the need to put every species on the same size sheet of paper). The world existed in the herbarium, but in a more manageable and orderly form than it existed in the field; every classifier played Adam's role, giving names to the plants God created, but in the manageable, miniature world of the herbarium, the systematist *was* God.[61]

However, the classifier's godlike role in the herbarium came at a cost; herbarium botanists had to work continuously to sustain the order of their world. Every new species meant a new sheet had to be created and kept track of, every revision in the classification meant relabeling and refiling sheets, and every specimen that arrived had to be compared with the hundreds of thousands of existing specimens to check what it was. For example, when a collection made by Michael Edgeworth, a Bengal-based civil servant, arrived from India, Hooker wrote despairingly to Bentham, "What shall be done about them? They are *most wretched scraps* for the greater part," in such poor condition that "they can hardly be called specimens." "The whole coll forms a bundle about a foot high of large newspaper sheets with one scrap or specimen as the case may be wandering over its acre of paper." Edgeworth's choice of newspaper would have been a problem: it was cheap paper, made mechanically from wood pulp, which was acidic and therefore yellowed and decayed much faster than the high-quality, handmade papers that metropolitan herbarium builders used. Moreover, single, isolated specimens made the herbarium naturalist's work harder, since they did not allow the variability of the species to be assessed, nor did they provide additional specimens for dissection. Through a lack of materials, skill, and care, Edgeworth had produced an almost valueless collection; as Hooker noted, "There are many marked 'new species' & they may be so, one could not tell."[62] Kew employed a Mr. Stevens as herbarium assistant, to sort through new collections and salvage what he could, but in the case of Edgeworth's "scraps," Hooker conceded, "I ought to go through them myself & so shall I think. My father grudges Stevens' time on such rubbish & so do I."[63] By contrast with such poor collections, every shipment of properly preserved and labeled specimens, free of unnecessary duplicates or misidentified plants, that arrived from expert collectors like Gunn and Colenso would have deepened Hooker's gratitude to them—and reminded him how rare and valuable such collectors were.

As he labored over his plants, Hooker complained to Bentham that "the jumble of sheets & specimens is frightful—I toil on & on & to little effect."[64] He later observed that "when species have been founded in error; this generally arises from their authors having imperfect specimens, or too limited a series of them."[65] I will discuss what a "series" was and its implications below, but we get a good sense of the problems caused by imperfect specimens from Hooker's letters to Bentham, which describe some of the day-to-day labor of classification. Hooker regularly asked for more specimens, to allow for dissection and perhaps soaking, either of which—as we have seen—would destroy the specimen. However, there were even more requests for *better* specimens when, for example, Hooker found that on some of his father's herbarium sheets there were so many specimens glued down together that, although they were numbered, "I can scarce say to which of the numerous specimens thus brought together the number belongs," and hence he could not relate the accompanying notes to the specimens and had to ask Bentham for better-labeled examples.[66]

As we have seen, creating and maintaining a large herbarium was an expensive business, requiring specialized equipment and skills. Not surprisingly, both were more common in the metropolis than in the colonies, which was one basis of Hooker's claim that naming could only be done by the metropolitan expert, who had a large enough herbarium to give him a broader vision of the plant world than any local naturalist could aspire to. However, this wider picture came at the expense of the details; the kinds of fine-grained classification practiced by some colonial naturalists could not have been attempted in the metropolitan herbarium. Even if the minor varieties were discernible in dried specimens and assuming such varieties had not already been destroyed by dissection or to make drawings, splitting numerous species into three or four—as Colenso constantly did—would have massively multiplied the number of herbarium sheets; even Kew could not have held the volume of specimens required. The practicalities of maintaining a large herbarium encouraged its owner to keep the number of species within manageable bounds; as a result, the keepers of large herbaria were usually prejudiced against any attempt to multiply species. In effect, the size of the collection shaped its owner's sense of what a species was.

Lumping and Splitting

Hooker was well aware of the differences between the kinds of classification practiced at the empire's center and periphery. In the *Flora Novae-Zelandiae* he had commented that "the local botanist looks closer, perceives sooner, and often appreciates better" than "the botanist occupied with those higher branches of the science." And he claimed, slightly disingenuously, that "there is no doubt but that the truth can only be arrived at through their joint labours; for a good observer is one thing, and the knowledge and experience required to make use of facts for purposes of generalization, another." Yet despite his tactful praise, Hooker nevertheless argued that local botanists' detailed knowledge led them to pay too much attention to "minute differences" that "when long dwelt upon . . . assume undue value." As a result, "the general botanist must always receive with distrust the conclusions deduced from a few species of a large genus, or from a few specimens of a widely distributed plant." He concluded that:

> I have been led to dwell at length upon this point, because I feel sure that the New Zealand student will at first find it difficult to agree with me in many cases, as for example on so protean a Fern as *Lomaria procera*, whose varieties (to an inexperienced eye) are more dissimilar than are other species of the same genus. In this (and in many similar cases) he must bear in mind that I have examined many hundred specimens of the plant, gathered in all parts of the south temperate hemisphere, and have found, after a most laborious comparison, that I could not define its characters with sufficient comprehensiveness from a study of its New Zealand phases alone.[67]

Since the New Zealand forms of this fern appeared "more dissimilar" from one another than some of the accepted species of the genus did from each other, New Zealand's botanists might reasonably conclude that the local varieties they knew so well should be classified as separate species. Hooker disagreed. By obtaining "many hundred specimens" from "all parts of the south temperate hemisphere" and subjecting them to "a most laborious comparison" he had uncovered a continuous series that united the apparently distinct New Zealand forms. This was what Hooker meant about erroneous species being founded on "too limited a series"; only a large herbarium could allow global comparisons.

Hooker explained how he worked in a letter to the entomologist Henry Walter Bates in which he noted the advantages of working with small specimens such as insects: "I was much struck in my small experience [of entomology] with the ample facility there was for working out questions of variation, owing to the readiness with which you can compare many specimens at one *coup d'oiel* [glance] & sort them in various [?diverging] series—with plants this is almost impossible—except by covering the floor or very large tables, which is the way I adopt."[68] In his herbarium, Hooker could see the plant world spread out at his feet. This ability to assess variation at a glance, by covering the floor or table with specimens, lay at the heart of the herbarium's power and demonstrates how central the practice of herbarium building was to Hooker's taxonomic method. Once nature had been reduced to mounted, filed, managed, and classified specimens, it lay open to the botanist's gaze, allowing him both to see patterns that were invisible to the field-worker and to ignore those details that threatened to obscure the perceived pattern.

Hooker's comments on *Lomaria procera* might have been intended for Colenso personally; when Hooker had classified the fern as a single species (fig. 5.2), Colenso had remarked that "I well knew that you would have difficulty with that very Protean fern, *Lomaria procera;* it has for years puzzled me. Notwithstanding, I believe, that there are several vars. of this sp.,—good, standard, well-marked & common vars., of which I have sent you specimens."[69] Colenso classified these "good, standard, well-marked" varieties as species in his unpublished *Glossarium Botanicum: Novae Zelandiae* (1834–41). Under "*Lomaria*" Colenso has sixteen species, with "W.C." (i.e., William Colenso) given as the author of four of them.[70] However, Hooker recorded just eleven species, with the others being reduced to minor varieties. And although Colenso was credited for his collecting efforts, none of his names were used by Hooker.[71] Here, as in other cases, Hooker suspected that the colonial botanists' ardor led them to inflate their floras; a decade earlier, while working on the *Flora Antarctica*, Hooker had confessed to Bentham that "I have a natural love of uniting old species which I may carry too far," but he persevered with this practice not least because he was "most anxious to avoid the stigma of having made species for the purpose of attaching undue importance to the Flora of the S. Pole."[72]

Given that Hooker required a large herbarium but a manageable number of species, he urged colonial naturalists to stop naming new species and instead search for the intermediate forms that would unite them; after rejecting Colenso's new species of *Phormium*, Hooker claimed that "if you endeavour to unite your *Ph.* [species] by intermediate forms you will do so too."[73] This endeavor was central to Hooker's approach to classification; even though the New Zealand forms appeared distinct, the apparent gap could be bridged using forms found elsewhere, and if it *could* be bridged, it *should* be, thus reducing the "frightful" jumble of sheets and specimens he groaned under.

When Colenso dissented from Hooker's classification of *Cardamine* and described four potential species instead, he added: "Your further observation upon making all into *one* 'by extensive Examination of hundreds of specimens from all parts of the world,' I have already expressed my dissent from, in my former Letter; perhaps however, neither philosophically, or satisfactorily." [74]

Hooker was a taxonomic "lumper," who constantly united numerous varieties by classifying them as a single species. By contrast, local naturalists were often "splitters," subdividing Hooker's broad species into more precisely defined ones, often by arguing that only living plants could reveal the numerous varieties of a species. Since Hooker relied on herbarium specimens to unite species, he could not possibly have agreed that only living plants revealed the intermediate forms.[75] In the case of *Cardamine*, for example, Hooker had told Darwin: "the little *Cardamine* or cress I prove by a comparison with about 50 states of it, running through the whole continent of S. Am., to be the same as the most common Europaean weed C. hirsuta." [76] Clearly, "50 states" from all over South America could only have been examined in the herbarium. Hooker justified his decision to unite various *Cardamine* species by claiming that they were linked "by intermediate forms of habit, of foliage, of inflorescence, and of pod." There was no break between the varying forms of each species sharp enough to warrant a new specific name, and so, Hooker firmly asserted, "it ranks according to my philosophy as a variety and not as a species." [77]

However, more things were dreamt of in Hooker's philosophy than abstract theoretical matters. In the introduction to the *Flora Indica*, Hooker had argued that it was "imperative, *on philosophical grounds as well as those of expediency*," to reduce the numbers of species.[78] A close examination of Hooker's classificatory practice reveals that often what was philosophically sound was also expedient. Philosophically, those forms of the plant that were not clearly distinguished from one another were varieties, not species; but this was also useful, since varieties did not get names and hence did not require a separate herbarium sheet, catalog number, or published description. Without examining classifying practices closely, it would be natural to assume that an individual's conception of a species determined his herbarium practices.[79] In reality, the reverse was regularly the case, and botanists' species concepts can often be correlated with geographical location and herbarium size; colonial botanists were more likely to be splitters, whereas metropolitan naturalists were usually lumpers.

Kew's herbarium made Hooker's classificatory method possible, allowing him to defend his broad definition of a species, but practical matters—such as the sheer size of the herbarium—undoubtedly encouraged the adoption of broadly defined species.[80] First, the quantity of specimens gave metropolitan botanists a practical reason for wanting to reduce species.

Second, the range of specimens revealed so much variety that it became impossible to identify unambiguous distinctions between certain forms, and therefore to differentiate species. As we have seen, these difficulties were exacerbated by the destruction of some of the living plant's characteristics, which ensured that some distinctions simply could not be made from dead materials. This last reason in particular suggests that Hooker's broad conception of species arose partly from making a virtue of necessity: since he could not use the characters of living plants, reasons had to be found to show they were not needed.

Hooker's predilection for lumping seems almost to have been forced upon him by the size of his herbarium. This might seem like hyperbole, but it is striking that other metropolitan naturalists, with large herbaria and similar ambitions, frequently shared his philosophy; while colonial naturalists and those who worked primarily with living plants tended to be splitters.

In 1855, Hooker's friend Asa Gray, who ran the Harvard University herbarium and was another lumper who approved of his colleague's efforts, told Darwin that "in boldly reducing nominal species Joe Hooker is doing good work."[81] Two years later, Arthur Henfrey (by this time professor of botany at King's College, London) argued that there were too many spurious fern species because "systematists have far overrated the distinctive marks"; if these were assessed properly, the number of species would be "much smaller than is usually set down in books." Like Hooker, Henfrey argued that those who create these false species often worked from "a limited number of specimens," but that if they were to compare their apparently distinct plants "with a complete collection containing all the intermediate types," they would soon realize their mistakes.[82]

In 1852, the *Westminster Review* claimed that "there are dullards, however, among botanists as among wise men of all categories," who were "slow to perceive essentials." Such "nigglers and hair-splitters" concentrated on minor differences "without concern about their relative value." To compound this problem, there were also what the reviewer called "fast" botanists, "anxious for the rather dubious glory of becoming the namegivers of new species at any price."[83] The *Westminster*'s piece was written by Edward Forbes (fig. 9.2), and the similarity between his views and those of Hooker and Henfrey had a lot to do with their common circumstances.[84] Overwork, ill health, and constant financial problems contributed to Forbes's difficulties in earning a living, but—like Hooker—the main problem he faced was the shortage of potential careers.[85]

Forbes contrasted the splitters with those who possessed "abundant knowledge and sound judgment." When these qualities were combined with the ability to distinguish "between important and unimportant details" and "a far-seeing spirit of generalization," "the systematist becomes a philosophic naturalist." He offered as an example of this rare beast the author of one of the books he was reviewing, "Dr. Joseph Hooker, a young, indefatigable, enthusiastic, truthful, and thoroughly-trained botanist," who despite his youth had achieved the ability "to perceive and develop new relations, and determine the true value of those that have long been noted." According to Forbes, Hooker had done so much because he was the "son of Sir William Hooker, the illustrious director of our national gardens at Kew" (which, as his readers would have recognized, was one of the country's few government-funded scientific institutions), where he had benefited from "constant association with all the ablest botanists of our time," combined with "the inappreciable facilities afforded by a library and herbarium of pre-eminent completeness." In addition, Hooker had "personal experience in the exploration of foreign countries"; many of the *Westminster*'s readers would have known that Hooker had gained that experience on Ross's celebrated voyage and had therefore been paid by the government to be a naturalist on a British naval vessel.[86] Forbes was not merely advancing his friend's career, at a time when Hooker still lacked a permanent position, but was also advancing the claims of all those, including himself, who wished to be philosophical at the public's expense. The campaigning tone is readily divined in the periodical reviews written by Forbes, Henfrey, and others, but it was also a goal of Hooker's essays; as we shall see, advocating a classificatory method that relied on large, well-funded collections was one reason for supporting a broad species concept, and the scorn Hooker and his fellow lumpers poured on the splitters was in part a demand for more publicly funded scientific positions.

Bentham, Lindley, Forbes, Hooker, and Henfrey were all metropolitan naturalists striving to pursue their calling as philosophical naturalists; all favored a broad species concept and condemned splitting. Their reviews and essays were part of their campaign for better-funded science, which included demands for more public institutions to house Britain's natural history collections, along with well-paid positions for systematists within them.[87] By emphasizing the shortcomings of existing species work—dominated by hairsplitters and fast botanists—the philosophic lumpers were arguing the need for broader comparisons, which could best be made using bigger, publicly funded collections.

"The Type Is a Phantom"

In the *Flora Nova-Zelandiae,* Hooker wrote that "I have proceeded on the assumption that species, however they originated or were created, have been handed down to us as such."[88] Nevertheless, he felt it necessary to "remind the New Zealand reader that the word [type] is often used in a vague and unphilosophical manner." These vague senses included the "individual of a species which was first cultivated, described, figured, or collected" or simply "that form which is most abundant in the neighbourhood of the writer"; in each case he was objecting to local knowledge, unsupported by the kind of broader survey that relied on a metropolitan herbarium. Hooker concluded his comments by observing that, "the fact is, we have no clue whatever to the originally created typical form of any plant," and so he announced that from a practical point of view, "the type is a phantom."[89]

When Hooker wrote these words, in 1853, he had known of Darwin's theory of natural selection for almost a decade, and it is more than likely that Darwin's ideas were on his mind when he wrote "however they [species] originated or were created." However, as we shall see in later chapters, the issue of *how* the typical form of each species had been "originally created" had no immediate impact on the way Hooker practiced classification. His main concern was always with how to define the phantom type in a way that made it stable enough to serve its traditional classificatory role.

Given that naturalists could not be sure what the "originally created typical form of any plant" might have been, Hooker argued that "for practical purposes we must assume the most common form to be the most typical." However, his sense of "common" was epitomized by Walter Fitch's pictures of plants or by a published diagnosis: the type was not an individual plant (certainly not the most common or best known in any specific neighborhood); it was a somewhat idealized composite of the most important features used in classification. And, of course, those features were the ones Hooker found most useful in his approach to classification. The type could—by definition—not be found in any local botanist's backyard, hence, as Hooker acknowledged, the "extreme difficulty in combating local prejudices." A "general botanist" must look at the wider picture, "but there are local observers who cannot be brought to see things in such a light." Despite their detailed knowledge and local expertise, such parochial observers take "the reduction of supposed local types to varieties of better known and wider spread plants, as little short of an insult to their understandings, and a slight upon the natural history of their village or island."[90]

Such comments illustrate the persuasive purpose of Hooker's introductory essays. They were written to cajole recalcitrant local naturalists into seeing things his way. Hooker's lumping reduced the size and diversity of New Zealand's flora, which, as we have seen, upset Colenso, who protested that "***seeking*** to unite 2 species" was a mistake; anyone could take the least characteristic forms of a species, "their aberrant . . . sportive, or hybrid forms," and use them to unite "almost any 2 (or more) species of any large genus." But such an approach, Colenso feared, would "certainly end in this—the breaking up of all species & genera." He quoted Alexander Pope's *Essay on Man* (1732–44) in his defense: "And middle natures, how they long to join, / Yet never pass th' insuperable line!" He did acknowledge, however, that such an argument, "you will say, is vastly more poetical than philosophical."[91]

In response to such complaints, Hooker argued that a New Zealand resident "may perhaps have to travel far beyond his own island to find the link I have found" (an option that was, of course, not available to someone like Colenso, whose position in the church kept him in New Zealand). Hooker concluded that the fact that varieties remain distinct in some locations but disappear in others "proves that . . . no deduction drawn from local observations on widely distributed plants can be considered conclusive. To the amateur these questions are perhaps of very trifling importance, but they are of great moment to the naturalist who regards accurately-defined floras as the means of investigating the great phenomena of vegetation."[92] The less-than-subtle implication being that those who insisted on using local knowledge to split species were unable to investigate the philosophical problems raised by "the great phenomena of vegetation."

Hooker's efforts to protect his notion of the phantom type from being undermined by the local splitters were often strident. He complained that, whenever an inexperienced botanist discovered an unfamiliar plant, "it rarely enters into his head to hesitate before proposing a new species": "Hence the difficulty of determining synonymy is now the greatest obstacle to the progress of systematic botany; and this incubus unfortunately increases from day to day, threatening at no very distant period so to encumber the science, that a violent effort will be necessary on the part of those who have its interests at heart, to relieve it of a load which materially retards its advancement."[93] A "violent effort" was exactly what Hooker and Thomas Thomson undertook when preparing the *Flora Indica*. Hooker described their labors to Bentham as "wild & exciting work, the species go smash smash every day," but although he was clearly enjoying "smashing" bad species, he assured Bentham that his broader species resulted "not so much from uniting varieties (which have led to founding bad species)" but from simply comparing a greater variety of specimens than the local botanists had access to, or, as he put it, "from putting side by side things that no one who ever saw these placed together would even have thought of distinguishing."[94]

This claim illustrates the complex relationship between Hooker's species concept and his interest in distribution: a broader species concept resulted in fewer species and fewer synonyms, which made it simpler to see how plants were distributed across the globe; yet, as Hooker's comments on *Lomaria* and *Cardamine* illustrate, surveying geographical variation was central to achieving that reduction.

Hooker was so passionate about combating the splitters that he claimed it was "my fate to destroy species as I go on & the more carefully I examine the more to fell," implying he was helpless in the face of the evidence.[95] But elsewhere he acknowledged a degree of subjectivity in how one classified. When Thomson was working on the genus *Ranunculus* (buttercups) for the *Flora Indica*, Hooker admitted to Bentham that, "if we three had independently to do that genus from these same materials we should have produced three monographs at total variance with one another."[96] What Hooker could confess to a fellow metropolitan gentleman like Bentham (but never to a colonial hairsplitter) was that, while some groups had clearly defined species, others did not. In highly variable groups, species definitions had to be made on some basis other than the unmistakable characteristics of the plants. As Hooker acknowledged when he defended his decision to lump *Cardamine*, it was "one of those plants of which you may make 20 species or one," and "if you make 2 you must make many more."[97] However, this does not imply that any arbitrary number of species would be equally valid; Hooker was asserting that, if one chose to pay attention to minor details, such as those visible in living plants, one ended up with twenty species, whereas his desire to tackle the philosophical problems of distribution meant he could only use herbarium materials and thus had to ignore some details, with the happy result that he ended up with just two species.

"For the Benefit of the Colonies"

Hooker regularly bemoaned the "chaos of synonymy which has been accumulated by thoughtless aspirants to the questionable honour of being the first to name a species."[98] His attacks on the splitters and the "incubus" of synonymy recur in many of his essays, a further reminder that he had to rely on more or less friendly persuasion to obtain compliance with his standards. Encouraging collectors to learn plant identification, and thus the rudiments of classification, evidently encouraged some of them to try naming species themselves; from Hooker's perspective that was preferable to the "dreadful work" of sorting through badly labeled duplicates.

Yet, even if it had been possible to entirely separate collecting and classifying, it is unlikely that Hooker would have done so, because teaching collectors how to identify plants was also his best opportunity to persuade them he was right about classification. The now-familiar exchanges of gifts, flattery, and friendship were part of this persuasion. Naming or not naming new species is the most obvious way in which Hooker tried to impose his species concept, but the process did not end merely because he had published (or not published) a name. A classification only works in practice, and Hooker had to induce collectors to use his, since they had the choice to either accept or reject his new names in their future collecting and labeling. Even if they accepted that classificatory standards could be set only in the metropolis, collectors could still choose whether or not to comply with them.

Each time Hooker sent one of his collectors a book, a journal, or the latest installment of his current flora, he was sending them his revised classificatory scheme and encouraging them to conform—by using it when they next collected. Colonial botanists could reject Hooker's approach to classification by, for example, sending further specimens of variant forms and labeling them as new species. However, in many cases, they simply accepted his judgments and tried to avoid repeating what he had called "mistakes" in their earlier collections; even the cantankerous Colenso accepted Hooker's verdict on most species. However reluctant they were to concede the right to classify, a collector who wished to be seen as competent and knowledgeable felt constrained to act in accordance with Hooker's guidance and would therefore take his published names as a guide as to which plants were new—and thus worth collecting—while ignoring those forms Hooker had dismissed as mere variants. In doing so, the colonial botanist had effectively accepted Hooker's species concept, his classifying method, and ultimately his authority.

Hooker's *Handbook of the New Zealand Flora* (1864–67), one of the "colonial floras" that Kew produced in the 1860s, was a particularly useful vehicle for exporting his metropolitan classification back to the colonies.[99] The *Handbook*'s classifications functioned as what the Victorians christened a "metrology," a system of universal standards and measurements that helped those in the metropolis impose their ideas on distant places and peoples.[100] Since the *Handbook* was much cheaper and thus more widely read than the original *Flora*, it was the ideal tool with which to distribute

Hooker's classification. Although these colonial floras were supposedly produced "for the benefit of the colonies," the *Handbook* had considerable benefits for Hooker as well, not least because his long-standing correspondent Charles Knight induced the New Zealand government to pay Hooker £600 to write the flora.[101] This subsidy allowed Lovell Reeve, Hooker's publisher, to sell the new work for just 16 shillings and—as intended—this helped the book sell well in the colony. By comparison, the *Flora Novae-Zelandiae* cost more than sixteen times as much as the *Handbook* (£13.2.6 colored, £9.5.0 plain). The 100 copies of the *Handbook*'s first volume, which the New Zealand government took in 1864, had all been sold before the second appeared in 1867, whereas Reeve only had sixty subscribers for the *Flora Antarctica*—ten short of the number he claimed to need to break even—and he had still not sold all 250 copies of the *Flora Novae-Zelandiae* eleven years after it had been completed.[102]

The low price of the colonial floras was one of their major attractions, as the *Natural History Review* had commented on an earlier volume in the series. Their reviewer noted that, when complete, the flora of the British West Indies (also published by Reeve) would "contain descriptions of no less than 3000 tropical plants, at the cost of 30*s*. This is undoubtedly by far the most inexpensive systematic botanical work ever published." [103] When the New Zealand *Handbook* appeared, Knight wrote to Hooker congratulating him on its success and told him he should be proud "to be, as you are, the first to lead the young people of a distant colony to the cultivation of an interesting and useful science." [104] And for Hooker, the *Handbook* also enabled him to "lead" the botanists of the "distant colony" toward his names and thus to his whole classificatory system. In addition, the *Handbook* contained detailed advice on how to collect, which helped ensure that future collectors would gather specimens that conformed to metropolitan needs. Hooker was thereby ensuring that, whatever the colonial botanists' intentions, their specimens would be made according to standards that did not permit hairsplitting.

In *On Nomenclature* (1883), Colenso noted with pride that he had been a member of the New Zealand House of Representatives when it commissioned Hooker to write the *Handbook*. Yet with hindsight, he felt that, "while its publication has been of service," it had been "not been unmixed with evil," since because of it cheapness and ready availability, "some in New Zealand have set themselves up for Botanists!" The *Handbook*'s clarity had convinced these newly minted botanists that "it is a very easy matter to name our N.Z. Ferns," a mistaken conviction which meant "our Cryptogamic Flora in particular—the chief botanical glory of New Zealand!—has suffered the most in its nomenclature." Colenso argued that these botanical *arrivistes* should pay more attention to the *Handbook*'s author, "Sir J.D. Hooker," who "warns his readers, and that frequently, against attempting great and new things, without, at all events, much study of those larger works and microscopical research, and a careful comparison of species with species,—those of New Zealand with those of foreign countries."[105] Although Colenso had still not accepted all of Hooker's opinions, he had accepted his geographical method. When Colenso's pamphlet arrived on Hooker's desk, what would surely have pleased the latter most was that Colenso had finally accepted that no local botanist could attempt "great and new things," at least not without "a careful comparison of species with species."[106] Despite all their arguments, forty years of actually using Hooker's classifications in the field had largely won over one of his most refractory collectors.

"Reducing Species Will Go on Apace"

At first glance, Hooker's decision to overrule Colenso's fern names seems to encapsulate an essential aspect of the colonial scientific relationship: the metropolitan expert using his position—both geographical and social—to overrule the distant colonial. However, as we have seen in earlier chapters, Hooker was not born with this ability, nor were men like Colenso passive in their dealings with the metropolis. Despite Colenso's respect for Hooker, he never conceded the right to name plants; as we shall see, in 1898, fifty years after that first meeting, the eighty-seven-year-old Colenso was still trying (and failing) to get his new names for ferns accepted. Clearly, no interpretation of how science is practiced in an imperial context can be adequate if it makes the periphery passive.[107]

Despite being isolated from Australasia's plants and having less first-hand knowledge of them, Hooker was gradually able to make himself central and the colonial collectors peripheral.[108] As his broad species definition was distributed via his publications and letters, it helped create both the center and the periphery; whether local botanists chose to accept or reject Hooker's classification, it eventually defined them as peripheral.

Attending to the ways in which the locations of center and periphery were negotiated makes it clear that, while Kew's herbarium was crucial to Hooker's eventual ability to overrule his colonial collectors, the accumulation of specimens alone does not explain its working. Specimens were constructed so as to ensure that only some of the living plant's features were preserved and thus could be used for classification; the features that were preserved were defined by the skills (including a training in preexisting classifications) of those who made them into specimens. Given the impoverished nature of herbarium specimens, Hooker could not have worked on Australasian plants without being a lumper. The paradox of what has sometimes been called "the tyranny of distance" is that it operated in both directions; the distance that separated colonists in New Zealand and Australia from Europe also "tyrannized" Europeans who wanted examples of Australia's unusual and largely unknown flora.[109] Hooker's isolation forced him to work in the herbarium with dried specimens, which helped turn him into a taxonomic lumper, while the practices of colonial botanists—not least their access to a limited number of living species—encouraged them to be splitters. Of course, there were other reasons why Hooker adopted his broad species concept, which will be examined in later chapters, but all are easier to understand once its material basis has been grasped. His preferred classificatory method—reducing the number of species by "condensing" the phantom type from a series of variations—was undoubtedly made *possible* by the wealth of specimens in his herbarium, but it was also made *necessary* by that same abundance.

As I noted earlier, when Hooker acknowledged the type to be a phantom, he was already familiar with Darwin's ideas and had spent many years pondering their implications for his science. During the late 1850s, while Darwin was writing what he called his "big species book," he regularly asked Hooker for help in analyzing plant classifications, partly in order to understand the significance of variability. Darwin's suspicion was that large genera, that is, those which contained many species, also contained species with many varieties. As he explained to Hooker, if this proved to be the rule, "it is very important for me; for it explains, as I think, all classification, ie the quasi-branching & sub-branching of forms, as if from one root, *big* genera increasing & splitting up &c &c, as you will perceive."[110] In other words, natural selection would act on species with many varieties (like *Lomaria procera*) and gradually develop them into multiple new species; the large size of such genera being evidence that this process had already occurred, and many species had already been created.

Hooker's reaction to his friend's claims was somewhat mixed, as I will discuss further in the conclusion. The possibility that natural selection provided a lawlike mechanism which "explains . . . all classification" was deeply appealing, since it would remove the element of idiosyncratic judgment from classification, providing a firm and unshakably philosophical basis for asserting the correctness of his classifications. However, the prospect that the same mechanism might justify Colenso's decision to split *Lomaria procera* into five extra species was much less attractive.

From Darwin's perspective, eager to prove that new species were being created, it was—as he told Hooker in a letter—"good to have hair-splitters and lumpers," since the former were carefully cataloging the minor varieties that Darwin needed to make his case.[111] This was an assertion Hooker would undoubtedly have disagreed with, if he had had the time (Frances had just given birth to their fourth child, Marie Elizabeth, in the week when this letter from Darwin arrived). A few months later, Darwin asked Hooker to consider a hypothetical large genus and imagine that four-fifths of its members had been destroyed: would botanists rank the surviving fragments of the genus as separate, albeit closely related genera? As Darwin put it, "are all the species in a gigantic genus kept together in that genus, because they are really so very closely similar as to be inseparable; or is it because no chasms or boundaries can be drawn separating the many species."[112]

Hooker's reply revealed the distance between his concerns and Darwin's. Although he wrote that "I will answer your query about big genera deliberately, in the affirmative," he went on to say that "many of the small genera still kept up would never have been made at all, had the whole of the Natural Order as now known been known when those genera were made," explaining that when a few isolated species of an unknown, large genus arrived in Europe, they were often so different to one another that botanists classified them into separate, new genera.[113] However, had the whole range of the genus been known from the outset, careful, herbarium-based comparisons would have allowed botanists to see that there were no clear gaps between these species and thus no need for separate genera. As a result, the genera could have been lumped together, just as Hooker lumped species.

Darwin's claim was that the distinct gaps that allowed species and genera to be clearly differentiated had been caused by previous extinctions, whereas in cases of closely related species, such as the local naturalists were studying, there had not been enough time to allow extinctions and hence no clear gaps existed. For Darwin, the numbers of species, genera, and ultimately families and orders were all gradually increasing. However, for Hooker, the main consequence of Darwin's claim was that it provided a firm basis for his predilection for lumping—common descent underwrote his claim that minor variations between species and genera were insignificant. Darwin constantly emphasized how slowly evolution progressed, from which Hooker concluded that many human lifetimes must pass before the minor varieties someone like Colenso wanted to classify would really become species.

Hooker's interpretation of natural selection as justifying his classificatory practice led him to become an early supporter of Darwin's views once the *Origin of Species* had been published (although, as I will show in the conclusion, this support was more qualified than has previously been realized). Darwin's views gave fresh impetus to Hooker's vehement denial of the validity of "local observations." Nevertheless, his unshakable preference for lumping illuminates the intimate connection between being philosophical, defining species broadly, and earning a living. When Bentham criticized him for lumping too enthusiastically, Hooker responded that "the tide will turn one day"; meanwhile, "reducing species will go on apace." The issue might seem unimportant, but (as we have seen) Hooker was one "who makes his bread by specific Botany," hence his comment that, while splitting might be "very pretty play to amateur Botanists," such unphilosophical shenanigans are "death to me." [114] Hooker's claim may have been exaggerated, but the obstacles on his path to a paid career were real enough. He had become engaged to Frances Henslow (daughter of the Cambridge professor of botany, John Stevens Henslow) in 1847, but they had to wait four years to get married, partly because Hooker lacked a steady income. So, the wedding was delayed while Hooker went on another career-building trip, this time to India.[115] This perhaps helps explain his intense hostility to the splitters, whose efforts he derided so vehemently: just after he married, Hooker mentioned in a letter to Bentham that Johann Klotzsch, who had worked in his father's herbarium in Glasgow, wanted to "make a frightful mess of the Rhododendrons" by splitting the genus into over twenty species, "it is dreadful—he wants me to be partner in his crimes." [116] In a letter to Gray, he wrote, "I scold you for calling Brackenridge's bad species 'all rubbish'—they are *crimes*—laughing at these things & people makes them worse." [117] The tone of these remarks suggests another implication of Hooker's comment that the numbers of species must be reduced "on philosophical grounds as well as those of expediency." Expediency referred not only to the practicalities of managing a large herbarium but also to those of earning a living; the splitters delayed Hooker's publications, held back his career, and, in effect, forced him to wait to marry Frances—no wonder he regarded their efforts as crimes.

This chapter once again highlights the importance of friendship in keeping the botanical networks running: Colenso not only overlooked his correspondent's high-handed attitude but continued to collect diligently,[118] just as Hooker put up with his colonial correspondent's "miscellaneous scraps" and the "dreadful work" of naming them, to say nothing of Colenso's complaints and increasingly idiosyncratic attitudes. By contrast, Gunn cared more for the friendship of a fellow gentleman than for the niceties of classification and generally accepted Hooker's judgments.

Hooker could not produce a stable classification without also producing compliant collectors. He tried to achieve this by urging them to use standard collecting equipment and techniques, to refer to standard illustrations, and to use standardized terminology to record notes on standardized labels. The practices of labeling and describing plants also restrained the species splitters. Just as with plant drawings, there was a cycle of imitation in which the end product served as the exemplar for the beginner to emulate. The outcome of the scientific practice, in this case the published description that fixes the name of a species, was the model for the label on a newly collected plant. Wherever possible, such a label should include an identification if the species was one that was already named, but even for an unknown species, the technical language—"*Involucre* marginal, scarious, continuous"—was used to define the characters of the plant as concisely as possible. Such terms collapsed the variability of a feature in a population of living plants into a single standardized term. Specimen-making techniques made taxonomic splitting harder; the techniques of seeing made it harder to depict variation; and standardized botanical language also promoted compliance with metropolitan standards.

CHAPTER SIX

Settling

When I commenced my Botanical Studies, the Linnean System was still the one followed in the Colleges, and if the Natural System was introduced at all, it was introduced, as it were by stealth, at the end of the course.

—Joseph Hooker, *Reply to the Toast "The Medallists"* (1887)

Joseph Hooker contributed substantially to the influx of natural history specimens that threatened to overwhelm him and his contemporaries; just a few years after returning from the Antarctic he was traveling again, in a further effort to gain experience and enhance his reputation. Between 1847 and 1849, he was in the Himalaya accompanied by Dr. Archibald Campbell, the British government's agent in the region. In November 1849, while in the Himalayan state of Sikkim, Hooker and Campbell crossed its northern border into Tibet in direct violation of the rajah of Sikkim's orders and were arrested and briefly imprisoned (the British government eventually secured their release by threatening to invade Sikkim). Hooker then spent 1850 traveling in eastern Bengal with his old university friend Thomas Thomson, and the two returned to England in 1851, laden with yet more specimens to be cataloged and named. They began work on a flora of India, the *Flora Indica* (fig. 8.2) and published the first volume privately in the hope of attracting patronage from the East India Company. However, no support was forthcoming and the work was never completed.[1]

Hooker spent the 1850s settling down: completing the floras of New Zealand and Tasmania; trying to get money for the *Flora Indica*; and looking for a secure, paid position that would support his new wife and family. He also struggled to achieve a stable set of botanical names, partly to rein in the colonial splitters but also to rescue botany from the rancorous debates about competing classificatory systems that were going on in the metropolis. The status of both Hooker and his discipline largely rested on his ability to settle these disputes and, in particular, to persuade his collectors to settle on his classification.

The disputes centered on the Linnaean, or sexual, system of classification, which remained in use in Britain well into the second half of the century (although some historians insist it had fallen out of use by the 1830s). The Linnaean system's survival, long after the rest of Europe had abandoned it, contributed to the perception that British botany was backward and unfit to take its place alongside either Continental botany or the physical sciences. However, those attacking the Linnaean classification were divided over what should replace it: the most widely used of its rivals was known as the natural system, founded in France in the late eighteenth century but only slowly taken up in Britain. As this system developed, its proponents found themselves disagreeing over its definitions and principles, thus fragmenting it. This led to destabilizing debates that were exacerbated by the existence of numerous other classificatory systems, all based on completely incompatible principles yet all claiming to be "natural." Both the debates within the natural system and the chaos of rival systems contributed to perceptions that botany was an unphilosophical study which lacked guiding principles.[2]

For Hooker, stabilizing classification involved winning over two different audiences, colonial and metropolitan. He had to persuade the former that his books and those he endorsed were the ones they should trust, while also establishing the correctness of his view in the eyes of his metropolitan colleagues. I shall discuss his efforts to win over the metropolitan audience in the next chapter; for the moment I want to focus on Hooker's efforts to demonstrate to his collectors that he was the metropolitan expert they should trust.

As we have seen, classificatory practices were in part determined by a hierarchy of deference and status; apparently trivial arguments over whether a particular plant should be classified as one or two species prove on closer examination to have been negotiations between rival pragmatic needs, contests over the relative importance of often-incompatible interests, and, perhaps most importantly, arguments over what gave one person rather than another the right to name a plant. As a result, the classificatory hierarchy itself—the names, groupings, and concepts used to classify—emerged from the same arguments, such as whether a colonial naturalist could name a new fern; these were, in effect, debates about what it meant to be philosophical.

The Linnaean "Incubus"

For hundreds of years, most European naturalists assumed that the lists of plants and animals they had inherited from the ancients—especially the Greeks—were more or less complete. That assumption was shattered by exploration and conquest; as Europeans traveled more widely, they realized they had no names or categories with which to describe the strange plants, animals, and peoples of Asia, the New World, Australasia, and Africa.

As European naturalists struggled to make sense of this flood of new things, many attempts were made to find new ways to classify them, to develop a systematic order that would allow naturalists to make sense of the world's diversity. Europeans gave new names to the new things, often in their own languages, which rapidly created a new Babel. These new names were not assigned on any consistent basis: some were (often wildly inaccurate) translations or transcriptions of indigenous names; others were long, descriptive names that were informative but hard to remember; or they were conferred in honor of a discoverer (or his patron) or to commemorate the place where the organism had first been found. Worst of all, the same plant or animal frequently received several different names, as numerous naturalists and explorers competed to cover themselves, their rulers, or their nations with the glory of a little piece of immortality.

By the eighteenth century, the need for reform was desperate, and the Swedish naturalist Carl von Linné, or Linnaeus, synthesized the achievements of several of his predecessors to produce a classification that used standardized names and—even more importantly—a standardized hierarchy of groups. Every creature belonged to a species, a group of almost-identical animals or plants; each species belonged to a slightly broader grouping called a genus (plural, "genera"); and so on up a hierarchy that culminated in the great kingdoms of animals and plants. Linnaeus created an orderly version of the two-part names (binomials), which were already in use by various botanists, to record the species and genus in a consistent form.[3] So, for example, in his *Species Plantarum* (1753), Linnaeus renamed the most common form of the herb bitter cress, which Hooker and Colenso were to argue over. The plant already had several common names (many of which, such as cuckoo flower, lady's-smock, and meadow-cress, are still in use) and gave it the two-part Latin name *Cardamine pratensis* (meaning "of a meadow").[4] At the same time, he took a closely related species of bitter cress (which rejoices in a huge variety of common names such as pepperweed, shotweed, and snapweed) and gave it the name *Cardamine hirsuta* (meaning "hairy").

Linnaeus's names were invaluable for botanists, firstly because they were short and standardized, unlike the numerous common names; *Cardamine pratensis* referred to just one, carefully defined species. The hierarchical classification was also invaluable because it saved time: all members of the genus *Cardamine* share common features, so a botanical book could now begin with a description of those common features and then, for each species in turn, simply add the much smaller set of characteristics that distinguished, say, *C. pratensis* from *C. hirsuta.*

However, when it came to the upper levels of the Linnaean hierarchy, things got much messier. It was clear to Linnaeus, and to most other botanists, that the similarities between plants went much further than those between genera. For example, *Cardamine* flowers have their four petals arranged in a cross, just like the flowers of the various kinds of cabbages (cabbage, cauliflower, kale, etc.) and the different plants from which mustard is made. These plants share other features as well, such as a rather harsh flavor. They seem to form a natural grouping, and Linnaeus gave them the name *Cruciformes* (derived from the Latin *crux,* or cross, in honor of the flowers).[5] Some other plants also share sufficient features to allow them to be placed into similar natural groupings; these patterns of what became known as affinity gave naturalists glimpses of an underlying order, an order embedded in nature itself. Many, including Linnaeus, assumed that the pattern which they dimly perceived was the plan God had used at the Creation. Eventually, once humans knew enough, they might be able to classify plants according to a truly natural system, one that reflected nature's own order.

Like his predecessors, Linnaeus hoped that, by taking the machinery of nature apart, he might recover the blueprint its creator had used; "this," as he wrote in the *Philosophia Botanica,* "is the beginning and the end of what is needed in botany."[6] Yet, by his own admission, he failed, offering his readers only "fragments" of the divine harmony. While a few groups, like the *Cruciformes,* were clear enough, there were many more plants where every classifier came to radically different conclusions; years, perhaps generations, of further work would, Linnaeus thought, be needed to sort out such confusion and complete the system.

To allow classification to proceed, Linnaeus devised his *methodus propria* (proper method) as a stopgap solution. It was designed to be simple and objective, obviating the disagreements between naturalists. To use it, all one had to do was count the plant's reproductive organs (which is why it became known as the sexual system): the number of stamens (the flower's male parts) determined its class, while the pistils, or carpels (the female parts), fixed its order; any plant that had two stamens belonged unambiguously to the class "Diandria," and if it also possessed one carpel, it fitted into the subsidiary order "Monogynia." It was, as Linnaeus himself readily acknowledged, an artificial system, in that its classification was based on a very small selection of the plant's features; therefore, the groups that resulted showed no hint of affinity, but at least they were clear-cut. A novice armed with one of Linnaeus's books could learn to identify plants in minutes, and the learned could fit every newly discovered plant into an existing order.[7]

The simplicity of the Linnaean sexual system made it both popular and influential, but the continued expansion of European knowledge of the world kept providing fresh challenges.[8] The French naturalist Michel Adanson decided in 1749 that the plants and animals of Senegal could not be fitted into the Linnaean categories. Adanson's call for a truly natural system, one that would be based on all the plant's features, was taken up by Bernard de Jussieu when he reorganized the French Jardin royal in 1759, and Bernard's nephew Antoine-Laurent continued his uncle's work when the garden became the Jardin des plantes following the French Revolution.[9] Antoine-Laurent de Jussieu took *Cardamine* and all the other plants that shared those cross-shaped flowers, such as the cabbages, and defined their natural group as a family that he renamed the Cruciferae (now also known as the Brassicaceae); this was one of a hundred natural families that he defined and that are still in use today. De Jussieu's work was continued and refined by the Swiss botanists Augustin-Pyramus and Alphonse de Candolle, and their joint efforts created a classificatory system that became known in the nineteenth century as the natural system.[10]

While the natural system was developing in France and across continental Europe, the British generally remained enthusiastic users of the sexual system, which played a large role in making botany popular both in Britain and in its colonies.[11] Britain's Linnaean enthusiasm partly resulted from Sir James E. Smith's purchase of Linnaeus's private collections; he brought them to London and devoted the rest of his life to preserving Linnaeus's legacy.[12] When the *Edinburgh Review* assessed Smith's *Memoirs and Correspondence,* the reviewer was "impressed with the great progress in systematic botany which has been made in England during the last fifty years." Behind this compliment, however, was a barely concealed attack on Smith, since the reviewer noted, which "is less flattering to our national vanity," that during this same period, "British botanists have been comparatively inactive in examining the structure of vegetable bodies,—in explaining their hidden functions."[13] Many of the *Edinburgh*'s readers would have understood the implication: Smith and the Linnean Society bore considerable responsibility for the British neglect of plant physiology in favor of lowly systematics.

By the 1830s, the perception that Britain was falling behind in botany's more philosophical branches led many botanists to question the usefulness of the Linnaean system. John Lindley described the Linnean Society as a "positive incubus upon science" that was holding back British botany by promoting a system that had been largely abandoned elsewhere in Europe.[14] And, as we have seen, in its comments on Smith, the *Edinburgh Review* noted: "Botany has hitherto been chiefly a science of observation," since it was dominated by the kind of "botanist who knows a plant only by its parts of fructification." It compared these Linnaean botanists to the kind of mineralogist who classified a rock "by throwing its lustre upon his eye, and by shaking it knowingly in his hand." Nevertheless, botanists need not despair, since mineralogy had now progressed to be one of science's "most interesting branches." The reviewer concluded: "Botany will, we doubt not, soon rise to the same dignity."[15] Lindley, who wrote for the quarterlies himself, would probably have known that the review was by David Brewster, who, as we have seen, argued that British science was in decline. Lindley shared many of his views and attacking the Linnaean system was a way of demonstrating botany's decline and arguing for increased funding.

Although de Jussieu's system had been introduced into Britain by Robert Brown in 1810, it took many years to become established.[16] Some historians have argued that it had largely triumphed by the 1830s, but there are good reasons to doubt this claim, not least because men like Lindley felt the need to continually attack the Linnaean system and stress the natural system's superiority.[17] In 1833, he reviewed an introductory volume whose author asserted that he was going to describe "the present state of Systematic Botany" but had in fact outlined the Linnaean system, which Lindley claimed "has been disused these forty years."[18] Lindley's claim was, as he well knew, disingenuous to say the least—the Linnaean system was discussed repeatedly in general periodicals throughout the 1830s, 1840s and 1850s. Long after professed botanists had largely agreed on the superiority of the natural system, the Linnaean one continued to be used in both medical textbooks and the general works aimed at gardeners, flower painters, part-time enthusiasts, and the myriad others with an interest in plants; the Linnaean system was not dropped from the *Botanical Magazine* until after Joseph Hooker took over its editorship, following his father's death in 1865.[19]

Like the Hookers, Lindley partly depended on selling books to these broader audiences for his income, so he could not afford to surrender them to the Linnaeans. Hence, he decided to write *Ladies' Botany*, which, as a reviewer noted, was "intended for the instruction of ladies in one of the most delightful of all the sciences."[20] Written using a popular genre of the period (as letters "addressed to a lady on the botanical education of her children"), it was nevertheless, as its subtitle proclaimed, *a Familiar Introduction to the Study of the Natural System of Botany.* Lindley's preface noted that "no one has, as yet, attempted to render the unscientific reader familiar with, what is called, the Natural System, to which the method of Linnaeus has universally given way among Botanists." He observed that few women managed to master it because "on all hands they are told of its difficulties; books, instead of removing those difficulties, only perplex the reader by multitudes of unknown words." While many introductions to the natural system had been written (including his own), "for those who would become acquainted with Botany as an amusement and a relaxation, they are far too difficult."[21]

The need to capture the popular and medical audiences for the natural system was one reason Lindley used both his own books and anonymous reviews in reforming journals like the *Athenaeum* to attack the Linnaean system. His hostility was so well known that in 1834 the *Monthly Review* commented that, "for a great number of years," Lindley had "unceasingly, not merely disparaged, but actually persecuted unto death the ancient Linnean system."[22] Almost twenty years later, the Tory *Quarterly Review* still defended the Linnaean method from Lindley's persecutions by claiming that its simplicity made it an ideal starting point for learning botany—even if the serious student must eventually graduate to the natural system. The reviewer quoted Lindley's claim that the sexual system was "well adapted indeed to captivate the superficial inquirer, but exercising so baneful an influence on botany, as to have rendered it doubtful whether it ever deserved a place among the sciences."[23] The *Quarterly* responded: "With all deference to the Doctor, we might rejoin that, if the Natural System were permitted entirely to extinguish the Linnaean, botany would soon deserve a place among the *mysteries* instead of the sciences. The 'superficial inquirer' is the very person who wants a clear and frank-minded guide that *will* show him what he wants, instead of letting him lose himself in a boggy maze where he can find no firm footing."[24] The reviewer claimed that the natural system was only suitable for the "practiced adept," having apparently been designed as "an excellent contrivance for fencing off the profane vulgar." By contrast, J. E. Smith's *English Botany* (which, of course,

6.1 The key to metropolitan authority; a large herbarium of specimens from all over the world. The Banksian Herbarium. Courtesy of the Natural History Museum.

used the Linnaean system) would allow the novice to identify a plant "in five minutes." The *Quarterly* therefore suggested that "the English student is advised to begin with Sir James Smith's works and end with Dr. Lindley's. The Knight should preside over the catalogue, the Professor over the herbarium" (fig. 6.1).[25] Ease of use was seen as more important than dubious philosophical claims made with the help of foreign (particularly French) classificatory systems.

Nevertheless, one might wonder why Lindley and others did not simply follow the *Quarterly*'s advice by retaining the Linnaean system in their popular books while using the natural one to "fence off the profane vulgar" from their more technical works. Instead of making de Jussieu's system accessible to ladies, why not allow its use to differentiate those of properly philosophical status? To some extent, that was what happened: William Hooker used both the Linnaean and natural systems in his books, using the former as an introductory key to the latter, and Joseph Hooker did the same in the *Flora Novae-Zelandiae* when he added a key "arranged according to the artificial or Linnaean system," in order to "facilitate the reference to many of the obscure genera to their proper places."[26] Hooker's decision might appear to be encouraging colonial naturalists to use the Linnaean system for identification, leaving the metropolitan naturalists to reclassify according to the natural system, but that was not his strategy—he stressed that the Linnaean system's use did not "obviate the necessity" for nor "supersede the use of a sound elementary acquaintance with the Natural Orders."[27] Even Lindley—despite supposedly having "persecuted unto death the ancient Linnean system"—gave a brief synopsis of its features in his books, conceding that "although now disused by men of science . . . many books have been arranged on its plan."[28] Far from erecting barriers to the "vulgar," the use of both systems was intended to provide a bridge to the natural system, across which the widest possible audience could be guided.

For would-be philosophical naturalists, the problems created by the persistence of the Linnaean system were exacerbated by the instability of its natural rival; despite many decades of work, nature's order had still failed to reveal itself. Not only did each proponent of the system have subtly different ideas about its application, some users even disagreed with themselves and were forced to continuously modify their version of the system in successive editions of their works. The *Athenaeum*'s review of Lindley's *Vegetable Kingdom* (1846) compared it to his earlier works on the natural system and observed that "the present arrangement differs materially from that in his former books": "This will, undoubtedly, be urged by many not only as an objection to the author's views on the ground of unsettledness, but, perhaps, amongst the remnants of the Linnean systematists left in this country, as an objection to the natural system itself." The *Athenaeum* attempted to refute this objection by noting that just as "every observation on the heavenly bodies influences the calculations of the astronomer, so does every observation on the tissues of plants, the discovery of every new form, tend to modify the views of the botanist." Yet such corrections do not alter "the fundamental principles on which that system has been constructed."[29]

The continuing need to shore up the unsettled natural system explains why, in 1850, the *Westminster Review* felt compelled to devote much of a review on "Natural Systems of Botany" to attacking the Linnaean system once again.[30] The reviewer claimed that the natural system was so widely accepted that "the appearance of a book whose professed object it is to depreciate the more rational principles of arrangement, and to advocate an unconditional return to the professedly artificial system of Linnaeus, could hardly have been anticipated at the present day." Nevertheless, that was what James L. Drummond, an Irish professor of anatomy and leading promoter of the Belfast botanic garden, had produced in his *Observations on Natural Systems of Botany,* and "it must be confessed," the reviewer wrote, "that Dr. Drummond's 'Observations' display far more ingenuity than could have been expected in the defence of a cause so hopeless and so ungrateful"—hence the need to devote eight pages to slaying a system that was supposed to be dead and buried.[31]

Like the Linnaean system's other advocates, Drummond defended its ease of use; he claimed that "the greatest possible recommendation of any system is the facility with which a person previously ignorant of all, save its merest rudiments, can master it and ascertain the name of the first plant he picks up."[32] The *Westminster*'s reviewer demurred, commenting that "we are no advocates for such a very superficial knowledge" (a comment that echoes Hooker's claim in the *Student's Flora* that analytical keys encouraged superficial knowledge). Fortunately, according to the *Westminster,* "a recurrence to the Linnaean system is impossible at the present day." Nevertheless, the reviewer felt obliged to condemn this impossible quest, in case, he explained, Drummond's book deterred beginners from studying botany.[33]

The real problem facing the promoters of the natural system was that it did not exist; there was no single "natural system." Instead, there were several incompatible ones, all of which claimed to be natural.[34] These were cited by Drummond as a reason for retaining the Linnaean system, and the *Westminster* had to admit that the advocates of the natural system "are far from being [unanimous] with regard to the grouping of the orders into cohorts, alliances, or whatever they may be called": "This circumstance is put forth by Dr. Drummond, as a proof of the want of agreement on first principles amongst modern botanists; and he has copied into his book a great part of the view, prefixed to Professor Lindley's 'Vegetable Kingdom,' of the various natural classifications, in order to show that his system 'is one chaotic mass of confusion and uncertainty.'"[35] Not surprisingly, the natural system's apparent "confusion and uncertainty" led many to remain loyal to the undeniably stable Linnaean system, but that was not the only alternative to the natural system on offer.

The *Westminster* disputed Drummond's claims, asserting that the disagreements were "far less than is generally supposed" and in any case were "of very secondary importance in practice, since the orders still remain, altogether, or for the most part, intact."[36] Yet, the *Westminster* then went on (presumably inadvertently) to provide good evidence for Drummond's argument, since its anonymous reviewer, George Luxford, devoted much of the rest of the review to explaining and defending a radical alternative to the natural system—the septenary system of classification.

"A Predilection for Favorite Numbers"

The septenary system had been devised by the Quaker naturalist, printer, and publisher Edward Newman, who tried to show that all living things could be classified in groups of seven, with affinities joining the members of the groups and analogies connecting the groups to one another.[37] His friend Luxford had introduced this obscure system into his review because he regarded it as having been "unintentionally confirmed by Professor Lindley," the latter having classified plants into seven main natural classes.[38]

Despite its unusual numerological basis, Luxford—like Newman—argued that the septenary system was a natural one: "most systematists have laboured hard to *invent* what they call a natural system, instead of contenting themselves with the endeavour to *discover* the system already existing in nature; and many of them, besides perversely following some certain track which they had previously marked out, have frequently encumbered themselves with a predilection for favorite numbers or cabalistic figures in which, as they imagined, lay the key to the mysteries of nature."[39] According to Newman, the pattern based on the number 7 was not a "predilection for favorite numbers" but simply emerged from a study of nature. In this, and in other regards, his system was similar to the quinarian one developed by William Sharp MacLeay in his *Horae Entomologicae* (1819–21), which used groups of five (with groups of three serving a subsidiary role).[40] The quinary system's most enthusiastic promoter, William Swainson, was a friend of Newman's and a contributor the *Entomological Magazine*, which Newman edited. MacLeay and Swainson's work may have inspired Newman to devise the septenary system. MacLeay had an a priori belief in God's designing power but also mentioned a range of independent facts to support the supposed naturalness of the system. Just as Luxford felt that septenary classification had been confirmed by Lindley, MacLeay believed quinarianism was confirmed because "different persons have respectively stumbled upon it in totally distinct departments of the creation."[41]

These numerological classifications seem strange now, but they were widely discussed in the early decades of the century and quinarianism in particular attracted many adherents. While aboard the *Erebus,* Hooker wrote to his father about his preliminary attempts to classify the Antarctic mosses, telling him "there are five groups I consider quite natural, and the three first [*sic*] of them abnormal; these are what McLeay's quinary system acknowledges." Hooker met MacLeay in Sydney (MacLeay had joined his father there in 1839) and discussed classification with him, but told William Hooker that "you must not think that I am led away by any system, for I formed this system before I saw McLeay's and before I understood his views."[42] Nevertheless, Hooker recorded that "I cannot, however, forget a remark he made, saying 'he was glad I had paid so much attention to the minute Orders and to Cryptogamic Botany, *for in them would be found the foundation of a truly natural system.*' Now, though I do not put any faith in the quinary arrangement, I believe that 5 *happens to be* the number of groups into which mosses most naturally divide themselves, and I am convinced of the truth of the circular system."[43]

Hooker quoted the work of several other botanists to support his opinion, mentioning in particular the Swedish botanist Elias Fries, director of the botanic gardens at Uppsala, who had classified the fungi into groups of five quite independently of MacLeay's work. After receiving a copy of Lindley's latest book (fig. 6.2), Hooker told his father that "Lindley's Elements seems a most valuable work to me and the very one I wanted, for I have a very high opinion of him as a Nat. Order man," and although he thought Lindley recognized too many orders, Hooker was impressed with his attempt to arrange them into a coherent plan: "A linear arrangement will never do, and Fries's Motto 'omnis ord. nat. circulum per se clausum exhibet' is daily gaining proof, Lindley's groups and alliances of plants . . . must be invaluable. I am no judge of the goodness of this arrangement of the groups, but it is the throwing the Nat. Orders into groups and showing the dependence of one group on another which impresses me."[44]

Fries's motto, "every natural order exhibits a self-enclosed circle," was precisely the point that MacLeay had seized on when he claimed that the Swede's independent discovery of circular groups proved that his system was indeed natural.[45] These apparent confirmations of the numerical systems by other naturalists led to a surge of interest in them, and both Hooker and Darwin were briefly interested in quinary classification, as were (at different times) Richard Owen, J. E. Gray, William B. Carpenter, Robert Chambers (anonymous author of the *Vestiges of the Natural History of Creation*), and Thomas Huxley (who also met MacLeay in Sydney, during the *Rattlesnake*'s voyage).[46] Quinarianism also gained acceptance in Britain because it was promoted by Nicholas Aylward Vigors and others

SYSTEMATICAL BOTANY. 127

22.	Calyx enlarged and irregular	DIPTERACEÆ.
	Calyx not enlarged	23
23.	Stamens monadelphous	MALVACEÆ.
	Stamens distinct	TILIACEÆ.
24.	Carpels disunited or solitary	25
	Carpels consolidated	30
25.	Carpels plunged in a tabular disk	NELUMBIACEÆ.
	Carpels clear of the disk	26
26.	Stamens perigynous	ROSACEÆ.
	Stamens hypogynous	27
27.	Carpel solitary	ANACARDIACEÆ.
	Carpels several	28
28.	Stamens polyadelphous	HYPERICACEÆ.
	Stamens free	29
29.	Herbs	RANUNCULACEÆ.
	Trees or Shrubs	ANONACEÆ.
30.	Placentas in the axis	31
	Placentas parietal	34
	Placentas dissepimental	
42.	Flowers unisexual	CUCURBITACEÆ.
	Flowers hermaphr. or polyg.	GROSSULACEÆ.
43.	Disk double	APIACEÆ.
	Disk simple	44
44.	Seeds few	45
	Seeds numerous	51
45.	Carpels solitary	46
	Carpels several	48
46.	Parasites on trees	LORANTHACEÆ.
	Root plants	47
47.	Leaves balsamic, acrid	ANACARDIACEÆ.
	Leaves insipid	COMBRETACEÆ.
48.	Calyx valvate	49
	Calyx imbricated	50
49.	Stamens opp. petals	RHAMNACEÆ.
	Stamens altern. petals	CORNACEÆ.
50.	Anthers curved downwards	MEMECYLACEÆ.
	Anthers erect	BRUNIACEÆ.

6.2 Analytic table, or key, which used a series of questions relying on technical botanical terminology to identify a plant (from Lindley, *Elements of Botany*). From the author's collection.

as uniquely British (despite its roots in German *Naturphilosophie*) and an alternative to the work of European systematists.[47]

Vigors, like MacLeay, was one of the young reformers in the Linnean Society's Zoological Club who sought to replace the existing Linnaean classification with a radical alternative but were unable to agree among themselves as to what the new scheme might be.[48] Adrian Haworth, another member of the Zoological Club, argued for a "binary," or dichotomous, classification, beginning with "Mind" and "Matter" and then proceeding by divisions of two until it encompassed all of nature—a system that had ancient precedents but whose immediate inspiration was the philosopher Jeremy Bentham (uncle of the botanist, George).[49] Even within the narrow confines of the Linnean Society's Zoological Club, there were several different post-Linnaean systems in use, while outside it were Newman, Luxford, and the septenary system.

Specifically botanical systems were also in circulation, of which the most unusual may have been Thomas Baskerville's *Arrangement of the System of Vegetable Affinities, on the Principle of a Sphere* (fig. 6.3). Baskerville, like the other naturalists discussed, claimed to have worked out a natural system; the preface to his book explained that the new system had grown out of his wish to "express upon paper the connexions of some of the more intricate portions of the vegetable kingdom, and the intention was by constructing circles of affinity to be able to concentrate attention on any required group of Plants." Although Baskerville's aims were similar to those of the quinary and septenary systems, he mentioned neither in his book but instead dedicated it to Lindley, with the latter's permission.[50] Like other prospective systematic reformers, Baskerville stressed the superiority of a natural system to the Linnaean one, because only a natural system allowed "the study of Affinities," which he hailed as "a new department of botanical science" since it was "by its aid alone the science has any claim to be philosophical."[51]

The impact of these metropolitan debates was surprisingly strong in the colonies. When Charles Knight's friend and fellow naturalist Robert Lynd died in New Zealand in 1851, Knight acquired his effects, including his botanical notebooks. He discovered that the self-taught Lynd "had completely wasted his time and abilities" by filling his notebooks with "the useless nomenclature and systematic arrangements of men like Swainson, Sowerby & Newman."[52] Lynd's notes suggest that the influence of both

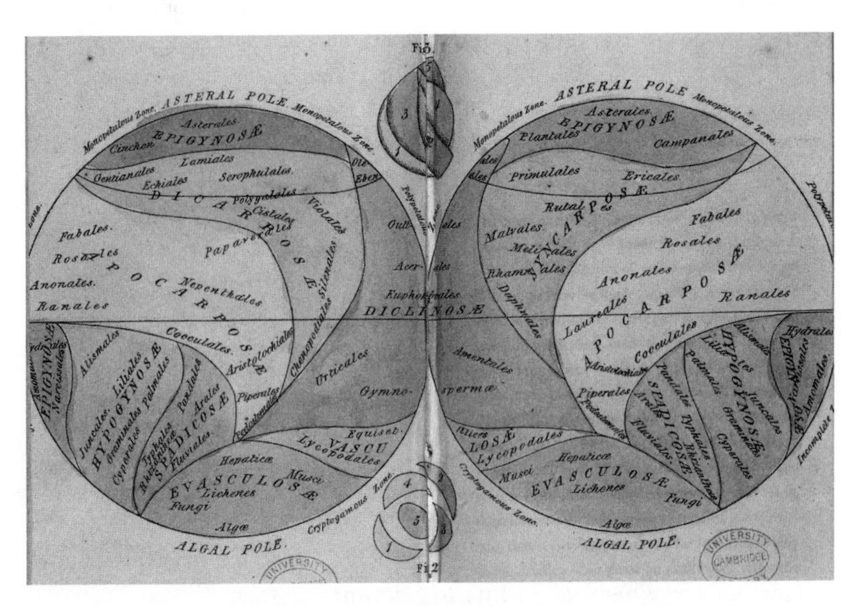

6.3 Thomas Baskerville's unique attempt at a natural system of plant classification: *An Arrangement of the System of Vegetable Affinities, on the Principle of a Sphere* (1839). Reproduced by kind permission of the Syndics of Cambridge University Library.

the Linnaean system (used in Sowerby's books) and the various rivals to the natural system persisted longer in the colonies than in the metropolis. The lasting impact of the quinary system in Australasia was undoubtedly because its founder, MacLeay, immigrated to Australia in 1839 but also owed something to the efforts of William Swainson, who immigrated to New Zealand in the following year. Swainson's cheap and widely available natural history books enthusiastically advocated the quinarian system.[53]

The lasting influence of the classification debates in the colonies, and the enthusiasm with which some colonial naturalists followed them, are evident from the work of Augustus Oldfield, one of Hooker's Australasian collectors, who wrote a book on classification sometime after 1865.[54] Although they remained unpublished, Oldfield's views give a clear sense of why Hooker attempted to dissuade colonial naturalists from speculating about classification; if similar opinions had become widely known and adopted in the colonies, they could have led colonial naturalists to use different classifications and names from those in use at Kew. Such speculations posed an even more serious threat to the prestige and authority of botany, and thus to Hooker's fragile status.

What would have made Oldfield's views particularly problematic for Hooker was that he used his wide reading and knowledge of the sciences to blend widely held opinions with unorthodox ones, creating a mixture that might have appealed to many colonial naturalists. For example, his view of the Linnaean system was one that many in the metropolis would have echoed: he argued that the sexual classification's very simplicity had turned botany into "a mere jargon of names," resulting in "the total neglect of the important study of the physiology of plants"—sentiments that Lindley himself might have voiced.[55] Similarly, Oldfield condemned the "lax notions" of naturalists who named new species on spurious grounds, adding that their "unphilosophical and groundless assumption that nature has plainly labelled all her various and multifarious productions" leads such men into "continually registering new species." He also argued that splitters—motivated by "their inordinate ambition to leave their names deeply engraven on the tablets of science"—were turning botany "into a perfect chaos of synonymy," to such an extent that "people begin to doubt the existence of species." Hooker would have particularly approved of Oldfield's claim that the real issue was the validity, "not of species, but of the methods by which we attempt to discriminate them."[56]

However, alongside these comments were others that showed the influence of Lorenz Oken and the German school of Romantic natural history known as *Naturphilosophie,* which Hooker would have found entirely unacceptable.[57] For example, Oldfield argued that a stable botanical classification relied on deciding what—if anything—actually exists. He argued that "if anything at all exists it is *mind*," and that "if there exists anything foreign to *mind*, it is *force*," from which he deduced that "the idea of *force* is necessarily accompanied by those of *space* and *time*." He proceeded in this mode, drawing pseudo-algebraic equations that related the universe (u) to the other four concepts: "$u = m + f + s + t$ . . . (!)" and so on for five, increasingly incomprehensible pages.[58] Oldfield also argued that there were no genuinely intermediate forms in nature, a claim that would have undermined Hooker's classificatory method.[59]

As Oldfield's book was never published, there is no way of knowing how colonial botanists would have reacted to it, but many of his ideas were not noticeably stranger than others in circulation in the mid–nineteenth century; indeed, its immediate inspirations appear to have been German *Naturphilosophie* and Adrian Haworth's binary system. Seen in the context of binary, quinary, septenary, and "Baskervillean" classifications, Oldfield's speculations—far from being bizarre—are clearly typical of the classificatory debates, not least because of his opposition to artificial systems and his focus on the search for affinities between plants. Had his ideas been published, there is every chance they would have been taken seriously.

"What Constitutes a Really Distinct Genus, or Species?"

While these unsettling debates rumbled on in the colonies, there were equally problematic ones going on in Britain, even though most professed botanists had adopted the natural system by midcentury. Despite the consensus that the de Jussieu/de Candolle system was the best, there were still sharp disagreements about *how* to use it. Between 1858 and 1862 three British floras appeared: George Bentham's *Handbook of the British Flora* (1858), Charles Cardale Babington's *Manual of British Botany* (5th ed., 1862), and William Hooker and George Walker-Arnott's (Walker-Arnott had succeeded William Hooker at Glasgow University) *British Flora* (8th ed., 1860). They all used the natural system, yet an anonymous reviewer noted that they came to wildly different conclusions as to how many plant species there were in Britain. As he observed, "While Mr. Babington's Manual (Ed. iv) contains 1708, Messrs. Hooker and Arnott have but 1571, and Mr. Bentham 1285"; in other words, Bentham's book contained only three-quarters of the number of species found in Babington's book.[60] How could botany be considered a philosophical discipline if no two botanists could agree on the simple issue of how many species there were in a well-botanized locality like Britain?[61]

Given such disagreements, it is hardly surprising to find William Colenso commenting that "I know not of any certain rule" for defining species since, whenever he opened a new book or journal, "I find the first Botanists of the day opposing one another in their speculations," arguing about classification and its principles, and either lumping together or splitting apart existing species, "laboriously undoing what their predecessors or compeers have toiled to rear."[62] The "first Botanists of the day" were supposed to set the standards with which collectors like Colenso were expected to comply, yet he found the British botanical community riven by arguments.

The instability of the standards coincided with Colenso's increasing confidence in his local knowledge and helped persuade the colonial botanist that he was as well placed as anyone to name his own plants and perhaps, eventually, to make his own contribution to the philosophy of system. Thanks to Hooker's having sent Colenso various publications and encouraged him to use the natural system, Colenso had mastered its terminology, as we can see from Colenso's response to some of the early parts of the *Flora Antarctica;* Colenso commented that "I cannot (at present) agree with you in believing *Aspidium venustum* to be identical with *A. pulcherrimum.*" His confident use of technical terms shows how much he had learned: he discussed whether a disputed fern possessed "a sub-coalescent caudex" or "coriaceous" fronds, the position of its sori and its habit ("always an epiphyte"), from all of which Colenso concluded, "I do not agree with you."[63]

Given his growing expertise, local knowledge, and the evident disorder of European classifications, Colenso was now ready to move on to the more philosophical aspects of classification. "Nature will not be bounded by *our* laws," he told Hooker, "and to the question,—*What constitutes a really distinct genus,* or *species*? I cannot give a satisfactory answer."[64] Although Colenso would have agreed with Hooker, Lindley, Bentham, and many other leading British botanists that "the" natural system was the only acceptable alternative to either the Linnaean system or its various numerological and other rivals, he knew that considerable individual expertise was needed to decide whether to create a new systematic group or abolish an existing one—expertise which he now felt he possessed.[65]

Colenso was not the only one who recognized the destabilizing effects of these disputes between classifiers; in 1856, the *Edinburgh Review* noted that they had created "the fashion of late years" for systematic botany to be held "in so much contempt." This was an urgent problem since classification was "the groundwork upon which the correctness of the speculation of the physiologist and geographical botanist must mainly depend."[66] The anonymous reviewer was Bentham, who admitted that the "contempt" in which his discipline was held was perhaps understandable—given that botanists could not even agree on how a species was to be defined, or even whether they existed in nature. The definition of basic categories, especially of species, was the most divisive issue for systematics. In his review, Bentham asked (just as Colenso did), "What is a genus?" noting that this was "a question which has led to as much controversy and difference of opinion as any other of the fundamental principles of botanical arrangement." Such differences over basic concepts meant that "many naturalists of the present day consider [a genus] to have no existence in nature, but to be a mere creature of the imagination, a kind of instrument to enable man to classify the infinity of forms exhibited by nature," while others believed that genera were not only real but "more natural than species themselves." Bentham asserted that "the truth probably lies between these two extremes," in that "all species may be arranged into groups indicated by nature," but it had to be accepted that "the precise limits and extent of each group,—call it class, order, family, tribe, genus, or section,—are purely arbitrary."[67]

Describing the precise limits of taxa as "purely arbitrary" was a dangerous admission; if they were merely the whims of individual botanists, why should a local botanist accept the judgment of a so-called expert who—as in the case of Bentham and the *Flora Australiensis*—had never even seen most of the plants he purported to describe?

Squabbles between rival experts were equally confusing for the wider botanical public. What, for example, would a botanical enthusiast who bought Hewett Cottrell Watson's *Cybele Britannica* in 1860, to learn about "British Plants and Their Geographical Relations" have made of its distinguished author's attack on botanical classification in general, and on Lindley in particular? Watson argued that "classification is sometimes erroneously supposed to require much ratiocinative capacity. It requires this in a very small degree only, as presently executed," adding that Lindley was "only a describer, very feebly a reasoner":

> After labouring on it during many years, *he has utterly failed to reason out a system*, properly so designated; and he has latterly even abandoned the word "system" as a book title. Through many changes, during which *the* natural system has become *a* natural system, and *a* natural system has sunk into *no* natural system, the learned Lindley has at

> last achieved a sort of mosaic classification of changeful pattern;—one much resembling Mrs Fanny Ficklemind's patchwork counterpanes; each new one different in its pattern, but each in its turn formed by ingeniously joining together some hundreds of pieces of all sizes and shapes.[68]

As Watson went on to observe, "much industry and skill, much labour time and tact, doubtless are required for nicely performing this sort of patchwork in botany; but it is not ratiocination. It is simply descriptive juxtaposition, nothing more."[69] If the natural system was perceived by some as a "changeful pattern" or "descriptive juxtaposition" rather than a stable classification built on rational principles, it is no wonder that displacing the Linnaean system and its competing successors proved so difficult. Such heated controversies also undermined botany's claims to scientific status; after all, individual physicists did not each have their own version of Newton's laws. Faced with such instability, it was hard for men like Hooker to claim that theirs was a genuinely philosophical science, and even harder to establish their authority over the collectors.

"The Knife and the Microscope"

Given the immense difficulty of creating a consensus around the natural system, one cannot help wondering why the philosophical botanists were so wedded to it; why not stick with the undeniably stable Linnaean one? Perhaps the most important reason was that the sexual system perpetuated the separation between botanical classification and its more prestigious branches, such as distribution and physiology. As we have seen, the Linnaean system used only the plant's reproductive organs, while the natural one required the use of all the plant's characters; as Hooker noted, a botanist who hoped to use it would find that even "an elementary acquaintance with the Natural Orders and Species of plants" required him to "commence with the knife and the microscope, tracing the development of important organs, however minute; and if he desire to obtain that knowledge of the affinities of plants which alone will enable him to prosecute other branches of the science, he can only do so by first making himself thoroughly acquainted with their comparative anatomy."[70]

The natural system made "the knife and the microscope" essential (fig. 2.6), and Hooker made it clear that he saw this marriage of anatomy and physiology with classification as the key to the "other branches of the science," such as distribution studies. For a systematic botanist like Hooker, the great attraction of the natural system was not that it made classification *one* of the philosophical studies but that it turned it into the key to the whole of botany, the central practice that connected the others. He was not alone in this view; the *Edinburgh*'s review of Smith's memoirs included the hope that "the botanists of the next age will apply themselves to these important objects": "and follow the example which has been set them by our countryman, Mr Brown (whom Humboldt has justly characterised as the *Botanicorum facile Princeps*) in making the microscope and the dissecting knife indispensable instruments of their science."[71] Brown—who had introduced the natural system into Britain—was Britain's foremost exponent of physiological plant studies, whose work raised botany above the mere counting of stamens; the *Edinburgh*'s review explicitly linked Brownian physiology with the higher-status sciences, noting that "the botanist has the same occasion for powerful microscopes that the astronomer has for telescopes of great penetrating and magnifying power." The Linnaean system prevented botany from reaching to the status of astronomy, and the reviewer explicitly recommended that "the young and aspiring student" should "consider systematic botany only as the means by which he is to attain higher objects," the necessary preliminary to physiology and distribution studies, and that he should "pursue those objects assiduously and ardently, with the Microscope in one hand, and the torch of Chemistry and Physics in the other."[72]

As we have seen, metropolitan classifications had to be sent back to the colonies and used by the collectors—hence Hooker's decision to use English for his colonial floras. His use of dual keys (to the Linnaean and natural orders) had a similar motivation, since many of the colonial naturalists had learned their botany from books that used the older classification—dual keys provided them with a way to cross over to the system Hooker wished them to use.[73] At the same time, the natural system's higher complexities helped re-create the distance between himself and the colonial collectors that was elsewhere being broken down. The separation between field and herbarium that Hooker was unable to maintain was replaced by one between collectors who used the natural system merely for identification and educated metropolitan gentlemen like himself who could describe new species in its technical Latin terminology.[74] For a would-be philosophical botanist, the complexities of the natural system gave classification an attractive degree of obscurity, more akin to the "mysteries" of the mathematical sciences.

"To Inculcate Caution"

Hooker's books were, as we have seen, partly intended to induce his colonial plant collectors to learn and use the natural system, thus absorbing his philosophy of classification and definition of a species and so reducing local speculation and species splitting. The books' pictures and descriptions of species were one way of doing this and the introductory essays were another; he told the botanist William Munro that the introductory essay to the *Flora Novae-Zelandiae* was "chiefly intended to open student's eyes to the great leading facts of the case and to inculcate caution, or they will have their Flora in a pretty mess."[75] He used almost the same words in the published essay: "I have further had a more practical object in view—the offering of theoretical reasons for inculcating caution on the future botanists of New Zealand; I have endeavoured to make it clear to those who may read these remarks, that systematic botany is a far more difficult and important object than is generally supposed."[76] It was, of course, primarily the local species splitters who, Hooker hoped, would become more cautious after reading his essays; emphasizing the difficulty and importance of systematic botany was a key strategy in inculcating caution. By explaining these complexities to his colonial audience, he emphasized that he was qualified to speculate but they were not. He described the *Flora*'s introduction to the American botanist Asa Gray as a "long-winded Essay" about "variation distribution & origin of species." Hooker claimed he had nothing "new to propose [nor] any dogma or theory to support," but "I think it high time for Systematists to take some decided stand upon such grounds; if every noodle that knows a cabbage from a Cabbage Palm is to set up as a describer of new species from every corner of [the] globe because he finds a difference in his specimens then [this] is an end of Systematic Botany."[77] The *Flora Indica* also claimed that the poor state of systematic botany resulted from a "misapprehension of the true aim and paramount importance of systematic botany," together with a similar neglect "of the proper mode of pursuing the study of the laws that govern the affinities of plants." Hooker and Thomson expressed the hope that their essay on botany's first principles would "fall into the hands of many beginners." By stressing the "true aim and paramount audience" of systematics and its "proper mode" of study, the essay's readers were placed in the role of "willing and anxious" students; the key point Hooker and Thomson wanted to establish was that systematic botany was more complex than it at first appeared.[78]

Given that Hooker needed colonial collections, it would clearly have been counterproductive to convince colonial readers that botany was too hard for them, so the *Flora Indica* essay stressed that "any observant person may readily acquire such a knowledge of external characters, as will in a short time enable him to refer a considerable number of plants to their natural orders." Nevertheless, referring plants to their natural orders and naming them was merely the first step; any botanist who sought "to go beyond this" needed to realize that "to develop the principles of classification, to refer new and obscure forms to their proper places in the system, to define natural groups and even species on philosophical grounds, and to express their relations by characters of real value and with a proper degree of precision, demand a knowledge of morphology, anatomy, and often of physiology."[79] Developing "the principles of classification" required a detailed knowledge of plant structure and development, which required long study and microscopic analysis (hence Hooker's joking proposal for a "tax on microscopes"). Colonial collectors had to learn how to refer plants to their natural orders, but to stop there and not try developing the principles of classification. Similar examples of this delicate balancing act, between encouraging the collectors (by emphasizing the "paramount importance" of the work they were contributing to) and yet at the same time stressing botany's difficulties (by persuading them of their role as students, "future botanists," not yet ready to set themselves up as systematists) can be found in all Hooker's essays.

Once the colonial audience had been convinced that they were unqualified to speculate, Hooker would be able to rein in their desire to multiply species and speculate as to their origins. He wrote that "it is very much to be wished that the local botanist should commence his studies upon a diametrically opposite principle to that upon which he now proceeds," and that he should try "to determine *how few*, not *how many* species are comprised in the flora of his district." As we have seen, standardized terminology and the conventions of visual depiction encouraged botanists to emphasize similarity rather than difference. Hooker added that "it is strange that local naturalists cannot see that the discovery of a form uniting two others they had previously thought distinct, is much more important than that of a totally new species."[80] It was, as he knew all too well, not in the least "strange" that local naturalists avoided lumping and preferred to name new species, enriching themselves (in terms of anything from cash to scientific reputation) and their floras in the process, but he nevertheless hoped his essays would gradually convince them to see the issue as he did.

Yet, despite the rhetorical and other resources brought to bear, Hooker's attempt to induce colonial readers that he had the right to overrule them was not entirely successful. The complexity of his goals created ambiguities in his essays that helped defeat his campaign. Given Hooker's limited control over his collectors, when a colonial reader refused to take the hint that he was one of the "local botanists" being complained of and instead chose to see himself as qualified to pronounce on species, there was little Hooker could do to disabuse him. Indeed, one of the ironies of Hooker's position was that the steps he took to persuade collectors into compliance often had the effect of encouraging them to speculate. At the heart of Hooker's dilemma was the natural system itself: using it required considerably more self-education and effort from the collectors.

By denying colonial naturalists the chance to name species, Hooker ran the risk of alienating them; to avoid this, he met as many of their demands as he felt able to. In 1883, Colenso sought his help in becoming a fellow of London's Royal Society, arguing that "I have often thought,—*I have won those spurs also*" through hard work in field and herbarium.[81] Since Colenso's service "in the battle-field of Science" had been not only "long" but "free"—having worked without pay—he saw the coveted "FRS" as being a scientific gentleman's proper reward. In a letter to Hector, Hooker confessed that "I am bothered with *Colenso* who . . . wants *me* to get him *made* F.R.S. which I have no *power* to do. Of course I should be most happy to forward his views in this matter provided that we could make out a case, but I do not think that his claims are strong enough."[82] Nevertheless, Hooker did assist Colenso by describing the admission procedures and explaining the need to get existing New Zealand–based fellows of the society to sponsor his admission.[83] Hooker must have known that helping Colenso join the Royal Society and the Linnean Society would add to Colenso's self-confidence, thus promoting the attempts at systematics that Hooker was anxious to discourage, yet—conscious of his debts to his correspondent—Hooker helped him join.[84] Colenso became a corresponding fellow of the Linnean Society in 1865 and got his FRS in 1886.

Confining Species

For colonial naturalists, compliance with metropolitan standards brought rewards, but at the potential cost of invalidating their prized local knowledge. Hooker's situation was in some ways the mirror image of theirs; the very steps he took to win their compliance encouraged them to be less obedient. And of course, it was Hooker's dependence on volunteer collectors that created this paradox.

In the *Flora Novae-Zelandiae*'s essay, Hooker referred to "all the individuals of a species (as I attempt to confine the term)"; "confining" species is an apt description of the practices involved in settling classification.[85] Surveying global variation was an attempt to confine species within a particular definition—his broad species concept. In the process, he restricted naming to the metropolitan herbarium, by confining colonial naturalists to a subordinate role as field collectors.

By "confining" species, Hooker hoped to establish his philosophical credentials while confining the local naturalists to the role of cautious students. Once again, Hooker's efforts were only partially successful. William Colenso, FLS, FRS, was not to be confined, arguing that "in our so-called *natural* system of classification" many of the taxa were "open to serious objections" since their advocates could not agree on the best characters to use when classifying. Local botanists could hardly be expected to comply with such uncertain metropolitan standards.

In Britain during the 1840s, the discord of competing classifications had threatened natural history's claims to be a mature science, especially when it was contrasted with geology, mineralogy, or chemistry, all of which were rapidly converging on internationally agreed, stable naming systems. Several metropolitan naturalists felt the need to bring stability by settling these arguments. Hugh Strickland was the most prominent zoological stabilizer, an opponent of quinarianism and other forms of classificatory radicalism.[86] As Gordon McOuat has shown, these competing systems prompted Strickland to try and establish the world's first formal rules of zoological nomenclature.[87] A committee of the BAAS, headed by Strickland, drew up proposed rules for zoological names, and although they were never formally adopted, Strickland succeeded in giving them quasi-official status by ensuring they were published in the association's *Report*.[88] Strickland's committee came up with an ingenious solution to the problems created by the competing systems; the rules they framed deliberately avoided the contentious issue of defining a species, especially since that might have led them into the explosive issue of transmutation. Instead, a pragmatic definition was used: species were whatever competent naturalists said they were.[89] As Strickland proclaimed, "Nature affords us no other test of the just limits of a genus (or indeed of any other group) than the estimate of its value which a competent and judicious naturalist may form." Competent naturalists were, of course, gentlemanly ones who exhibited their competence by joining metropolitan societies and publishing in their journals. By tacitly defining who was entitled to pronounce rather than defining the basis on which pronouncements were to be made, Strickland's rules limited the range of speculators and speculations by disbarring local naturalists from the classificatory process.[90]

From the perspective of practitioners like Hooker, there were many reasons why the Linnaean system had to go, yet it cast a long shadow over British botany and played a major role in establishing its low status. Perhaps the decisive, symbolic nail in its coffin came in 1863, when the third edition of *English Botany* appeared. Although the original plates had been retained, Smith's text—and with it the Linnaean classification—was finally gone, replaced by a natural classification. The *Athenaeum* welcomed this, commenting that since the 1840s "no general account of our native plants *written by men of note* has appeared which did not reject the Linnean classification."[91] The author of a flora that used the Linnaean system was, by definition, not "a man of note," whereas one who used the natural system was; Lindley had effectively made the same move when he declared that the Linnaean system was "disused by men of science." The natural system, and its attendant philosophical practices of anatomy and distribution, helped to more clearly define a science of botany.

Hooker's solution to the problem of instability in botanical systematics had some similarities with Strickland's approach.[92] It might be assumed that the prestige and authority of Kew would be enough to impose de facto nomenclatural rules.[93] However, while Kew eventually built up collections and a reputation that allowed its systematists to impose order, during the middle decades of the century it was still struggling to be recognized as a scientific institution rather than as a royal park (a topic I shall return to). Since Hooker's institutional position was not strong enough to impose a definition of philosophical naturalist, he used the material practices of collecting and classifying to persuade colonial botanists to accept his authority. In the next chapter, I shall turn from Hooker the herbarium builder to Hooker the essay writer, to see how he used his introductory essays, not merely to settle classificatory disputes in the colonies, but to engage in speculations that he hoped would establish his reputation in the metropolis.

CHAPTER SEVEN

Publishing

The abolition of the Stamp Duty on Newspapers and the consequent reduction of prices, enlargement of size, and wider diffusion of the daily and weekly press mark an era in our progress from the old region of enormous expenditure and fiscal restraint to an age when journalism may be said to be as universal as air or light.
—"Cheap Literature," *British Quarterly Review* (1859)

The massive proliferation of print in Victorian Britain created something of a crisis: with so many books, magazines, and pamphlets available, how could readers decide which ones to read and, more importantly, which ones to trust? The men of letters formed a growing and increasingly diverse group; from aristocrats and others who had no need of additional income, to penny-a-line hacks, who eked out a precarious living on Grub Street. Some men of letters were socially accepted and "lionised," but there were often intense discussions as to whether or not writers were "gentlemen"; since this term's fluid meanings still included the connotation of someone who did *not* work for a living, the social standing of even the most successful writers could be compromised by their association with trade.[1] Since the social standing of authors was uncertain, so too was their credibility, and as writing was often an important source of income for the men of science, they became caught up in these problems: how could they establish their status and trustworthiness?

Given the lack of scientific careers, many men of science—such as John Lindley, Edward Forbes, and William Hooker—wrote to supplement their incomes. They and their contemporaries soon found themselves churning out everything from scholarly tomes to encyclopedia entries, from practical manuals for the amateur chemist or astronomer to luxurious, illustrated monographs for the genteel connoisseur. Their words filled both the columns of the highbrow quarterlies and the cheap, steam-printed pages of organizations like the Society for the Diffusion of Useful Knowledge or the Religious Tract Society. As new technologies and sciences (including, of course, the steam engines that powered the presses and distributed their products) affected every aspect of Victorian life, the public and the publishers turned to the men of science—a miscellaneous group of often self-appointed experts—for enlightenment and guidance. As a result, the influence of science and those who wrote about it extended well beyond strictly scientific articles or publications, but as the opportunities for promoting science (and earning a living from it) multiplied, so too did the cacophony of contradictory opinions.[2]

As book historians have shown, trustworthiness is not an inherent property of printed matter; at a very early stage in the history of printing, it became apparent that errors and forgeries were preserved and distributed just as readily as accurate facts. Questions of reliability pressed especially hard on the natural history sciences because they were particularly reliant on such genres as published accounts of travels and of the organisms travelers discovered. The authority of a particular book was ultimately established through a complex negotiation between authors, editors, publishers, printers, booksellers, reviewers, and—of course—readers.[3] For example, a publication's credibility was partly established by where and by whom it was published: works published in the colonies were, unsurprisingly, usually less well regarded than those that emerged from the metropolis. However, standards of credibility, like the other kinds of standards I have discussed, varied from time to time and place to place; publication in New Zealand might enhance the reliability of a book for a New Zealand audience, even as it undermined it for a British one.[4] The language employed in a book was also crucial: using correct technical terms enhanced an author's trustworthiness while intemperate or abusive language undermined it. Finally, the plausibility of any specific claim was partly determined by the claims that accompanied it: a highly speculative tone might undermine the believability of some of the book's factual observations. However, a speculative tone also established the work's genre, indicating the kind of audience the writer hoped to address. From a more or less explicit consideration of all these factors, the reader formed an impression of the author. Readers could also draw on anything that might be known about the writer (their previous works, formal qualifications that might be listed on the book's title page, and so forth) in making a decision on what kind of trust, if any, they were going to place in the book. By considering the issues of a book's publisher, its language, and its tone and genre in turn, we can build up an idea of the kind of author someone like Hooker hoped to be—and whether he succeeded.

"Observers of All Ranks and Classes"

With growing literacy came a growing market for publications and, within that, an increasing diversification that reflected—and, of course, partly created—a parallel diversification of audiences. As we have seen in earlier chapters, within the very broadly defined market for "scientific" works, there were several more specialized and overlapping markets for books and periodicals that dealt mainly with flowers and plants—from expensive floras to cheap, practical works on gardening. However, there was no recognizable genre of, or market for, "philosophical botany." When the *Quarterly Review* discussed John Lindley's *Introduction to the Natural System of Botany,* its comments appeared under the running headline "gardening," since Lindley's book had been reviewed alongside such works as Joseph Paxton's *The Cottager's Calendar of Garden Operations* and the *Miscellaneous Writings of John Evelyn, Esq.* After dealing with the question of when to plant specific crops, their reviewer observed that "at the present epoch . . . the amateur gardener can hardly get on with satisfaction to himself . . . without acquiring some knowledge of botanical arrangement," which allowed a smooth segue into his criticisms of Lindley. Once the philosophy of classification had been dealt with, the reviewer turned to practical tips for raising old, long-neglected varieties of salad plants that were then coming back into fashion, followed by a lengthy discussion of salad dressings.[5]

Such genre-bending juxtapositions were commonplace in the reviews and often reflected the similarly diverse contents of botanical books and periodicals. There was widespread interest in natural history across Britain, and several journals were founded in an effort to take advantage of it—with varying degrees of success.[6] When *The Gardeners' Magazine of Botany, Horticulture, Floriculture, and Natural Science* began publication in 1850, its catchall title was indicative of the wide audience it needed to attract if it was to survive. The magazine aimed to include everything from new plants published in English and foreign magazines to such plans for gardens, greenhouses, and similar structures "as may assist amateurs and others to comprehend these matters" and to build them themselves. The publishers also boasted: "The pages of the Magazine will, moreover, be open to communications and criticisms on all branches of Natural

History, Gardening, or Botany; and Observers of all ranks and classes, Naturalists and Cultivators, Gardeners and Garden amateurs, are invited to record the results of their investigations." The publishers, W. S. Orr and Co., intended their magazine to replace the bankrupt *Paxton's Magazine of Botany* and clearly hoped to build a commercially viable readership by attracting "Observers of all ranks and classes." Orr and Co. published everything from periodicals such as *The Magazine of Domestic Economy* to books on a broad range of natural history topics (from Jane Loudon's *The Ladies' Flower-Garden of Ornamental Bulbous Plants* to Robert Patterson's *Letters on the Natural History of Insects, Mentioned in Shakespeare's Plays*), but the number of such works was equaled—and perhaps surpassed—by a miscellaneous range of commercial works, from *The Comic Album: A Book for Every Table* to the rather more forbidding *China and the Chinese . . . the Evils Arising from the Opium Trade*, by Henry Charles Sirr.

Given their varied list, Orr and Co. were determined to ensure that no sort of plant enthusiast would be overlooked; the magazine was to be "published in MONTHLY PARTS only, price at 2s. 6d. each., containing Five Plates and Thirty-two pages of Letterpress, interspersed with Engravings on Wood." Using both text and images, it would "teach Lady or Gentlemen Amateurs the art of designing, laying out, selecting, and planting gardens and pleasure-grounds" as well as communicate "all additions to the Physiological department of Botany." It was therefore hoped that "*The Gardeners' Magazine of Botany* is thus designed to become a Miscellany of useful and interesting information on every branch of horticultural science. 'Science with Practice,' is to be its motto." This was no empty slogan; alongside the new flowers and garden plans, the editors proudly noted that the "scientific portions of Systematic and Physiological Botany, have been entrusted to A. Henfrey, Esq."[7]

At this time, Arthur Henfrey was a poorly paid botanical lecturer at St. George's Hospital, London, but he had philosophical ambitions to match those of Lindley or Joseph Hooker; he wrote numerous scientific articles and eleven books and was an influential translator (fig. 7.1) who played a key role in bringing the new Continental botany to English-speaking readers.[8] Nevertheless, he could not afford to turn down the chance to increase his income by writing for as broad an audience as possible.

As we have seen, William Hooker was in a similar position, and when the *Monthly Review* assessed the latest issue of Hooker's *Botanical Miscellany* in 1831, it noted that "this is one of the few scientific journals that are published in this country . . . for the support of which . . . no

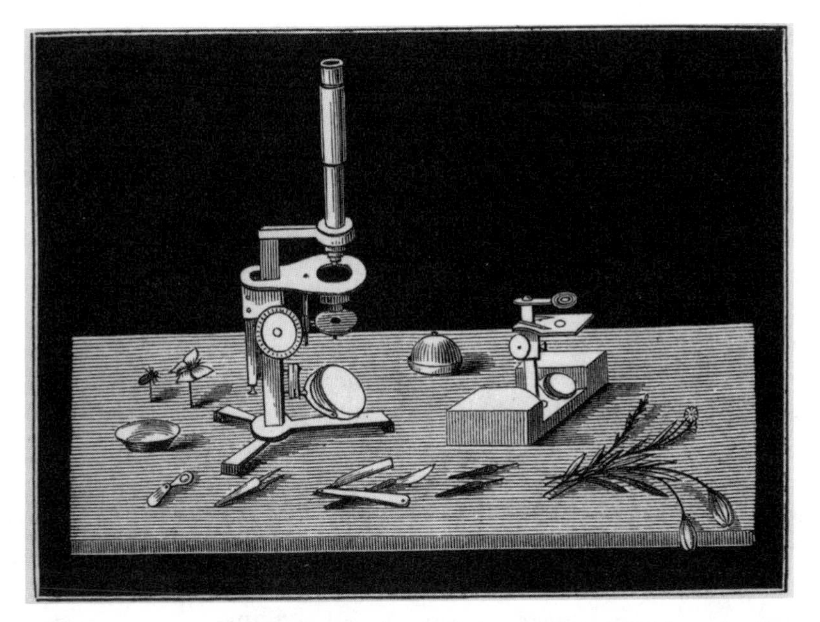

7.1 The knife and the microscope, the key tools for a philosophical botanist. From Schleiden's *The Plant*, translated by Arthur Henfrey. By kind permission of the Trustees of the Royal Botanic Gardens, Kew.

adequate patronage has yet been obtained among those classes who can best afford it"—hence the necessarily commercial nature of Hooker's undertaking.[9] Color plates were essential to the success of many of Hooker's publications, of which the best example was the *Botanical Magazine*, which had been founded by the Quaker naturalist William Curtis in 1787 to bring him an income which his large, lavishly illustrated *Flora Londinensis* (1777–87) had not. As Curtis said, the latter brought him praise, but the magazine brought "pudding." As previously mentioned, when Hooker took control, he worked to make it more scientific, yet it was clearly still intended to reach a broad audience, particularly the growing army of amateur Victorian gardeners. The *Botanical Magazine* described new species in technical detail but also gave handy hints about their cultivation; it brought the Hooker family "pudding" for many years. As the suburban garden was popularized as a place for healthy, educational exercise—for self-improvement as much as for horticultural improvement—numerous publishers took advantage of the new breakthroughs in printing and publishing technology to produce popular books and magazines for gardeners. John Claudius Loudon founded the first, the *Gardener's Magazine*, in 1826. Its success inspired competition, and by the 1840s, it had been joined by the *Gardener's Gazette,* the *Horticultural Magazine,* and the *Cottage Gardener.*[10]

Joseph Hooker, as I have noted before, generally avoided producing popular works, perhaps because, as he explained to Ronald Gunn in 1844, there was not "the slightest chance of making money now by Nat. Hist. publications, they are at a discount altogether, the market is glutted."[11] According to William Hooker, convincing Lovell Reeve to take on the *Botany of the Antarctic Voyage* had not been easy; the book was accepted only on condition that not one copy was be given away to anyone who might otherwise buy it.[12] Such strictures reflected Reeve's relatively uncertain finances, which partly resulted from the fact that, by contrast with a firm like Orr and Co., Reeve's catalog consisted almost entirely of natural history books. Reeve himself was an enthusiastic conchologist and a fellow of both the Linnean Society and the Geological Society, although the "taint of trade" blocked his attempt to join the Royal Society, despite Darwin's support.[13] In addition to Hooker's books, Reeve published other technical and systematic works, such as George Bentham's *Handbook of the British Flora* and his *Flora Hongkongensis,* as well as William Harvey's *Phycologia Australica; or, a History of Australian Seaweeds.* However, the profitability of such works was unpredictable, and so Reeve supplemented them with a series of "popular" histories, which ranged from *Popular British Conchology* and *Popular History of Birds* to *Popular Greenhouse Botany.* The success of the popular gardening magazines led Reeve to launch his own, the *Floral Magazine,* in May 1860, with illustrations by Walter Fitch (William Hooker was understandably annoyed at having his artist poached by a rival to his own *Botanical Magazine*).[14] The fact that Reeve would run the risk of alienating such an influential figure as Hooker suggests his business must have been quite strongly in need of additional income.

It was difficult for a serious publisher to make a living without producing popular works, and those who hoped to become philosophical botanists frequently found themselves in the same position. So, despite the doubts he may have had about popular works, Joseph Hooker continued to edit the *Botanical Magazine* for many years after his father's death and, as we have seen, Hooker's *Student's Flora* was also intended to make money, albeit from a rather narrower audience.

While periodicals multiplied in Britain, the English-speaking, educated populations of colonies like Australia and New Zealand remained too small to support many local scientific publications. Nevertheless, the changing economics of printing did permit a few to start up, such as the *Tasmanian*

Journal of Natural Science, Agriculture, Statistics, &c., which Ronald Gunn edited during its brief existence.[15] In 1844 the journal's eighth issue included "a paper on some New Ferns by Colenso."[16] When Hooker complained about Colenso multiplying the numbers of fern species, the latter reminded him of this "little V.D.L. publication," in which he had tried to *reduce* the number of published species: "to you both personally & by letter, I remarked, that *Asplenium polyodon*, & *lucidum*, & *obligum*, were only *one* sp.—All this at the outset! & against such first rate authority as your own dear & honored father."[17] Of the more than one hundred papers that Colenso published on every kind of scientific subject, most appeared in the *Transactions of the New Zealand Institute*.[18] In his later years, when he found himself constantly frustrated by Hooker's refusal to use his names, Colenso took to publishing the descriptions of his many new species in this journal, but all were ignored by Hooker.[19]

Colenso's papers in *Transactions of the New Zealand Institute* are invariably concerned with creating new species, which might explain Hooker's decision to ignore them, but his refusal to take note of Colenso's 1844 paper, which reduced the numbers of fern species, suggests that he had other goals in addition to reining in the species splitters. Whenever possible, Hooker would ignore any publication in a colonial journal, such as many of the names published by the Australian naturalist Ferdinand von Mueller.[20] In private letters, Hooker complained constantly about Mueller's systematic labors, telling Bentham that the Australian was "vomiting forth new genera & species with the lack of judgement of a steam dredging machine."[21] However, despite the huge numbers of species Mueller published, he was—unlike Colenso—a lumper, not a splitter; in fact, his species were often even broader than Hooker's.[22] Clearly, Mueller's species concept was not the issue; his perceived "lack of judgement" was a factor, but his real offense was refusing to be brought under Kew's control—publishing in colonial journals was just a symptom of his recalcitrance.

The common factor that united Colenso and Mueller was that by publishing in the colonies they were trying to evade Kew's supervisory role. Hooker argued in the *Flora Indica* "that the time is rapidly approaching, when the difficulty of obtaining access to the necessary periodicals must render the effectual study of botany impossible." Even with Kew's library at their disposal, he and Thomas Thomson had been unable to examine "several journals of local or ephemeral interest"; they suggested that "it would be well if isolated naturalists paused before they sought to establish such." Instead of a proliferation of local journals, Hooker promoted "centralization within reasonable limits," by urging such "isolated naturalists" to send their contributions to "the Transactions of well-established Associations for the furtherance of natural science." One benefit of this, for a metropolitan naturalist at least, was that the papers sent to such journals were refereed before publication, "which ensures their being worthy of it."[23] Little imagination is needed to guess whose papers Hooker thought would prove *un*worthy of publication.

This proliferation of colonial and foreign journals led Bentham and Hooker to devise the "Kew Rule" as a way of consolidating their botanical empire and limiting the role of colonial naturalists.[24] Their informal law concerned priority in naming, and one of its conventions was that commonly used names should be allowed to stand, even if they were not the first names to have been given. Bentham (who had a legal training) argued "that long-established custom amounts to prescription" and so "could and may justify the maintenance of names," even when those names were "exceptions to those laws which should be strictly adhered to in naming new names."[25] The Kew Rule reduced the time Bentham and Hooker had to spend in searching "journals of local or ephemeral interest" and was the equivalent in the library of lumping in the herbarium.[26] And, as with lumping, the Kew Rule had the not-unwelcome effect of preventing colonial botanists from naming species by justifying the metropolitan naturalists' decision to ignore names published in the colonies.[27]

When Hooker ignored Colenso's papers in the *Transactions of the New Zealand Institute*, Colenso (fig. 7.2) tried sending a paper describing some new ferns to the Linnean Society of London in December 1884, but it was rejected.[28] He wrote immediately to Benjamin Daydon Jackson, the society's secretary: "I received your letter of 9th April,—informing me of the fate of my Botanical Paper through 'my specimens being quite identical with specimens at Kew, previously sent thither by myself.'—Never man was more surprised than I on receiving your letter! and although 10 days have since elapsed I have not yet forgotten it—or, more plainly, recovered the shock it gave me."[29] He apologized for having sent Daydon Jackson on an unnecessary trip to Kew, but "I never supposed that such was requisite on your part"; Colenso had probably hoped that no recourse to Kew would be involved. Once again, colonial species were simply being "disallowed" by Kew, but Colenso, ever the champion of local knowledge, still would not bow to metropolitan authorities, convinced that they would agree with him if they "had seen them living and growing and in plenty as I have."[30] Inevitably, his complaints did him no good; by this time, Kew's reach had extended to the point where it could prevent names appearing, even when they had been sent elsewhere to be published.

7.2 William Colenso in later years. This hand-colored photo was sent as a gift to Joseph Hooker. By kind permission of the Trustees of the Royal Botanic Gardens, Kew.

"Correct Botanical Language"

As Colenso found, simply getting into print could be difficult for an aspiring botanist, but it was only the beginning of the process if one wished to produce publications that would enhance one's reputation. A reader's decision to purchase a specific book or journal might depend on its publisher's reputation, but his assessment of the author's language, the tone and style in which he wrote, would invariably give a stronger sense of an author's persona and intentions.

As we have already seen with botanical images, the conventions followed helped define the genre to which the picture belonged, and the same was true of botanical writings. For example, using the correct technical terms enhanced an author's credibility. John Lindley (fig. 7.3), ever mindful of the need to earn extra income, published a one-shilling guide to *Descriptive Botany* in 1858, which he described as intended "For Self-instruction, and the use of schools." As Lindley noted in his preface, botanical description had become part of botanical examination at the universities of

7.3 John Lindley, reading the galley proofs of one of his many publications. Drawing by Sarah Crease (née Lindley). British Columbia Archives (PDP02938).

Cambridge, London, and elsewhere, so he had produced his primer to assist students. As he argued, "the art, for it is an art, of drawing up descriptions of plants in correct Botanical language, should be rendered so familiar that all who have mastered the rudiments of Botany . . . may understand it." And, to broaden his market as much as possible, he also recommended his volume to "Gentlemen who mainly rely upon self-instruction [who] cannot dispense with a guide to such technical subjects as descriptive Botany."

The most obvious feature of botanical language was, as we have already seen, the use of Latin, and Lindley included many of Linnaeus's maxims in their original Latin. He assured "those who are unacquainted with that language" that the English text would still be invaluable, but urged his readers "to accustom themselves to describing plants in Latin, which is much better suited for the purpose than English." Lindley argued that common names and words lacked precision, and as a result, "popular" descriptions, "like the drawings of plants made by persons who are not botanists, are more calculated to mislead than inform."[31]

One effect of using "the technical and precise language of the science" was to achieve the brevity that was regularly urged on collectors. Lindley gave an example of a lengthy description of a tree in everyday language, which remained vague, whereas had the writer used technical terms "instead of the vague and uninstructive phraseology of mere conversation, he would have conveyed a distinct idea to the mind of the reader, at the expense of only about a dozen words instead of sixty-eight."[32] Lindley concluded: "In drawing up a description care must be taken that every term is used in its exact sense; that all is perspicuous and free from ambiguity, and that nothing superfluous is introduced."[33] This injunction was followed by many pages of terminology and illustrations plus ten pages of examples. These standard terms had, of course, been coined by European botanists to describe European plants, so by encouraging collectors to use them, metropolitan experts were also encouraging the colonists to look for those features of their local plants that were already familiar. In effect, local naturalists were being encouraged to look for similarity, not difference.

For colonial naturalists in particular, the use of indigenous languages was also an issue. The use of anything other than Greek or Latin was generally condemned by Linnaeus and most of the European naturalists who followed him, but—as we have seen—for someone like Colenso, who was an expert in the Maori tongue, it was a source of considerable pride to be able to record and use indigenous plant names. However, Maori had a deeper significance for Colenso, as is obvious from his pamphlet *On Nomenclature*, which ranged across both the uses and abuses of Maori place-names to a critique of changing fashions in scientific naming. This pamphlet also reveals that it was not just in Britain that Colenso had difficulty getting published: *On Nomenclature* was one of three pamphlets that he had privately printed after they had been, as he put it, "omitted" from the *Transactions of the New Zealand Institute*; he deliberately chose to get them printed in the same format as the *Transactions*, so that they "could be bound up with them."[34]

In his pamphlet, Colenso noted the numerous errors that the "ignorance of the *pakeha* (= foreigner)" had inflicted on Maori place-names; he argued that such errors could not be left uncorrected, since "Language adheres to the soil, when the lips which spoke it are resolved into dust. Mountains repeat, and rivers murmur, the voices of nations denationalized or extirpated in their own land."[35] These are unusual sentiments for the nineteenth century; it was more common for *pakeha* to describe those "extirpated in their own land" as victims of the inevitable superiority of the white races. Colenso's feeling for the Maori is unusual; he wrote that "strange thoughts arise at times within me," when he contrasted "the uncivilised unlettered Maori carefully handing down the names of places and things obtained from his forefathers from time immemorial, without error or change," with the supposedly "civilised lettered European," who, while claiming to use those same names, "neither speaks nor writes them correctly, and, worse still, does not care about doing so!"[36]

As evidence of the "copious, fluent, flexible, and euphonious Maori language," Colenso gave examples of "their genius for Nomenclature, and apt and fertile invention," such as *wiwi,* "Frenchman," "from their own manner of saying *Oui.*" Colenso believed that, sadly, the period of inventiveness was largely over, because of the rise of a patois, which he derided as barbarous.[37] He argued that "had it not been for their obtaining a written language through the Church-of-England Missionaries, and also had books printed in correct Maori by them, the Maori language would have soon become irretrievably lost."[38] He also attacked the policy of government documents being "printed or written in 'Official' Maori," which used many of the "barbarous *patois* words and phrases," which he felt were ugly, unnecessary and often ill-understood.[39] Ironically, however, Colenso had contributed to the impoverishment he complained of by bringing the first printing press to New Zealand (fig. 7.4). The missionaries had indeed preserved aspects of the Maori language and culture, but at the cost of standardization. The Maori had no written language prior to European settlement, so the missionaries created one for them; the effect was to preserve just one of the many Maori dialects—along with a smattering of neologisms and simplified terms—which then became fixed as standardized Maori.[40]

Colenso's discussion of names and naming moved from Maori words to botanical names, and he urged the importance of giving apt and meaningful names. He gave examples of Latin names that met these criteria, such as *Metrosideros,* from the Greek *metra* (heart) and *sideron* (iron), and so literally means "iron-heart" because of the tree's hard wood. Such names, he noted, had been extremely helpful to him when he first began studying botany.[41] Colenso therefore urged that New Zealand's plants be given names that were derived from their Maori names, as long as the Maori words were apt, properly understood, and correctly spelled.[42] This made perfect sense for naturalists like himself, who were based in the colony and knew some Maori, but Linnaeus had condemned what he called "barbaric" names (those not derived from Greek or Latin) because "they are from languages not understood by the learned," such as English, and so were a barrier to international communication.[43] Colenso was well aware of the learned Swede's condemnation, but he cited Lindley's observations on the benefits of retaining names that were familiar to local residents and

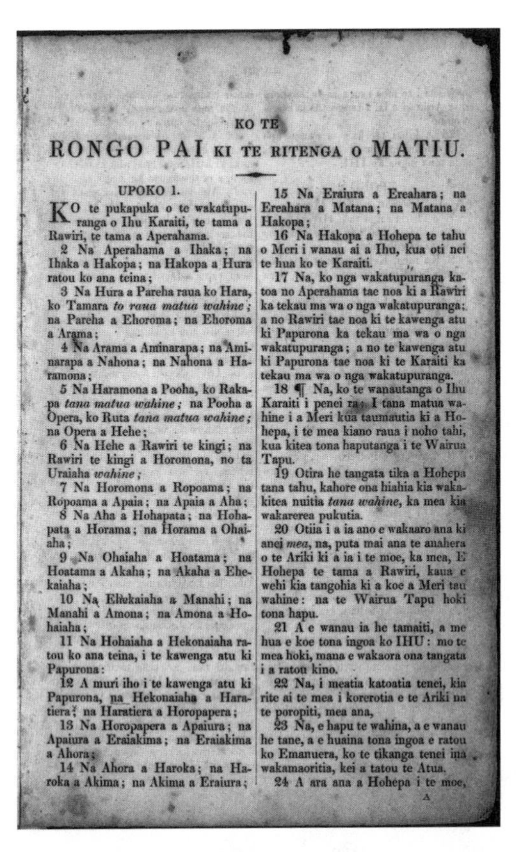

KO TE

RONGO PAI KI TE RITENGA O MATIU.

UPOKO 1.

KO te pukapuka o te wakatupuranga o Ihu Karaiti, te tama a Rawiri, te tama a Aperahama.

2 Na Aperahama a Ihaka; na Ihaka a Hakopa; na Hakopa a Hura ratou ko ana teina;

3 Na Hura a Pareha raua ko Hara, ko Tamara *to raua matua wahine;* na Pareha a Ehoroma; na Ehoroma a Arama;

4 Na Arama a Aminarapa; na Aminarapa a Nahona; na Nahona a Haramona;

5 Na Haramona a Pooha, ko Rakapa *tana matua wahine;* na Pooha a Opera, ko Ruta *tana matua wahine;* na Opera a Hehe;

6 Na Hehe a Rawiri te kingi; na Rawiri te kingi a Horomona, no ta Uraiaha *wahine;*

7 Na Horomona a Ropoama; na Ropoama a Apaia; na Apaia a Aha;

8 Na Aha a Hohapata; na Hohapata a Horama; na Horama a Ohaiaha;

9 Na Ohaiaha a Hoatama; na Hoatama a Akaha; na Akaha a Ehekaiaha;

10 Na Ehekaiaha a Manahi; na Manahi a Amona; na Amona a Hohaiaha;

11 Na Hohaiaha a Hekonaiaha ratou ko ana teina, i te kawenga atu ki Papurona:

12 A muri iho i te kawenga atu ki Papurona, na Hekonaiaha a Haratiera; na Haratiera a Horopapera;

13 Na Horopapera a Apaiura; na Apaiura a Eraiakima; na Eraiakima a Ahora;

14 Na Ahora a Haroka; na Haroka a Akima; na Akima a Eraiura;

15 Na Eraiura a Ereahara; na Ereahara a Matana; na Matana a Hakopa;

16 Na Hakopa a Hohepa te tahu o Meri i wanau ai a Ihu, kua oti nei te hua ko te Karaiti.

17 Na, ko nga wakatupuranga katoa no Aperahama tae noa ki a Rawiri ka tekau ma wa o nga wakatupuranga; a no Rawiri tae noa ki te kawenga atu ki Papurona ka tekau ma wa o nga wakatupuranga; a no te kawenga atu ki Papurona tae noa ki te Karaiti ka tekau ma wa o nga wakatupuranga.

18 ¶ Na, ko te wanautanga o Ihu Karaiti i penei ra: I tana matua wahine i a Meri kua taumautia ki a Hohepa, i te mea kiano raua i noho tahi, kua kitea tona haputanga i te Wairua Tapu.

19 Otira he tangata tika a Hohepa tana tahu, kahore ona hiahia kia wakakitea nuitia *tana wahine*, ka mea kia wakarerea pukutia.

20 Otiia i a ia ano e wakaaro ana ki anei *mea*, na, puta mai ana te anahera o te Ariki ki a ia i te moe, ka mea, E Hohepa te tama a Rawiri, kaua e wehi kia tangohia ki a koe a Meri tau wahine: na te Wairua Tapu hoki tona hapu.

21 A e wanau ia he tamaiti, a me hua e koe tona ingoa ko IHU: mo te mea hoki, mana e wakaora ona tangata i a ratou kino.

22 Na, i meatia katoatia tenei, kia rite ai te mea i korerotia e te Ariki na te poropiti, mea ana,

23 Na, e hapu te wahina, a e wanau he tane, a e huaina tona ingoa e ratou ko Emanuera, ko te tikanga tenei ina wakamaoritia, kei a tatou te Atua.

24 A ara ana a Hohepa i te moe,

A

7.4 First page of the Gospel according to St. Matthew, from Colenso's Maori-language Bible. By permission of the Alexander Turnbull Library, Wellington, NZ.

so recommended preserving indigenous names by converting them into scientific ones "by adding a Lsatin termination to them." [44] However, botanists like Hooker who shared Linnaeus's concern with global knowledge generally ignored such names, although they were still used by those like Colenso who wished above all to record their flora's unique names, habitats, and endemic species.

The processes Colenso complained of with regard to Maori—the standardization of a limited language designed for imperial purposes and then preserved and disseminated through printing—were strikingly similar to the impact of botanical terminology on colonial naturalists. The technical terms that European botanists attempted to impose on colonial ones were ostensibly designed to make the colonists' reports more reliable, partly by allowing them to understand European botanical textbooks. In a similar way, written Maori had been designed to give the indigenous people access to the Bible and other kinds of European knowledge. So, like the Maori he preached to, Colenso adopted Europe's botanical terms, abandoning a rich local language in an effort to be taken seriously by Europe's men of science. However, the effect of his compliance was that he lost the ability to express or record much of the local knowledge he valued; Maori names, like their plant lore and place-names for obscure local habitats, were invariably missing from the metropolitan publications in which Colenso hoped to see his name.

The credibility of a man of science's writing was assessed by his use of not only the correct terminology but also a courteous tone. However, while these elements were necessary to establish one's scientific credentials, they were not sufficient to show readers that one was genuinely philosophical; achieving that distinction depended on engaging in speculations that went beyond mere cataloging and recording. As Hooker had told Darwin, his delight at being able to address Darwin's questions on "philosophical" botany was that addressing such issues was "the best means of keeping alive a due interest in these subjects. I indulge vague hopes of treating of them some day."[45] Nonetheless, speculation was fraught with potential danger, and the would-be philosophical naturalist needed to ensure that he engaged in the right kind of speculation for the audience he wished to address.

"Perhaps a Fanciful Idea"

Because of the enormous range of metropolitan and colonial publications, men like Joseph Hooker had to put considerable effort into distinguishing the work they produced from the other kinds of botanical publishing. The increasingly long and detailed essays that accompanied Hooker's successive publications were declarations of intent, intended to distinguish himself and his science from the many other kinds of botanical writers and their products. This was a goal in which he generally succeeded; the independently wealthy naturalist and traveler Charles James Fox Bunbury described the *Flora Antarctica* as "an admirable book, and gives [Hooker] I think the very first place among the botanists of the new generation."[46] More importantly still, the essays were intended to establish his reputation as a respectable philosophical speculator in the eyes of his fellow metropolitan gentlemen of science. Once again, it is worth recalling that the de facto definition of "philosophical" that Hooker and his peers shared was only one of many possible definitions, even among metropolitan gentlemen; his essays served both to exemplify his conception of what was properly philosophical and, in the process, to persuade a specific metropolitan audience that he was one of them. Moreover, that audience had to be convinced that Hooker's view was not merely the consensual one but was founded on clearly stated, philosophical principles, thus establishing his science as mature and stable; in particular, his broad species concept had to be accepted as central to good systematic practice.

Given Hooker's desire to restrict colonists to collecting and discourage them even from naming plants, much less speculating about the philosophy of classification, one might expect the *Flora Novae-Zelandiae*'s introduction to eschew hypotheses and focus instead on the minutiae of collecting and recording data. However, it included Hooker's most speculative theory. He noted that the tiny isolated sub-Antarctic islands, despite being "few in species" like all insular floras at such latitudes, nevertheless had some characteristics that were shared with New Zealand, Australia, and the larger landmasses. It was as if, he argued, they were "the vestiges of a flora which belonged to another epoch, and which is passing away." He speculated that the islands of the Southern Ocean were home to "the remains of a flora that had once spread over a larger and more continuous tract of land than now exists in that ocean; and that the peculiar Antarctic genera and species may be the vestiges of a flora characterized by the predominance of plants that are now scattered throughout the southern islands."[47] The relationships among the plants of the southern oceans suggested that the islands they grew on were the remains of a sunken southern continent. He admitted that "this is perhaps a fanciful idea," but it was, nonetheless, "one which I believe to contain the germ of truth."[48]

Hooker's hypothetical missing continent was his contribution to a wider debate about the distribution of plants, the flora and fauna of islands, and the potential role played by geological processes in explaining the observed distribution. I will return to the details of this debate and its implications in the next chapter, but for the moment, I want to examine what Hooker might have hoped to achieve by prefacing a fairly straightforward catalog of species with such "a fanciful idea."

Two years after the *Flora Novae-Zelandiae*'s introduction appeared, in July 1855, Hooker asked Bentham, who was then in Paris, how the French botanists were reacting to the even longer and more wide-ranging introduction that he had written to the *Flora Indica*, adding: "I have frightened them here out of their wits, and some of them thank me for the presentation copy with a frigidity that delights me." He claimed that "Henfrey is shot, and proposes altering his whole system of Botanical instruction at

King's College!" and that "My cher confrères the geologists shrug their shoulders and do not half like it." Hooker anticipated that the strong reactions that the essay would provoke would advance his career: "Hitherto Botany has been dull work to me, little pay; no quarrels; an utter disbelief in the stability of my own genera and species; no startling discoveries; no grand principles evolved, and so I have a sort of wicked satisfaction in seeing the fuse burn that is I hope to spring a mine under the feet of my chers confrères, and though I expect a precious kick from the recoil and to get my face blackened too, I cannot help finding my little pleasure in the meanwhile."[49] Hooker hoped his essay would end the period of "dull work" and "little pay" by provoking quarrels over "grand principles" that would establish him as a philosophical speculator, thus giving him the prestige to transform "utter disbelief" in the stability of his genera and species into acceptance. The 250 privately printed copies of the *Flora Indica* were to be a mine, detonated under the feet of his colleagues, that would bring him to the attention of the whole natural history world.

The ambition behind Hooker's introductory essays helps explain their relatively unusual character. Most mid-nineteenth-century floras were baldly descriptive; their introductions simply outlined the scope of the work and the history of botanizing in the region described, rather than expounding a philosophy of systematics or distribution. Such concerns were more typical of a slightly earlier genre, typified by Alexander von Humboldt's *Prolegomena* to his *De Distributione Geographica Plantarum* (1817) or Brown's *Prodromus* to the *Florae Novae Hollandiae et Insulae Van Diemen* (1810). In these cases, a largely descriptive work had been prefaced by an essay that surveyed broader philosophic questions; when Hooker decided to structure his own books along similar lines, he was allying his work with that of his most illustrious predecessors rather than with the descriptive empiricism that characterized botany in his day.[50] The novelty of Hooker's writing was evident to Darwin, who gently scolded his friend for "the way you sometimes speak of other people having philosophical minds, as if you had nothing of the sort," and told him to "compare any half-dozen-pages of the Antarctic Flora with other systematic works, dull as they generally are, & you will never have the face to depreciate yourself again."[51] Hooker's efforts to create a new botanical genre, which married the form of older essays with the most recent theories of classification and distribution, were intended to further his ambitions, partly by conveying subtly different messages to his colonial and metropolitan readers. I have already discussed the colonial audience, so now I want to consider how Hooker intended to "spring a mine" under the feet of his metropolitan contemporaries.

Hooker and Thomson gave 120 copies of the *Flora Indica* to various prominent men of science, including Darwin and William Whewell, master of Trinity College, Cambridge, in the hope that its emphasis on classification's philosophical precision would demonstrate botany's maturity.[52] When the essay's authors argued that "in the present unsatisfactory state of systematic botany it is the duty of each systematist to explain the principles upon which he proceeds," they were claiming to be equipped to tackle the philosophical issues on which the progress of botany depended.[53] Hooker hoped his emphasis on the degraded state of systematics would dissuade local botanists from creating further synonymy; meanwhile, the same words were intended to be read in Britain (particularly by the directors of the East India Company, whose patronage Hooker and Thomson were hoping to secure) as a plea for better funding, new publications, larger collections, and more extensive libraries—the resources with which they could bring order to the chaos.[54]

It is in the context of essay writing that "philosophical" has its most obvious, familiar meaning: like most of his contemporaries, Hooker used the term to refer to the business of identifying causal mechanisms, using the methodological rules of induction expounded by Whewell, John Herschel, and others.[55] Inductivism held a central place in nineteenth-century British philosophy of science, in part because the language of the inductive method worked as a tool with which those who wished to could try and improve the standing of their science.[56]

Geology provided naturalists with a clear example of how an appropriate philosophical approach could raise the prestige of a collecting science. In the 1820s and 1830s, geology's standing was constantly threatened by religious and other controversies, brought on by what James Secord has called its "philosophical promiscuity." This led to a focus on stratigraphy, which provided safe, relatively neutral ground, but at the cost of making the study seem trivial. (There are evident parallels with the consequences of collecting and classifying as unifying botanical practices.) The threat of controversy formed the background to Charles Lyell's attempt to establish the "philosophical foundations of geology," and Secord notes that "making science philosophical was a way of making it respectable."[57] Although few adopted Lyell's views in full, there was widespread agreement that his book put geology on a properly philosophical footing.[58] The increasing respect in which geology was held, combined with its practical importance for mining, led to its being better funded.

Before botanists could begin to ascend the path the geologists had marked out, they had to agree on a stable classification; everything else depended on this. The first step was to settle on a system that grouped plants in fewer categories and made the objective basis of those categories clear. As the *Athenaeum* commented in a botanical review, "The tendency of modern investigations and discovery has been rather to reduce than enlarge the number of Natural Orders."[59] Hooker expressed similar views when he claimed that the more experienced the botanist, the broader his species concept would be, because each botanist "begins by examining a few individuals of many extremely different kinds of species, which are to him fixed ideas, and the relationship of which he only discovers by patient observation; he then distributes them into Genera, Orders, and Classes, the process usually being that of reducing a great number of dissimilar ideas under a few successively higher general conceptions." This reduction of observations under "higher general conceptions" was, of course, precisely what Whewell thought had enabled sciences such as astronomy to progress. Hooker argued that, in the botanist's search for higher conceptions, "the abstract consideration of the species itself is generally lost sight of."[60] The higher taxa—genera and orders—become the botanist's main focus; lumping was both pragmatically useful and philosophically virtuous, and in both roles it concentrated authority in Hooker's hands. He berated those who created spurious species by overestimating minor variations: "To this tendency there can be no limit, when the philosophy of system is not understood," adding that "students are not taught to systematize on broad grounds and sound principles."[61] Broadly defined species were one of the "sound principles" upon which Hooker's philosophy of system was built.

Hooker cited Whewell's *History of the Inductive Sciences* in the introduction to the *Flora Indica*, and the two men soon became friends; Whewell consulted Hooker over revisions to the *History* and acknowledged his assistance in the third edition.[62] By comparison with Herschel, Whewell put a greater stress on the creative role of the theorist, which helps explain why Forbes's and Lyell's speculative ideas about geological processes (which, as we shall see, lay behind Hooker's missing-continent speculation) became central to Hooker's distribution ideas.[63] Hooker argued that "some perfectly original course of study must be adopted" before botanical geography could advance; it needed "some such bold original ideas as led Lyell first to conceive and then to prove that species may be older than the lands which they now inhabit, and that led Edward Forbes to seek in the distribution of the fossil remains of existing British species a key to their present diffusion."[64] Hooker expected ideas like the geological ones to overcome botany's weaknesses: "the vagueness of its principles, the inexactness of its methods, the puzzling complexity of its phenomena, and the purely speculative character of those hypotheses upon which all inquirers base their efforts to explain its facts and develope its laws."[65] Whoever came up with the bold idea that transformed distribution studies would do for botany what Lyell had done for geology; Whewell had celebrated Lyell as geology's Galileo, whose "geological dynamics" would eventually transform the subject into "a Science, and not a promiscuous assemblage of desultory essays."[66]

As we have seen, in the 1830s and early 1840s, philosophical natural history mainly referred to work that was either German, zoological, or anatomical (often all three); the BAAS's neglect of botany relative to zoology was one result of this de facto definition. George Bentham, Arthur Henfrey, Joseph Hooker, John Lindley, Edward Newman, and Hewett Cottrell Watson were among the botanists engaged in an attempt to make botany philosophical in its own terms—to have their systematic and distributional work recognized as philosophical, rather than accepting zoology's anatomical and physiological hegemony. I will argue that it is useful to think of Hooker and some of his contemporaries as hoping to expand the definition of "philosophical" to include French, botanical, and systematic.[67] These are not, of course, mutually exclusive categories, but it is noticeable that Hooker was as uninterested in much of German botany (such as Mathias Jakob Schleiden's work) as he was in German zoology.[68]

French, Botanical, Systematic

There was, of course, a German botanical tradition that focused on geographical matters and that ran from Johann and George Forster (Captain Cook's naturalists), via Karl Ludwig Willdenow, to Humboldt, and this was work Hooker and his contemporaries respected and cited.[69] However, the German zoologists were generally better known in Britain, and for Hooker and his colleagues, French or Francophone examples were at the heart of their idea of philosophical botany, particularly—as we have seen—the work of Michel Adanson, Antoine-Laurent de Jussieu, and Augustin-Pyramus and Alphonse de Candolle (and, every bit as important, the Jardin des plantes itself as a model of state-financed botany).

However, putting botany in its proper place required turning what was still known as "French botany" into an indigenous tradition; John Ray was a key historical figure in this attempt, but it was the living example of Robert Brown that was most important (fig. 7.5).[70] In 1831 the *Athenaeum* noted with satisfaction that botany was finally becoming a respectable,

7.5 Robert Brown: *Botanicorum facile princeps, Britanniae gloria et ornamentum.* State Library of New South Wales, Sydney.

well-regarded science. The anonymous author was Lindley, and as was so often the case, his comments reflected his aspirations rather than the actual state of his science. Nevertheless, he contrasted the present state of botany with that "twenty years since" when "the first really, we will not say philosophical, but logical, work upon the subject, made its appearance." This book "was so far either misunderstood, or not understood, or undervalued, as to have been finally withdrawn from public circulation by the author, after a sale of about 70 copies in two years." The unnamed volume was Robert Brown's *Prodromus Florae Novae Hollandiae et Insulae Van Diemen* (1810), which introduced the natural system to Britain. Hence Lindley's comment that: "At that time, such botany as is now taught was considered scientific heresy. The science (?) was confined within what the followers of Linnaeus chose to call the bounds defined by that celebrated naturalist; and all who dared, if any such there were, to overstep those limits, were accounted without the pale of science."[71] The Linnean Society refused to countenance any challenge to the sexual system, and the situation was no better at the English universities, where "what was called French Botany, was proscribed with more than political rancour." However, these days were past and "philosophical botany is now taught by the Professors in Cambridge, Edinburgh, Glasgow, and London."[72] This change had come about because a few botanists (the review's anonymity allowed Lindley to include himself among their number) had had the courage to defy the Linnaean "clique of English botanists." William Hooker was one of these, but the most important was "Mr. Robert Brown," who produced "many papers indicating deep philosophic research."[73] Brown, the Scottish curator of botany at the British Museum and de facto heir to Sir Joseph Banks, had introduced the natural system to Britain and pioneered both microscopic plant anatomy and distribution studies using botanical arithmetic. He was a mentor to Joseph Hooker and several of the other young botanists previously mentioned, not least because he gave "French botany" an aura of indigenous, British respectability.

In the mid-1840s, the *Athenaeum* described the state of British botany when Lindley had begun his career (i.e., c. 1820): "when botany could hardly be said to be cultivated as a science" because the "collecting of strange plants, and sticking them on white pieces of paper with a hard name attached to them, were all that was considered necessary to make a man a botanist." It was true that "Robert Brown had also lived and observed,—but the facts and principles of that greatest of botanists were regarded only as ingenious theories." Plant anatomy and physiology were neglected, and Lindley soon recognized "the insufficiency—not to say absurdity—of the artificial system, as used by the followers of Linnaeus in this country." He therefore rejected the Linnean approach in favor of "that method of studying plants, which had first been pursued by his own great countryman, Ray, successfully developed by Adanson and Jussieu on the continent, and rendered so much more philosophical and accurate by the profound genius of Robert Brown."[74] Tracing the natural system back to Ray helped defuse its French nature, especially when it had been made "more philosophical and accurate" by another Briton. However, insiders like Lindley (who reviewed for the *Athenaeum* himself) and Hooker would have known that the anonymous reviewer was Edwin Lankester, a doctor and lecturer at the Grosvenor Place medical school, who—as we shall see below—had a concept of philosophical natural history that was rather different from theirs, despite their common aversion to the Linnean system. Although Lankester was respected enough to be secretary to both the Ray Society and Section D of the BAAS, he was a controversial character, a champion of educational and medical reform, including phrenology, and even of Chartism.[75] These were precisely the kinds of radical views that were traditionally associated with all things French, which the "Brownian" botanists needed to distance themselves from.[76] Lankester was "convinced, that Brown has found in Dr. Lindley an ardent disciple, and one who has done more to extend his fame and apply his principles of classification than any other European writer." The natural system was crucial to this genealogy—Lindley inherited the task Brown had begun—and the review concluded "that, in our present state of knowledge, this is at once the most philosophical and practical method of arrangement."[77]

If the correct philosophy of system was one part of being a Brownian botanist, an appropriate philosophy of distribution was the other; indeed, as we shall see in the next chapter, the two were so closely related as to be almost indistinguishable. Hooker said of geographical distribution that this "most interesting branch of botany, has made very little real progress of late years, owing to the confused state of Systematic Botany; for we do not consider rudely cataloguing the ill-defined species of limited areas, or loosely defining geographical regions . . . as calculated to advance directly the philosophy of distribution." However, he was forced to admit that the data were still inadequate to "pursue these interesting subjects"; classification had to come first.[78]

German, Zoological, Anatomical

German naturalists of this period can be divided into two main camps: the Kantian-inspired group Timothy Lenoir has christened "teleomechanists," such as Ernst von Baer; and the transcendental anatomists influenced by *Naturphilosophie*, including Lorenz Oken.[79] The latter group were better known in Britain, partly because Richard Owen was an exponent of this approach.[80] Both groups were dominated by zoologists but included a few botanists: Schleiden among the former group and Christian Gottfried Nees von Esenbeck among the latter.

In 1833, Lindley published a paper entitled "On the Principal Questions at Present Debated in the Philosophy of Botany," in which he discussed only physiology and anatomy, fields in which he admitted Britain had long been overshadowed by Europeans. Fortunately, Brown and one or two others had recently "entered into competition with the anatomists of Germany and France," which allowed him to hope that Britain might contribute to this important field.[81] The same equation of philosophical with anatomical was apparent in the *Westminster Review* in 1850, which acknowledged the close relationship between the natural system and "the geography of plants" but nevertheless insisted that "the theory of morphology, that great principle, . . . owing to the labours of Göthe, Oken, Carus,

Owen, and others is now universally acknowledged as the basis of all philosophical researches in natural history."[82] Goethe was the only botanist on the list, and his work was sometimes viewed as being of doubtful scientific status.[83]

Many British inductivists were repelled by any hint of German a priori thinking—whether influenced by *Naturphilosophie* or Kant; such idealism seemed to subordinate nature to preconceived schemes, instead of building up from facts to theories. Under Lindley's editorship, the *Gardeners' Chronicle* campaigned vigorously against Oken's *Naturphilosophie*, commenting: "In England, where men prefer fact to fiction, and expect all philosophy to be founded on intelligible principles, the dreamy mysticism of Oken will find little favour."[84] On another occasion, the *Chronicle* noted with relief that, in the book they were reviewing, "the philosophy of Oken has appeared in so very unattractive a form . . . that our young botanists are not at present likely to be Okenised."[85]

Hooker shared Lindley's low opinion of German theorizing and tended to "lump" all Germans together. In a letter to Huxley, he mentioned a translation of Alexander Braun's *Rejuvenescence of Plants* and asked, "when is this German rubbish to end?" Adding that "I protest boldly against such work as Oken, Braun, Schleiden, Meyer, and others, being given to the British Public without one word of explanation," he suggested that they all required an introduction "pointing out what can be understood from what can not."[86] After a visit to Zurich in 1862, Hooker commented that he was not surprised to find the place dominated by the "Okeno-Schleidenists," given the "dull and dreary" lectures of the traditional botany teacher.[87] Hooker's coinage of "Okeno-Schleidenists" reveals how little he knew about German thinking: Schleiden was a Kantian and a vigorous critic of the *Naturphilosophie* Oken espoused.[88] Nevertheless, behind this ignorance and the rhetoric about "German rubbish" and "fanciful" theorizing was a concern that botany would be adjudged a second-class science if German influences dominated, since they made systematic work secondary to physiology.

Oken's name became synonymous with unacceptable speculation among those British botanists who, despite their differences, shared the goal of making their discipline philosophical without subordinating it to zoology. Edward Newman ridiculed mesmerism by equating it with "Okenism."[89] Forbes made a similar linkage in his review of the geologist Adam Sedgwick's *Discourse on the Studies of the University of Cambridge*, where he noted that "Atheism, Pantheism, Vestigianism, Okenism, Hegelism [*sic*], Straussism, and Puseyism, alternately and intermingled, are splintered or rather pounded by the Woodwardian hammer; the 'Physio-philosophy,' the 'Vestiges of Creation,' and the 'Tracts for the Times,' are submitted to a process of critical cleavage."[90] This catalog of unacceptable ideas is similar to those listed by the liberal weekly the *Manchester Spectator*, which attacked Oken for his "pantheism" and described his supporters as "half-witted atheists. Who have got a smattering of chemistry, and a smattering of astronomy, and a smattering of phrenology," all of which was connected with "communism and sundry other *isms*."[91]

By contrast, those like Lankester who supported the Germans argued that the British classificatory tradition would never lead to genuinely philosophical natural history. In a review in the *Athenaeum* of Mathias Schleiden's *The Plant* (fig. 7.1), Lankester argued that the Linnaean system's ease of use had "induced men to search the world for new forms; and ship-loads of dried vegetables were deposited in the museums of Europe." Not only could "these haystacks" never supply "a true knowledge" of the vegetable kingdom, but the collecting mania "has given to the science of botany a repulsive character; and the sound common sense of the world has felt that a man could be no better for knowing the number of stamens or pistils in a buttercup or calling chickweed by a hard name." By contrast, Schleiden's work promised "to render botany not only an amusing but a useful and instructive science" and so, "as an introduction to the philosophical study of the vegetable kingdom, we can cordially recommend this volume."[92]

Rival definitions of "philosophical" often led to heated debate between naturalists. Historians rather enjoy describing arguments, because they often reveal underlying commitments, but we should be careful not to assume that the protagonists also enjoyed them, not least because for botanists trying to make their studies philosophical—and thus respectable—such arguments could be dangerous. It is no accident that Hooker's comments about "German rubbish" and "Okeno-Schleidenists" came in private letters, not in the columns of a periodical; he believed that such matters were better resolved between specialists in private; public dispute could bring science into disrepute. When Darwin became involved in Hugh Strickland's controversial campaign to reform zoological nomenclature, Hooker wrote from India to advise him to "drop the battle" because the subject was one "so closely touching the honor of some, the pride of many & the conflicting interests of all, that you will get into hot water." He warned his friend that "naturalists are of the genus irritabili" and "they will be down on you like Sikhs if you do not look out."[93] Despite his own obsession with stable nomenclature, Hooker saw no benefit to Strickland's high-profile attempt to enact a nomenclatural code and enforce it through the BAAS; as we have seen, Hooker's strategy, whether of persuasion, coercion, or flattery, was conducted almost entirely through private letters; when he did venture into print on controversial topics, it was in his relatively expensive botanical monographs, not in the popular quarterlies or weeklies. As we shall see, the same approach was apparent in his refusal to engage in religious controversy. Given this caution, Hooker's desire to "spring a mine under the feet" of his fellow botanists seems contradictory, but I think it can be understood partly by considering where the "mine" was placed, in a privately printed book of very limited circulation among the elite, and also by recognizing that Hooker was sensitive to the way in which disputes were conducted; they must not be marred by abuse or ungentlemanly language, as we shall see below.

"The Scientific Hive"

For both Whewell and Hooker, formulating hypotheses was essential to demonstrating one's right to be treated as one of science's "master spirits," and thus entitled to manage its untrained observers.[94] The anomaly of Hooker's publishing his "Forbesian" speculation about a missing Antarctic continent in the same volume as his attempts to warn his collectors off speculation is only apparent: speculating about lost continents was a way to demonstrate that he was no mere cataloger. Hooker's reliance on colonial collectors was closely paralleled by Whewell's use of untrained observers in his work on tides, which helps explain why Whewell was wary of a simplistic Baconianism that stressed facts at the expense of theories; devising appropriate theories was a first step toward validating the position of elite gentlemen like himself as supervisors of networks of observers.

The role inductivist philosophy might play in helping manage collecting networks is apparent from Forbes's 1843 "Inaugural Lecture on Botany" at King's College, London: "The scientific systematist, surrounded by the stores of his herbarium," should remember that his specimens had often been collected "by adventurous and earnest men," whose role was as important in its way as his, for "in the scientific hive as in the apiary there must be working-bees and neuters as well as queens and drones: it is necessary for the economy of the commonwealth."[95] This invocation of science's "working-bees" was a conventional Baconian trope (Forbes had quoted Bacon's *Novum Organum* on the preceding page): for those who saw themselves as ruling "the scientific hive," another reason to use the language of induction was that it necessitated a division of labor between themselves and the "drones" in the field—between, for example, Hooker and his colonial collectors.[96]

Forbes argued that botany, like all sciences, must pass through three phases, "OBSERVATION, CLASSIFICATION AND PHILOSOPHICAL INVESTIGATION," because a botanist needed "a correct acquaintance with the forms and structures of plants," and then "with the various families, orders and classes into which the genera and species of plants are grouped," "he is prepared to enter upon the philosophy of the science." In Forbes's view, botany was ready "to enter on the third aera of its existence, that of *philosophical investigation.*" This era was characterized by "the observation of facts, not so much for their own sakes as for the illustrations they afford of the laws of the science." Together with coordinated "experimental inquiries into the phaenomena of vegetation" and "minute anatomical investigation under the microscope," there must also be "the construction of local floras and local catalogues," "not with the limited view of assisting the inhabitants of a province to a knowledge of their vegetable compatriots, or with the pardonable vanity of showing how many fine plants grow in the author's country, but in order that the great laws of distribution of organized beings on the surface of the globe may be discovered and developed." These comments closely parallel Hooker's own on the vanity and parochialism of the local botanists. He would surely have agreed with Forbes that floras were to be used in "the construction of systematic arrangements, not framed solely for the ascertaining of the natural alliances of families, important as such an object is, but also with the view of discovering the great laws which doubtless regulate those alliances equally in the animal and vegetable kingdoms."[97] Forbes's lecture captured the way he and his colleagues hoped to use induction as a tool that would elevate them and their facts, via generalizations, to prestigious definers of laws.[98]

Timothy Alborn has shown that British scientific reformers—especially Whewell, Herschel, and Babbage—used Adam Smith's notion of the division of labor as their reforming principle. Smith had argued that such division allowed "the most dissimilar of geniuses" to be "of use to one another," a claim the scientific reformers seized on to argue that Britain's problem was not lack of native genius but poor organization and lack of job opportunities. Men like Charles Babbage and Herschel argued for free competition within a "deregulated" economy of science, while Whewell preferred to see individual geniuses in command of the scientific economy—a vision with obvious attractions for the autocratic, opinionated Hooker.[99] However, if men like himself were to command the scientific economy, they had to invest their intellectual capital—the authority they had accumulated—wisely; wild speculations would squander it by inviting the hive's worker bees to usurp the speculator's role. The master spirits also had to demonstrate that they were morally qualified for leadership.

"A Model Which May Well Be Adopted"

Imagine yourself as a Victorian reader, picking up one of Hooker's books, such as the *Flora Tasmaniae*—how would you assess its author's standing and the book's trustworthiness? First, there were the physical aspects of the book: as previously noted, it was a large quarto, which cost £9.5.0 or, for a version with hand-colored plates, £13.2.6; this would have been a very substantial sum relative to average earnings, between £500 and £800 ($900–$1,500) in today's terms. That limited the audience, of course, not merely to those who could afford the book but to those predisposed to take its author seriously. The publisher, Lovell Reeve, was impeccably scientific and respectable, and the book was, of course, published in London, not in some obscure colonial town.

Upon opening the book, the title page announced the author as "J. D. Hooker, M.D., R.N., F.R.S., F.L.S., &c. &c.," an impressive list of qualifications and memberships, including his medical degree—the closest thing to a botanical degree then available. A reader might well have known of Hooker's previous works, both technical ones like the earlier volumes of the *Botany of the Antarctic Voyage* (fig. 1.4) and more popular ones, particularly his *Himalayan Journals* (1854), which—while they did not rival Darwin's *Journal of Researches* (*The Voyage of the Beagle*)—brought Hooker's name to much wider attention than his specialized books would ever have done.

Its introduction began with the words "The 'Flora of Tasmania' completes the series of works on the Botany of the 'Antarctic Expedition,' with the publication of which I was entrusted by Her Majesty's Lord Commissioners of the Admiralty in 1843"; this, along with Hooker's naval affiliation ("R.N.") proclaimed on the title page, would have made it clear to every reader that the volume he or she held was the product of government patronage.[100]

And once the buyer started to read the book, he or she would be impressed, and perhaps overwhelmed, by the precision and brevity of its descriptions, the learned Latin, and the hard technical terms. With the aid of official support, Hooker had produced not merely a catalog of some of the empire's potentially valuable plants but a speculative work that dealt with the historical causes of plant distribution, including the role of geology and—rather more controversially—the transmutation of species. Fifteen years earlier, the anonymous best-seller *Vestiges of the Natural History of Creation* had been widely pilloried for venturing to discuss this deeply disturbing idea, but all the details of the *Flora Tasmaniae*—its publisher, price, patron, and, above all, its language—helped create a respectable authorial persona for Hooker, making the book a safe place in which to engage in such potentially tendentious speculations.

Publishing was yet another practice that created and maintained the balance between center and periphery: labeling had to be modeled on metropolitan practice, and the published diagnoses were exported to the colonies to serve as models of terse description. But if brevity was prescribed for the collectors, the speculator was allowed—indeed required—to be more prolix.

Hooker's essays were intended to complement his classificatory method, including its labeling and descriptive practices, by explaining the principles on which it was based: as he argued, botanists had to generalize about their results before they could be presented "in that systematic form which can alone render them available for the purposes of science." However, for that to happen, "it becomes necessary for the generalizer to proceed upon some determinate principle." [101] His essays expounded these "determinate" principles as a way of making his pronouncements on classification more persuasive. However, as I have stressed, stabilizing classification in the colonies was only one-half of his battle; he faced equally complex problems in the metropolis, where his growing authority was potentially being undermined by unstable classifying systems, which affected his attempts to persuade the colonial collectors to comply with his classification.

Hooker's two audiences—metropolitan and colonial—meant that he had conflicting needs: to inculcate caution in one audience while springing a mine under the other. This conflict created some confusion among reviewers of his essays. He was disappointed by the reaction to the *Flora Indica*, which did not create the explosion he had hoped for; he complained to Asa Gray: "With the exception of the Kew Journal, not one single journal, literary or scientific has so far as I am aware made the smallest allusion to its existence." [102] When the *Gardeners' Chronicle* did finally review it, the anonymous reviewer (either Lindley or Miles Joseph Berkeley)[103] felt it would be "of infinite service," since it attacked the prevalent "loose notions . . . on the limits of species" and criticized "the reckless way in which multitudes of useless names are daily inflicted on science." By contrast, the *Flora Indica*'s approach to classification should be treated as "a model which may well be adopted by future botanists." However, it is a mark of the novelty of the introductory essay's genre that the *Chronicle*'s readers were particularly encouraged to buy the book because its information on India's climate would be useful to every "intelligent cultivator" trying to grow exotics, especially rhododendrons.[104] It was impossible, even in the most philosophical of essays, for the professed botanist to entirely disentangle himself from the low-status world of the gardener.

However, if the *Gardeners' Chronicle* did not quite understand Hooker's essays, Darwin certainly did. After reading the introduction to the *Flora Tasmaniae,* he wrote to congratulate Hooker on having written "by far the grandest & most interesting Essay on subjects of the nature discussed I have ever read," pronouncing it even "superior . . . to the famous one of Brown."[105] Darwin was especially delighted by the fact that he could "look at your conclusions with the philosophic abstraction" that a mathematician brought to bear on his equations. "I hardly know which parts have interested me most," Darwin continued, "for over & over again I exclaimed this beats all."[106] Darwin, of course, did not need to be impressed, but his reaction was shared by some of the wider community of naturalists; in his letters Charles Bunbury described the essay as "exceedingly able and curious," commenting that he had read it with "great admiration. It is masterly, like most of his writings."[107] William B. Carpenter, registrar of University College London, pronounced Hooker "a truly philosophical systematist" on the strength of the essay, and Hewett Watson praised the essay's "great knowledge and generous sincerity."[108]

For Hooker, the need to impress London's men of science was as important as the need to curb the enthusiasm of some of those in the colony; until the full range of his complex, and occasionally contradictory, objectives are borne in mind, his essays seem ambiguous, especially—as we shall see—on the topic of transmutation. However, for Hooker the pressing issue throughout his career was not the origins of species but their classification. His never-ending war with the species splitters was the paramount concern. He described the *Flora Indica*'s introduction as an attempt "to illustrate our great argument, the imperative necessity of checking the addition of new species on insufficient grounds, and the importance of treating scientifically those already known." With this end in mind, he reiterated the classificatory principle enunciated in the *Flora Novae-Zelandiae* that "for all practical purposes species be regarded as definite creations, the offspring each of but one parent or pair," endowed with "great powers of migration," from which he could urge his readers "to use every effort to reduce the vast bulk of forms we have to deal with in the Indian Flora to as few species as we can." [109] This was the "great argument"—ending the disorder that undermined philosophical botany's standing. Only when that is understood does the full significance of Hooker's essays become clear.

CHAPTER EIGHT

Charting

> Oceanic islands are, in fact, to the naturalist, what comets and meteorites are to the astronomer; and even that pregnant doctrine of the origin and succession of life which we owe to Darwin, and which is to us what the spectrum analysis is to the physicist, has not proved sufficient to unravel the tangled phenomena they present.
> —J. D. Hooker, "Insular Floras" (1867)

The empire's growth involved many naturalists in surveying and inventorying the curious productions and natural wealth of newly acquired territories. As these activities gathered pace, naturalists began to search for order amid the accumulating specimens; classification alone could not resolve the puzzles of distribution. How did the same species come to be growing in widely isolated spots? Why were some species missing from places whose climate should have suited them or were replaced by similar but unrelated species? Such questions produced divisive debates within the natural history community, especially between proponents of "single" or "multiple" centers of creation. The former argued that each species had been created in one place and at one time only, from where its descendants had migrated out to occupy their present range, while supporters of multiple centers were convinced that species found in widely separated localities had been created where they were now found.[1] There were many different factors at work in these debates, but it is important to realize at the outset that neither view was necessarily connected with a particular opinion about *how* species had been created. There were natural and supernatural mechanisms being posited on both sides of the argument, and as a result one sometimes finds transmutationists and their opponents united, conventional Christians sharing a view with outright materialists, and similar—often uncomfortable—alliances between philosophical idealists and inductivists or between land-bridge theorists and believers in accidental migration. These debates cut right across the more familiar but often anachronistic categories of "pro-" and "anti-evolution"; the fact that Darwin used distribution as a key argument in favor of his theory does not prove that his contemporaries were investigating the same species question; for Hooker—and many other naturalists—matters such as distribution's economic benefits (both for him and for the empire) were probably more important.

Since speculating was a key practice for a philosophical naturalist, the debates over distribution were crucial to Hooker's scientific standing, but they also shaped—and were shaped by—his material collecting practices. Surveying geographical variation was integral to both his species concept and his classifying method, but global surveys were about more than creating stable names. As with other kinds of speculation, Hooker's theories were partly intended to deter colonial naturalists from joining the debates; such colonial speculations were potentially even more disruptive than those over classification because distribution was a more philosophical topic.

"We Want Better Collections"

Before Hooker could chart the patterns of distribution, much less speculate about their causes, he needed comprehensive plant collections, which depended on encouraging colonial collectors to search for specific plant families and genera. He wrote to Ronald Gunn in 1846 to insist that examples of the genus *Forstera* "*must* & do exist though you *wont find* them": "My dear Gunn you must sprawl on your hallowed [?belly] on the top of the Mts. & pick little things out of the ground for I still want analogies to the genera of L^{d}. Auckland group & Fuegia which your Mts. must & shall produce."[2] Hooker's conviction that Tasmania "must & shall" contain the plants was based in part on his knowledge of those in Robert Brown's herbarium (even though Brown's legendary meanness with duplicate specimens forced Hooker to obtain his own).[3] However, Hooker's confidence was also derived from his distribution theorizing, especially botanical arithmetic—the calculation of the ratios between different plant groups at given locations. These ratios gave numerical values for the typical composition of, for example, the floras of islands (insular floras), which were characterized by smaller genera (i.e., those containing fewer species) than those of a continental flora at the same latitude. Hooker assumed that if the ratios between groups of plants gathered in Tasmania did not match those for similarly sized islands at comparable latitudes, there were still plants waiting to be found.

Hooker's desire for what he called "analogies" referred to the idea of representative species, those which were similar but not identical to others growing in geographically separated areas.[4] Naturalists recognized that species were fitted for the particular place they inhabited; those that grew in deserts could deal with water shortages, but although every American desert held cacti, each desert held slightly different ones. The problem of representative species was that these distinct places in the economy of nature could have been filled by the same species—as, indeed, they often were—so what accounted for these seemingly pointless variations on the cactus theme?[5] The concept of representative species meant that Hooker could predict that Tasmania would hold plants similar to (i.e., "representatives" of) those he knew from Lord Auckland's islands and Tierra del Fuego (based on his knowledge of these other southern mountain floras; fig. 8.1). As Hooker told Gunn, "The representation of the New Zealand & Cape Horn Flora on the W. Coast of your Island is a most wonderful & valuable fact, which suggests many ideas to me," but before these ideas could be developed, "we want better collections."[6]

Better collections required standardization of collecting practices, particularly of labeling to ensure locations were recorded appropriately. Collectors were encouraged to add such details as the altitude at which specimens were gathered (a key datum according to Humboldt's theories), so Hooker recommended "the plan of carefully determining elevations by the temperature of boiling water, as amply sufficient for all botanical purposes."[7] He urged the need for accurate observations on altitude, climate, and similar matters in his letters and wrote about them in the *Gardeners' Chronicle,* which had a broad readership among plant enthusiasts of all sorts, including those in the colonies.[8] Hooker noted that observers' carelessness had previously impeded progress in distribution studies, but that techniques such as botanical arithmetic could make even a gardener's observations accurate.[9]

Standard specimen tickets were intended not only to encourage brevity but also to ensure that an appropriate level of geographical information was included. Hooker complained that he was "often perplexed by collectors sending as localities the names of insignificant hamlets or streams, which are not to be found in attainable maps, and convey no meaning whatever."[10] And—as in so many other cases—William Colenso objected,

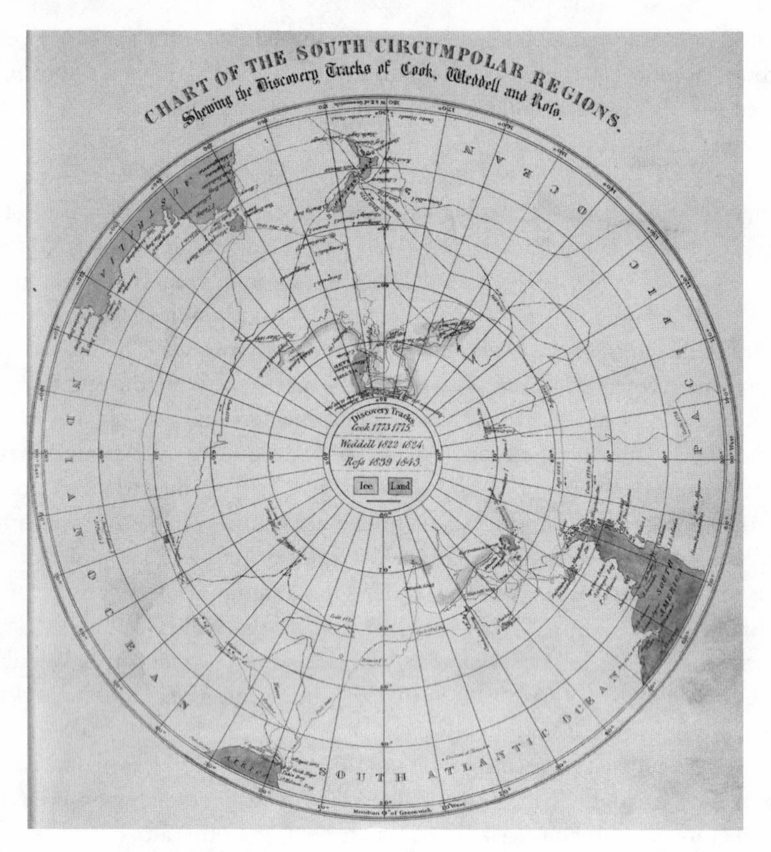

8.1 A chart of the circumpolar regions, showing the voyage of the Ross expedition. From J. D. Hooker, *Flora Antarctica* (1844–47). Reproduced by kind permission of the Syndics of Cambridge University Library.

asserting that Hooker had "certainly erred in not publishing the precise habitats of several of my plants": "You have, it is true, excused your not doing so, by pleading, that, my habitats are not to be found in any Map or Gazetteer! It would indeed be a curious thing if the all but unknown hills, streams, and hamlets of this new country were to be found in any Map or Gazetteer."[11] Precise locations were an aspect of topographic knowledge, but colonial collectors had no more luck persuading Hooker of the value of such knowledge than they did with its endemic and indigenous forms.

Calculating arithmetic ratios required standardized classifications; as Hooker told Darwin, it was clear that some floras were "peculiar" in that they contained unusual species that were not found elsewhere, but he bemoaned the fact that "our ideas of peculiarity are most loose, we have no standard." The first step in establishing a flora must be to "know the absolute numerical amount of peculiar species," which of course required accurate classifications. For Hooker this "must ever be the primary point," and until it was established, there was no possibility of discovering the causes of such peculiarity. Hooker continued: "Except Brown & Humboldt, no one has attempted this, all seem to dread the making Bot. Geog. too exact a science, they find it far easier to speculate than to employ the inductive process. The first steps to tracing the progress of the creation of vegetation is to know the proportions in which the groups appear in different localities, & more particularly the relation which exists between the floras of the localities, a relation which must be expressed in numbers to be at all tangible."[12] Botanical arithmetic was the key to making "botanical geography," or the study of plant distribution, into an "exact" science. As Hooker argued, this required botanists to "employ the inductive process," to gather specimens, name and classify them, and then analyze the ratios between species and genera; all this work had to be done before anyone could speculate about the processes which had *caused* the observed patterns of distribution.

However, proper classifications were only the starting point; it was also essential that every specimen have an appropriate location recorded (i.e., neither too narrow nor too broad). In the introduction to the *Flora Indica*, Hooker defined "physico-geographical or geographico-botanical districts" intended to provide "intelligible localities to the species in the body of the Flora, and such as may be easily committed to memory, or found with little trouble on the map."[13] The accompanying map illustrated the proposed divisions (fig. 8.2), to guide future collectors into providing specimens labeled with standardized locations; this was yet another attempt to get colonial collectors to meet the terms of metropolitan standards.

"A Mass of Most Valuable Information"

Emphasizing the importance of the work collectors were contributing to was a way of inspiring them to greater efforts. The introduction to the *Flora Novae-Zelandiae* stressed: "Of all the branches of Botany there is none whose elucidation demands so much preparatory study, or so extensive an acquaintance with plants and their affinities, as that of geographical distribution."[14] This was read by Augustus Oldfield, who told William Hooker that reading it "has considerably modified my views with respect to sending specimens," since it had convinced him that "I might throw some light on the distribution of Tasmanian plants." With this in mind, he had "be-

8.2 The "Botanical Provinces" described by Hooker in the Flora Indica, which he hoped would serve as a guide to future collectors. Detail from J. D. Hooker and T. Thomson, *Flora Indica* (1855). By permission of the librarian, Trinity College, Cambridge.

gun to cull from my herbal; 1st all plants of which I Know not the place either with regard to order, genus or species. 2nd such plants as I have found here which none of my authorities refer to as indigenous to this Island. 3rd some which I think doubtfully named."[15] The thought of contributing to such an important project evidently inspired Oldfield, and presumably his initial response—such as removing badly labeled specimens—was precisely the one Hooker was hoping for.

However, from Hooker's perspective Oldfield's desire to "throw some light on the distribution of Tasmanian plants" also had unwelcome consequences, since it led him to read widely on distribution issues. As we have seen, his unpublished manuscripts reveal an extensive knowledge of both classificatory and distribution theories; for example, he discussed—and criticized—several of Darwin's, including his claims about waterbirds spreading plant seeds via the mud adhering to their feet.[16] Oldfield thought the latter distinctly implausible, whereas Gunn was more confident about the effects of accidental migrations. The latter noted in a letter to Hooker that there were many water and marsh plants that were common to Europe and Tasmania and suggested that "aquatic birds which migrate [might] carry the seeds about the plumage &c" and that "currents of air might waft spores of Mosses, Fungi, &c. but not the seeds of water plants."[17] Colenso was also interested in distribution; when he told Hooker that New Zealand's flora could "compete with the Botany of Islands of a similar size and parallel," he was demonstrating an awareness of the same distribution theories that guided Hooker's instructions to Gunn.[18] Once again, Colenso owed his interest to Allan Cunningham, who had assured the missionary that he would eventually "acquire a mass of most valuable information, in regard to the Botany of Islands, daily becoming more and more important in the Eyes of Europe, in this Age of Colonizn. and immigration."[19] The "Botany of Islands," the character of insular floras, was central to plant distribution theories, and Colenso was apparently enthused by Cunningham's promise that gathering "valuable information" would eventually make him "important in the Eyes of Europe."

From Hooker's perspective, the immediate problem presented by colonial botanists who took too close an interest in distribution was that, like Gunn and Oldfield, they might arrive at different conclusions, which might encourage different collecting practices (Gunn might, for example, have decided to search for new aquatic plants only along birds' migration routes, while Oldfield would have ignored such a possibility). However, that was a minor problem compared to what would happen once their speculations led them to dispute Hooker's theories and authority, as Oldfield's did. Oldfield's manuscripts discuss what he called "the numerous specious theories that have been invented to account for the distribution of organic life" and asked "this simple question: are they necessary?" His short answer was "no": he argued that in the case of most animals and plants, no theory was needed to explain their distribution.[20]

Had Hooker read Oldfield's manuscripts, they would have reinforced his desire to dissuade his colonial correspondents from speculating, since they contain ideas that were typical of the current of speculation in Britain. Once again, botany's more philosophical topics created a dilemma for Hooker: as in the case of the natural system itself, persuading his collectors of the importance of distribution—because he needed their data—could also lead them into speculation.

"The Vestiges of a Flora"

As we have seen, Hooker's introduction to the *Flora Novae-Zelandiae* included the speculation that the plants of the Southern Hemisphere were "the vestiges of a flora" of a sunken continent. His theory illustrates the multiple purposes and audiences of his essays, demonstrating that he was qualified to speculate while establishing his philosophical credentials in the eyes of his fellow men of science. As he worked to become a distinguished member of the metropolitan scientific community, he also had to be acknowledged in the colonies as a person whose advice and guidance must be sought and followed. Once again, these conflicting goals help explain why, despite his avowed lack of theories, he decided to advance the "general theoretical views on the origin, variation and dispersion of species" with which he prefaced the *Flora Novae-Zelandiae.*[21]

Although both Hooker's audiences, metropolitan and colonial, needed to read his discussion of speculative topics, he had to communicate different messages to each—hence his equivocation and the contradictory elements in his writing. At one moment he was arguing for the importance of theorizing, yet a few pages later he claimed: "It is no part of my intention to discuss the theoretical views that have been entertained on these obscure subjects: my aim is to draw attention to a few leading questions of great practical importance, which ought not to be overlooked."[22]

While curbing colonial speculations was an important motive for Hooker's essays, he was even more concerned with his status and that of his discipline. The essays were to stress both botany's economic importance and its ability to tackle abstract questions.

When, in 1833, the *Edinburgh Review* had claimed that among the activities that defined a "philosophical botanist" was that he "develops the laws of [plants'] geographical distribution," their reviewer, David Brewster, stressed that this was an activity that was "at least as important to society" as astronomy's application "to the purposes of navigation and of commerce."[23] For Brewster, stressing the commercial benefits of an apparently abstract science like astronomy (particularly for Britain's maritime trading empire) was central to his declinist argument; tangible benefits helped justify higher government funding. He extended his argument to botany: the efforts of "the philosophical botanist," while perhaps not quite as "elevated" as those of an astronomer, were nevertheless just as important to society. But in both cases, this importance was a combination of the abstract value attached to discovering laws *and* botany's contribution to mundane matters like "diet, medicine, and the arts."[24] Even high-status sciences had to bring practical, commercial benefits if they expected government patronage, a point Hooker would not have needed reminding of—the Admiralty had funded the Magnetic Crusade because of its value to navigation and commerce rather than to philosophy. The perception that distribution held the key to advancing the standing of the natural history sciences was a widespread one (as the campaign for the Magnetic Crusade demonstrated, both John Herschel and Roderick Murchison shared Brewster's and Hooker's view; see chapter 1), but in discussing its importance we must always keep in mind the crucial link between high scientific standing and a high income.

Economic botany was at the forefront of Hooker's mind in 1842, when the *Erebus* arrived in the Falkland Islands. He immediately began surveying the islands' plants and sketched penguins as they waddled among labyrinths of a grass that grew so high as to resemble small palm trees. This was *Poa flabellata,* the Falklands tussock grass, a favorite food of the local wild cattle that, Hooker noted, grew even on the poorest soils. Inspired by his son's reports, William Hooker immediately sent Wardian cases to the Falklands to obtain living plants in the hope that they might provide fodder on otherwise-worthless land.[25]

The Hookers' attempt to introduce the tussock grass to Britain was one of numerous similar attempts at getting plants to grow in new countries. As Richard Drayton has shown, this process began in 1787, when Sir Joseph Banks expressed the hope that the Royal Botanic Gardens at Kew might become "a great botanical exchange house for the empire"; like the Hookers after him, Banks argued that a network of botanic gardens would aid in the discovery of new, economically beneficial crops and in the transplantation of both old and newfound crops to British colonies where they could be grown profitably.[26] Banks used this imperial role to secure Admiralty patronage for naturalists, including, most importantly, the policy of continuing to pay them after their voyages so that they could publish their results. One of the first to benefit was Robert Brown.[27]

Economic concerns form the primary context within which plant distribution studies have to be considered; while the philosophical issues were at least as important to the botanists, they were distinctly secondary from the government's perspective. Demonstrating Kew's usefulness was an essential part of William Hooker's strategy of maintaining government funding. Kew's Museum of Economic Botany was intended to complement Kew's living collections: merchants, manufacturers, travelers, and colonists could all come to see preserved specimens of plants and the products they supplied.[28]

In the Admiralty's *Manual of Scientific Enquiry* (1849), William Hooker had listed regions whose plants were still comparatively unknown and included specific questions about them. For example, under "Asia," Hooker asked travelers to look out for "*Sagapenum*—a gum-resin: its source? It is said to come from Persia, and to be derived from a *Ferula.* Specimens of the plant with the gum-resin which it affords are desirable."[29] In his various collecting instructions, he always urged travelers to make notes of plants' "uses and properties" and to send "ample collections" home.[30] Joseph Hooker had received similar advice when the *Erebus* set sail, having been told to "collect everything" since even "such scraps as are useless for other purposes may yet, so long as they exhibit the Natural Order to which they belong, prove of service in illustrating the geography of plants."[31] This was advice Joseph followed throughout his life: Kew's collections include dozens of plant-based products that Hooker collected during his travels (fig. 10.1).

Botany's claims to economic importance depended on successfully transplanting crops, but results were unpredictable. The tussock grass proved a disappointment; its seeds were hard to germinate and it was too slow growing to be a useful crop.[32] One of Kew's supposed successes was cinchona (the tree from whose bark quinine is made). When the *Edinburgh Review* surveyed books on cinchona, it claimed: "To transplant a vegetable or a tree from the soil where it is indigenous to some other region fitted to receive it, is to extend the realm of Nature herself."[33] Yet this much-vaunted transplantation was a failure—the varieties the British acquired and planted in India proved to have a low quinine content, while the Dutch trees in Java proved much more productive.[34]

Other plant transfers—such as rubber, tea, and cocoa—produced huge profits. For the botanists, these successes and failures were a mystery. Why did the tussock grass grow so fast and high in the Falklands but so poorly in Britain? Why did the plants in some Wardian cases die, while others thrived? Why did the Dutch cinchona trees produce so much more quinine than the British ones? The answer to the last question might be the particular species or variety of cinchona, but the *Poa flabellata* brought to Britain was the same species that grew in the Falklands. The atmospheric moisture or soil type might make a difference, or the average temperature and rainfall, or the latitude or altitude at which the plant grew. The economic practicalities of moving crops made it urgent to understand the relationship between geography, meteorology, and botany.

"Chartism"

Hooker's time aboard the *Erebus* provided him with an ideal opportunity to contribute to botanical geography, and at the same time, his active involvement in the day-to-day work of the magnetic survey and in the other record-keeping activities of his ship reinforced his earlier training in both medical and botanical record-keeping. While the *Erebus* was in Hobart harbor, Tasmania, Hooker had to cancel a botanizing trip ashore with Gunn planned for the following Sunday because, Hooker explained, that was to be "the great day for simultaneous experiments on Magnetism all over the world. . . . all the officers are ashore at the observatory but the non-Executives, who, (including myself) have volunteered to lighten our shipmates labor by keeping the meteorological Register & Log on board."[35] In many regards, the Magnetic Crusade's global survey became Hooker's ideal of science; long after his return to England, he promoted global surveys of vegetation as the key to uncovering botanical geography's laws.[36]

Despite the crusade's prestige, its results were somewhat disappointing. The magnetic observatory mentioned by Hooker was one of a network whose work continued in the years following the Ross expedition (fig. 8.3), but as the results poured in, no less a figure than Herschel expressed disappointment at the astronomer and explorer Edward Sabine's obsession with merely compiling data and creating maps. Sabine, who was a general secretary of the BAAS and de facto head of the Magnetic Crusade, published more than a dozen "Contributions to terrestrial magnetism" in the Royal Society's *Philosophical Transactions*, but his papers did little more than summarize the raw data coming in from the magnetic observatories.[37] The maps used Humboldt's isoline technique (joining points of equal declination) in an effort to plot the relation between magnetism and other phenomena, but Herschel was not impressed and dubbed the project "chartism," arguing that data collection alone would never lead to a proper theory.[38]

Herschel used his presidential address to the BAAS in 1845 to discuss the results of the Ross expedition and mention his doubts about the "chartist" approach: "Do we call for *facts*?" he asked rhetorically, "they are poured upon us in such profusion as for a time to overwhelm us." "Witness the piles of unreduced meteorological observations which load our shelves and archives; witness the immense and admirably arranged catalogues of stars which have been and still are pouring in from all quarters upon our astronomy. . . . What we now want is *thought*, steadily directed to single objects." "Reducing" observations to comprehensible generalizations was

8.3 The Rossbank Observatory (Hobart, Tasmania) established during the Magnetic Crusade. From J. C. Ross, *A Voyage of Discovery and Research in the Southern and Antarctic Regions,* 1847. Reproduced by kind permission of the Syndics of Cambridge University Library.

more urgent than gathering more data; observational sciences like astronomy required data gathering to be followed by "a constant collateral endeavour to concentrate the records of that observation into empirical laws in the first instance, and to ascend from those laws to theories."[39]

So, if distribution studies were to raise botany's reputation, it was vital for botanists to move beyond simply gathering and classifying plants, which produced unordered observations. Like many of his contemporaries, Hooker turned to Brown's work to help convert his raw observations into higher laws. Although botanical arithmetic was associated with Humboldt's name, it had been pioneered by Brown, whose *Botany of Terra Australis* (1812–14) included mathematical ratios between different families of plants at various latitudes, providing the first statistical treatment of such issues.[40] Humboldt acknowledged Brown's priority in the field and bestowed upon him the epithet *Botanicorum facile princeps, Britanniae gloria et ornamentum* ("Easily the prince of botanists, glory and ornament to Britain"), which was often quoted proudly by the British press.[41]

Brown's name became a byword for high botanical attainment; in 1835, the *Edinburgh Review* praised the BAAS annual meetings for providing "a privilege of immeasurable value," which is "to discuss the topics of abstract or natural science . . . with the Daltons, the Aragos, the Herschels, the Faradays, the Browns, the Airys, of modern science"—Brown being the only naturalist considered worthy of a place alongside the physicists and astronomers.[42]

Given Brown's status as both national hero and a pioneer in the study of the Australian flora, it is not surprising that Joseph Hooker admired him deeply. During the *Erebus* voyage he told his father that if he succeeded in pursuing a botanical career, he hoped to be "like Brown," albeit "without his genius."[43] He read Brown's work onboard, understood the importance of botanical arithmetic within it, and took it as a model for his own work, telling his father that he was "anxious to know the proportions that the Nat[ural]. Orders bear to themselves at different Antarctic Longitudes and to themselves in each locality, as an object of primary importance to the elucidation of Bot[anical]. Geog[raphy]." With these lofty goals in mind he was pleased to discover that "several of the tabular results I have drawn out show a delightful accordance" with distributional laws. His charts revealed, for example, that the grasses "rose in the scale of importance, beating even Brown's published ideas," and he added, "nothing so satisfies me, that I have observed carefully in any Island, as to find these laws to hold good."[44]

However, while Brown's numerically expressed generalizations captured the patterns of distribution, they did not explain what *caused* these patterns—and causes mattered more than generalizations in the hierarchy of scientific knowledge. Hence, Hooker utilized geology—and the works of Charles Lyell and Edward Forbes, in particular—in formulating his own distribution speculations, such as his missing southern continent; by associating botany with the most sophisticated geological theorizing, Hooker could establish its philosophical credentials—and his own. Distribution studies would provide a chance to show he was more than a classifier.

In 1858, the *Edinburgh Review* commented in an article entitled "The Progress and Spirit of Physical Science": "It has been justly said by Sir J. Herschel that *number, weight,* and *measure* are the foundations of all exact science."[45] Botanical arithmetic provided the clearest example of botanists attempting to apply such ideas to their own science. Calculating the ratios of different taxa expressed "some numerical relation between two quantities dependent on each other," which was Herschel's definition of the "simplest stages" of the inductive ascent to higher-level laws.[46] Hence, Darwin could comment that botany had "more a philosophical spirit than

Zoology," so much so "that I scarcely ever like to trust any general remark in Zoology, without I find that Botanists concur."[47] The numerical basis of botany's distribution data was an important reason for this preference, as Darwin explained in a letter to Hooker when he noted there is "no little ambiguity in the mere assertion of 'wide ranges,' (for zoologist seldom go into strict & disagreeable arithmetic, like you Botanists so wisely do)."[48]

Numbers were expected to bring simplicity and clarity to the "piles of unreduced observations." However, Hooker's use of botanical arithmetic, like his utilization of Lyell's and Forbes's theories, grew out of his charting practices, particularly his need to standardize the work of his network of unpaid assistants.

Just as Herschel had complained about the data generated by the Magnetic Crusade, Hooker grumbled about the "unreduced observations" he had to tackle: the mountains of plant specimens that were "pouring in from all quarters" to Kew. Botanical arithmetic provided him with the tools to tackle them, but as Janet Browne has argued, the collection of plant statistics was beginning to be seen as a pointless exercise in botanical bureaucracy (it had been partly modeled on the demographic census-taking of various European governments); like the magnetic surveyors, the botanists were engaged in "chartism."[49] The situation was not helped by the fact that the British tradition of botanical geography was dominated by local naturalists, such as Nathaniel John Winch, whose *Essay on the Geographical Distribution of Plants, through the Counties of Northumberland, Cumberland, and Durham* (1819) was simply a compendium of detailed information about the plants of those counties.[50] As discussed in the previous chapter, distribution theorists needed "some such bold original ideas" to help them make sense of their data.[51]

On the "Connexion" between Geology and Causes

When Roderick Murchison praised British naturalists for becoming "more philosophical," he cited Edward Forbes's work as evidence of this trend.[52] Like Hooker, Forbes lacked an independent income and was forced to combine appointments in order to make ends meet. It was while he was paleontologist to the Geological Survey that he published his most influential paper: "On the Connexion between the Distribution of the Existing Fauna and Flora of the British Isles, and the Geological Changes Which Have Affected Their Area, Especially during the Epoch of the Northern Drift."[53] This was the focus of Murchison's praise and it lifted Forbes's reputation to the point where he was appointed president of the Geological Society in

1853 and then to the chair of natural history at Edinburgh.[54] Hooker felt that Forbes's ideas supplied answers to some of the most pressing questions in distribution studies, but before examining them, it is important to understand the questions he was asking, which focused on the issue of single versus multiple centers of creation.[55]

Naturalists had long recognized that each part of the world had its own distinctive flora and fauna, but as European knowledge of the world grew, they discovered things were not so simple. Augustin-Pyramus de Candolle built on the work of the French naturalist Georges-Louis Leclerc de Buffon and of the German Karl Ludwig Willdenow for his pioneering 1820 essay "Géographie botanique," in which he argued that if such factors as temperature and moisture were the sole determinants of vegetation, every region with the same climate would have the same plants, yet they do not.[56] He drew on Buffon's thesis that each region of the earth is characterized by its own distinctive flora and fauna to argue that these distinctive assemblages could only be explained by reference to historical patterns of migration, often across an earth whose physical features were very different from those we now observe.[57]

Like many early-nineteenth-century naturalists, de Candolle believed that all the members of a species were descended from a single pair: Christian naturalists (like Forbes) believed that the original pair had been directly created by God, but many transmutationists (like Darwin) assumed that each species had evolved only once. For those committed to a single origin (regardless of how they thought species had been created), present distributions had to be explained by assuming that the original pair had multiplied and spread from their point of origin. Given these suppositions it was puzzling to discover "sporadic species," members of a single species growing in widely separated areas, such as isolated islands, with no obvious means by which they could have migrated between the places where they were now found.[58]

Alongside the problem of sporadic species (i.e., identical species in disconnected localities) was that of representative ones (similar but non-identical species with comparable distributions). Some naturalists tried to resolve these anomalies by positing multiple centers of creation for each species. Since the machinery of creation (however it operated) could work in one time and place, many naturalists assumed it could work twice or, indeed, as many times as were necessary to solve the problems of distribution. Once the idea of more than one center of creation had been admitted, there seemed no clear reason why centers should not proliferate ad infinitum, a fresh one being invoked to explain each fresh problem.

Brown was unsure whether the 150 species of European plants he had found in Australia were immigrants or had been created there, as he assumed the Australian ones had been—and if more than one center was needed for the Australian native flora, why not two (or more) for the European?[59]

Many prominent European naturalists shared Brown's belief in multiple centers, as did some naturalists in the colonies.[60] Oldfield claimed that European botanists failed to solve distribution puzzles because they were wedded to the mistaken assumption "that all forms have originated in one locality only." Once this was dropped, Oldfield argued, "there is little room, and still less necessity, for the working of any theory of distribution." His alternative was to assume multiple creations by natural laws: "All organic life originated from the combination of the organic elements in definite proportions"; that is, organisms were spontaneously generated according to the available raw materials and their precise local conditions (temperature, moisture, and so on). He felt that his view "affords an adequate explanation of every anomaly in the distribution of organic life in time or space."[61] Oldfield's theory was intended to explain what his local knowledge of endemic species had revealed and what he felt other theories could not: Australia's endemic species resulted from Australia's unique local conditions. It also explained both sporadic and representative species: "It may be admitted as an axiom that 'Identical causes produce identical effects,' to which may be added as a necessary corollary, 'Similar causes produce similar effects.'"[62] Identical species in separated locations were the result of identical physical conditions, while similar but closely related ones were produced by similar conditions.

The New Zealand missionary Richard Taylor also discussed contemporary distribution theories—including Forbes's and Hooker's—in his book *The Age of New Zealand* (1866).[63] Perhaps not surprisingly, given his religious vocation, Taylor argued that "animal and botanic centres are to be viewed as epochs of creation, originally of the widest distribution"—in other words, that plants and animals were created in an instant and that whole floras and faunas came into being at once.[64] Taylor added that "these so-called centres preserve the peculiar condition of each, which, by subsequent convulsions, have been lost in other parts," which explained why the flora and fauna of New Zealand and Australia were so distinctive: they were much older than those of Europe and were in effect "living fossils" akin to the Northern Hemisphere's Carboniferous flora and fauna.[65]

Hooker was vehemently opposed to the doctrine of multiple centers of creation whether it appeared in the metropolis or colonies. Once again, his enmity was largely prompted by the need to restrain the splitters. He told

Asa Gray that he opposed multiplying centers "more upon principle than upon facts" since, if it were to be admitted, "the flood gates are opened to species-mongers, and it is cast in your teeth every moment." In the hands of local botanists, multiple centers could be used "as an argument for making every slight difference, if only accompanied with geographical segregation, of specific value."[66] He was not alone in his opposition to multiple centers; geologists also opposed them because they undermined the rationale behind their key tool, stratigraphy. Fossils were used to correlate different strata on the assumption that similar fossils were the same age—an assumption rendered invalid by multiple creations. Multiple creations also threatened comparative anatomical studies: how could past and present organisms be compared if species could be created at several times? Some naturalists were troubled by the way the doctrine conflicted with literal readings of the Bible, while others were disturbed by the materialism implicit in the claim that living things were somehow directly produced by their physical surroundings.[67]

For those naturalists who rejected the idea of multiple centers of creation, for whatever reason, geology seemed to offer an alternative. A.-P. de Candolle had argued that distributional anomalies could be explained by assuming that "habitations are probably determined in part by geological causes that no longer exist today"; among the many naturalists who read his essay was Charles Lyell, who cited it in the second volume of the *Principles of Geology.*[68] Forbes was inspired by both de Candolle and Lyell to attempt his own explanation of the distribution of the British flora by reference to geological causes. His conviction that each species began with a single pair was partly formed by his religious views but also because—as paleontologist to the Geological Survey—he could not contemplate the destruction of the vital tool that stratigraphy provided.[69] In the "Connexion" essay, he argued that single specific centers "must be taken for granted, if the idea of a species (as most naturalists hold) involves the idea of the relationship of all the individuals composing it, and their consequent descent from a single progenitor." He also argued that "the abandonment of this doctrine would place in a very dubious position all evidence the palaeontologist could offer to the geologist towards the comparison and identification of strata, and the determination of the epoch of their formation." His essay therefore aimed to resolve the "apparent anomalies and difficulties in distribution" that seemed to require "more than one point of origin for a single species."[70] His solution, drawing on Lyellian geology, was to focus on what he felt was the neglected historic dimension of the distribution problem.[71]

As Forbes noted, the botany of the British Isles included several sporadic species—in particular, several species of saxifrage that were found in northern Spain and Portugal and in the southwest of Ireland but nowhere in between. His solution was to suggest that, "at an ancient period, an epoch anterior to that of any of the floras we have already considered, there was a geological union or close approximation of the west of Ireland with the north of Spain; that the flora of the intermediate land was a continuation of the flora of the Peninsula." He admitted that his theory was "a startling proposition, and demands great geological operations to bring about the required phenomena," but he nevertheless argued that—according to Lyell's theories—such operations were possible.[72] This hypothetical intermediate land inevitably became known as "Atlantis," and Forbes's theory became known as the "Atlantis Theory" (an epithet that was often intended to express incredulity).[73]

Hooker's missing southern continent was, as he acknowledged, a "Forbesian" conjecture.[74] Other botanists greeted the Atlantis theory with similar enthusiasm; Arthur Henfrey adopted it and credited Forbes with having been the first to apply geology to the problems of distribution.[75] In 1856, Alphonse de Candolle expressed a degree of support, although Hooker felt that "M. de Candolle here perhaps hardly recognizes sufficiently the real value of Forbes's Essay."[76] And in the same year, the *Edinburgh Review* noted: "A new turn in the right direction has of late years been given to these enquiries by geologists. The principle first laid down by Sir Charles Lyell, and practically applied to a particular district by the late lamented Professor Edward Forbes, of the dispersion of plants during a state of the globe anterior to the present geological period, has greatly enlarged our views of the subject."[77] Forbes's ideas were crucial to moving British distribution debates beyond "chartism" during the middle decades of the century. As the *Edinburgh* noted, in early works on botanical geography, "we find little more than numerical statements of the number of species contained in particular districts, of the relative proportions of botanical genera, classes, or orders."[78] But such surveys, even when presented arithmetically, did not bring botanists closer to understanding the causes of distribution.

The *Edinburgh*'s reviewer, George Bentham, noted the many points of agreement between de Candolle's *Géographie botanique raisonée* and Hooker's *Flora Novae-Zelandiae,* despite the fact that the latter was published too late for de Candolle to refer to it.[79] Bentham was particularly enthusiastic about the way de Candolle and Hooker had shifted distribution from mere "numerical statements" to "research into cause and effect."

However, he agreed with Hooker that the progress of distribution studies relied on better classifications.[80]

The Debates between Hooker and Darwin

As Janet Browne has shown, while Hooker was an eager enthusiast for Forbes's ideas, Darwin was not, commenting, "I must still think it a bold step, (perhaps a very true one) to sink into depths of *ocean, within the period of* **existing species,** so large a tract of surface. But there is no amount or extent of change of level, which I am not fully prepared to admit, but I must say I sh[d]. like better evidence, than the identity of a few plants, which *possibly* (I do not say probably) might have been otherwise transported."[81] Nevertheless, despite his doubts about speculative continent building, Darwin shared Forbes's opposition to multiple creations and his belief that species originated as single pairs at a particular location.[82]

As is well known, Darwin preferred to explain distributional anomalies as resulting from accidental processes, such as the transportation of seeds by winds and waves, and he spent many years testing possible methods and proving their plausibility.[83] Hooker was often surprised by his friend's results and gradually came to concede a larger role for migration than he had originally done. However, it took him many decades to come round to Darwin's view, and I am interested in understanding why.

Hooker and Darwin agreed that insular floras were best explained by assuming that isolated islands had indeed been populated by the haphazard transport of a random assortment of species, which had washed up and taken root.[84] However, Hooker reminded Darwin that "it is much more difficult so to wash seeds up that they shall grow, than to transport them"—the lessons of the Falklands tussock grass had not been forgotten.[85] More importantly, Hooker rejected Darwin's accidental migrations in the case of the Australian and New Zealand floras because the floras were not insular in character. Somehow whole communities of species and higher taxa had reached them, a fact that he felt demanded some more consistent migratory route, such as one of Forbes's land bridges.

Hooker's skepticism about migration was probably formed during the *Erebus* voyage. He noted that the remoteness of the Antarctic islands, "together with their inaccessibility, precludes the idea of their being tenanted, even in a single instance, by plants that have migrated from other countries."[86] However, his doubts mainly resulted from his focus on higher taxa and on communities of plants. In the *Flora Novae-Zelandiae,* for example, he noted that floral relationships do not consist of isolated plants, but that "there are upwards of 100 genera, subgenera, or other well-marked groups of plants entirely or nearly confined to New Zealand, Australia, and extra-tropical South America."[87] Hooker felt accidental transport could not account for these large-scale patterns of affinity; if Darwin's mechanisms were effective enough to spread whole plant communities, why were there Indian plants in Australia, but no Australian plants in India? Why were some common European plants absent in Britain, if the wind could spread seeds?[88] And when discussing the relationship between the Australian and New Zealand floras, he noted: "It is impossible in the present state of science to reconcile the fact of *Acacia, Eucalyptus, Casuarina, Callitris,* etc., being absent in New Zealand, with any theory of transoceanic migration that may be adopted to explain the presence of other Australian plants in New Zealand."[89]

Hooker was always more interested in the general character of a country's vegetation than in the individual species of which it was composed.[90] As his disagreements with Darwin make clear, the migration of individual species (by whatever means) was not the main issue; he was perplexed by communities of vegetation, which he felt Darwin's accidentally transported seeds could not explain.

Hooker's concern with economic botany also shaped his initial rejection of Darwin's ideas on migration. His interest in the roles plants played in the economy of nature did not displace his concern with their potential roles in the human economy of imperial trade. Distribution studies might eventually allow botany to become a predictive science like geology, which could tell miners where to look; as we have seen, Murchison's apparent ability to predict gold deposits led to publicly funded geological surveys in many colonies.[91] If tools like botanical arithmetic could turn botany into a similarly valuable enterprise, the government might found a national—or even international—"Botanical Survey." However, Darwin's haphazard migrations obviated such possibilities; if, as he implied, the character of a country's vegetation was ultimately the product of blind chance, botanical prediction would surely fail.

"The *Bête Noire* of Botanists"

In 1866, Hooker addressed an audience of over two thousand people at the Nottingham meeting of the BAAS. He took geographical distribution and the importance of insular floras as his subject and illustrated his talk with a huge map, which summarized the topic's complexities and anomalies. He concluded with "the most difficult part of my task," which was "to discuss in a brief space of time the hypotheses that have been invented by naturalists" to explain these fascinating floras. There were many theories, all "as yet unverified and insufficient," and none of them "have yet helped us to a complete solution to this problem, which is at present the *bête noire* of botanists." However, Hooker told his audience—in the words that appear at the head of this chapter—that the "botany of islands" held the key.

Although I have discussed classification and distribution in separate chapters for analytic reasons, they were all but inseparable in Hooker's practice. As a result, there are strong analogies between Hooker's systematic and distributional philosophies: he saw correctly grouping species together into higher taxa as more important than precisely delimiting the boundaries of individual species—the big picture of patterns of affinity was more important. No image of the map Hooker used at Nottingham survives, but it is likely it would have had affinities with geological maps, which grouped individual strata into "units" or "systems," subordinating individual rocks and fossils (and collectors) to grand imperial designs. The mapping of Murchison's celebrated Silurian system was, literally, spreading the standardized British system of broadly defined patterns of geological strata across the world.[92] No doubt Hooker hoped his map of India might produce similar results, but—as we have seen in the case of classification—there is some element here of making a virtue of necessity: without more extensive networks of compliant collectors, Hooker could not hope to chart the locations of every species.

Hooker used global botanical surveys to abolish local names and lump species; a major motivation for this practice was that it simplified the identification of the plant community that characterized a particular region. There is a strong circularity in these practices: a broad species concept was used to assist in surveying vegetation, but surveys of vegetation were used to maintain the broad species concept. This is unsurprising, given his subordination of both systematics and distribution to the larger issue of botany's philosophical standing.

Distribution studies provided a further reason for Hooker's broad conception of species. Given the conception of natural law that he (and most of his contemporaries) worked with, the yet-to-be-discovered global laws of vegetation could not be dependent on resolving the minutiae of such matters as whether or not the New Zealand flax genus *Phormium* was composed of one species or two. Broad generalizations, ideally expressed in mathematical terms, would provide the first step on the inductive ladder from facts to laws. Unsurprisingly, Hooker saw such facts as the overall ratio of species to genera within a particular flora as more important than the precise numbers of either species or genera: one or two more species would not alter the big picture that Hooker was trying to build up. He was charting vegetation (rather than individual plants), and that entailed mapping families and genera rather than individual species.

While charting vegetation with a broad species concept solved some problems, it created others. For Darwin, the accidents of wind or waves were adequate for his purposes, but Hooker could not use them to explain the relationship between islands and complex overlapping patterns of continental floras. This conviction forced Hooker to accept Forbes's continental extensions; only now-vanished continuous land could have allowed whole vegetational communities to migrate across the globe. Only then could botanical arithmetic come into use.[93] However, before specimens could be treated as data for the calculations of botanical arithmetic, they had to be stripped of both the plants' and the collectors' idiosyncrasies. Certain kinds of information had to be removed to amplify those Hooker was really interested in; his lumping ultimately made it impossible for him to decide whether a particular plant variety was or was not found in a particular place.

For Hooker's mentors Lyell and Forbes, multiple centers of creation threatened the very foundations of geology. For transmutationists like Darwin, the threat was even starker; multiple creations meant that modern forms were not the descendants of earlier ones, whatever comparative anatomy or embryology might say. While Hooker would have appreciated the force of both these objections, multiple centers posed an even greater threat to him: as Oldfield had seen so clearly, without the assumption of single centers, there was no need for distribution theories, and without distribution theories, botany lost its main chance to become properly philosophical.

As Janet Browne has argued, the threat posed by the alternatives makes it easy to understand Hooker's adherence to Forbes's "doctrine of specific centres."[94] However, his commitment makes even more sense when seen in the light of his classificatory philosophy. Hooker's broad conception of species meant he could deny that parallel places in nature were occupied by closely related species; supposedly "representative" species were merely varieties of a single species, as in the case of *Lomaria*. By denying the validity of geographical isolation as a classificatory character, Hooker abolished much of the evidence for troublesome representative species. He commented in the *Flora Tasmaniae* that the distribution of Australian plants seemed to some as "an argument in favour of there being many centres of creation for each vegetable form," but he disagreed and assumed "that variation will account for change of species and genera; that the force of variation being a centrifugal one tends to diversity of forms and opposes reversion."[95] Hooker's belief that once a species varied it kept on varying, departing further and further from the original type, meant in practice that the more geographically isolated a species was, the more justified he was in lumping it.[96] Centrifugal variation meant that denying widespread endemism had the advantage of abolishing the favorite evidence used by those like Colenso and Oldfield who claimed the vital importance of local knowledge: fewer "peculiarly local" plants meant less peculiarly local knowledge.

Just as Hooker's rejection of the use of living plant characters was both an effect and a cause of his species concept, certain fine-grained geographical differentiations were impossible without more comprehensive collecting networks. However, far from trying to build up more data, Hooker actively discarded some of the information he had, such as Colenso's overly detailed localities. Instead, he used his collecting networks to reduce living forests and fern-tree scrubs to a representative group of dried specimens with locations; these were then reduced to fewer species by lumping; then they were reduced to written descriptions in the published flora; in the introductory essay they were reduced to lists of species; and, finally, botanical arithmetic was used to reduce them to ratios.[97] Only by ignoring such things as collectors' topographic knowledge could Hooker see the plant world as he wished, spread out at his feet.

While their philosophical implications were important, we must not lose sight of the original economic context of Hooker's distribution studies. The early-nineteenth-century networks of colonial botanic gardens, government officials, missionaries, traders, and travelers were an essential tool for Hooker, just as they were for Darwin's vast fact-gathering project, but we should avoid the Panglossian assumption that their usefulness to metropolitan theorists was the reason they existed. As Drayton has reminded us, these networks were created to serve colonial agriculture, medicine, and trade. He comments that, "by the late 1850s, . . . there were many convenient overlaps between the 'internal' technical ambitions of scientists, and those of the commercial and religious allies of imperial expansion."[98] However, this way of describing the situation runs the risk of creating an anachronistic distinction between the scientific and economic; for Hooker and his contemporaries, economic and philosophical questions were aspects of a single set of concerns.

Browne and others have noted the early-nineteenth-century fascination with ideas about distribution, on the assumption that distribution was being studied in order to shed light on the origin of species.[99] She argues that it was these debates "that led Hooker to concentrate on representative forms and the districts in which they appeared."[100] However, it is important not to assume that Darwin's interests were identical to those of Hooker or the majority of his contemporaries. For them, distribution theories were intended to have practical applications; like the related practice of founding colonial botanic gardens, distribution was about the acclimatization and transplantation of economically valuable plant crops. Understanding the laws that allowed a given plant to grow in a particular place had practical consequences: if local botanists like Colenso and Oldfield were right, local plants grew best in their local habitations, which meant their metropolitan colleagues would have to either defer to local knowledge or travel to New Zealand. More importantly, valuable new crops would have to be grown in their native countries and could rarely be transplanted. However, Hooker's view that many local species were merely forms of highly variable cosmopolitan species implied that plants could be transplanted and acclimatized more readily. Distribution theories could raise botany's philosophical standing, but that was not the only, or indeed the main, reason for formulating them.

On 30 November 1887, Hooker told the gentlemen of the Royal Society that after one of his childhood botanizing trips with his father: "I built up by a heap of stones a representation of one of the mountains I had ascended, and stuck upon it specimens of the mosses I had collected on it, at heights relative to those at which I had gathered them. This was the dawn of my love for Geographical Botany."[101] This anecdote is part of the myth of Hooker the "born botanist," of which he and his first biographer were so fond. In reality, formulating a distribution theory was far more than an innate love or childhood passion—it was both a goal of Hooker's ambitions and a tool with which to advance them. Distribution theory's economic importance justified higher expenditure on botany and raised its status. And it was a tool: both in the obvious sense that it helped him achieve these goals but also because it helped manage and organize collectors. It is no coincidence that more than thirty years after Brewster had compared botanists and astronomers, Hooker would declare: "Oceanic islands are, in fact, to the naturalist, what comets and meteorites are to the astronomer"; astronomy had originally been "subservient to the purposes of navigation and of commerce," as botany was in Hooker's day, but the study of distribution might raise it up alongside the queen of the sciences.

CHAPTER NINE

Associating

> Although these remarks will not be sufficient in themselves to *make* you a *gentleman*, yet they will enable you to avoid any glaring impropriety, and do much to render you easy and confident in society.
> —Charles W. Day, *Hints on Etiquette* (1836)

Hints on Etiquette went through twenty-six editions between 1834 and 1849, and its companion, *Etiquette for the Ladies*, had appeared in thirty-three editions by 1846.[1] The middle decades of the century saw a boom in etiquette books, an indication of how many people were anxious to "avoid any glaring impropriety," but as the anonymous author of *Hints on Etiquette* acknowledged, avoiding mistakes was not sufficient "to *make* you a *gentleman*." The difficulty of knowing what made one a gentleman provided the publishers who were eager to satisfy the demand for etiquette books with another profitable market—satires of those who struggled to join a class they clearly were not born to. Charles Dickens's *Sketches of Young Gentlemen* was typical of works that mocked those who aped a gentility they did not possess.[2]

A fear of social embarrassment was one aspect of life in the growing and anonymous cities, but even more significant was the erosion of trust that came from not really knowing who your neighbors were. Anonymity also contributed to increasing social mobility—a new career in a new town or colony allowed you to shed past embarrassments. These developments helped incite the widespread concern, especially among urban Victorians, over whom to trust and how to establish oneself as trustworthy. The belief that gentlemen were, by definition, trustworthy, fueled the desire both to acquire the right manners and to expose those who were not genuine gentlemen.

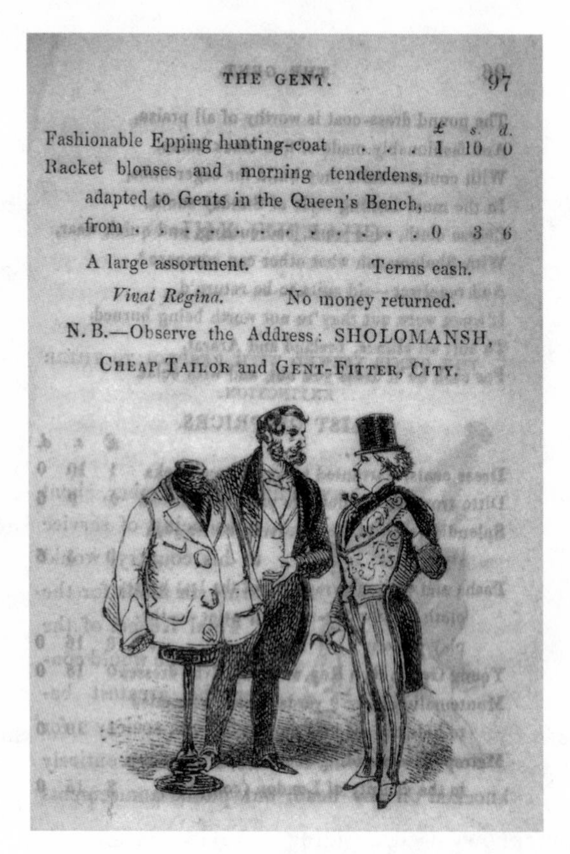

THE GENT. 97

	£	s.	d.
Fashionable Epping hunting-coat . . .	1	10	0
Racket blouses and morning tenderdens, adapted to Gents in the Queen's Bench, from	0	3	6

A large assortment. Terms cash.

Vivat Regina. No money returned.

N. B.—Observe the Address: SHOLOMANSH, CHEAP TAILOR and GENT-FITTER, CITY.

9.1 One way to recognize the "Gent," or false gentleman, was by his fondness for cheap tailoring. From A. R. Smith, *The Natural History of the Gent* (1847). Reproduced by kind permission of the Syndics of Cambridge University Library.

In 1847, purchasers of *The Natural History of the Gent* were offered guidance that would help them identify and avoid a "most unbearable" species of the human race, the "Gent," or fraudulent gentleman, who "is evidently the result of our present condition of society—that constant wearing struggle to appear something more than we in reality are, which now characterizes every body" (fig. 9.1). The author noted: "The Gent is of comparatively late creation. He has sprung from the original rude untutored man by combinations of chance and cultivation, in the manner as the later varieties of fancy pippins have been produced by the devices of artfful [*sic*] market-gardeners."[3] The satirists were not the only Victorians concerned that new varieties of person might be produced, like new varieties of apple, "by combinations of chance and cultivation."[4] As I noted earlier, Francis Galton's contrast between "nature" and "nurture" grew out of a concern that traditional distinctions between social groups were being eroded by groups of "comparatively late creation," whose breeding—and thus qualities—were uncertain. It is no coincidence that one of Galton's first uses of the phrase "nature and nurture" was in the subtitle of his book on *English Men of Science*.[5] They were one of the many new varieties of Victorian whom "combinations of chance and cultivation" had brought to prominence; even if one agreed with Galton's claim that men of science, like other kinds of genius, owed their prominence to inherited qualities, that did not help the public decide which men of science, if any, were to be trusted.

Galton's fellow men of science shared the concerns of their fellow citizens to establish themselves as trustworthy while exposing those whose claims to scientific respectability were illegitimate. As we have seen, the ideal of a disinterested pursuit of knowledge, preferably conducted by independently wealthy gentlemen, was a key factor in making the term "philosophical," rather than "professional," into the highest accolade one could bestow on a scientific man. However, the importance of not equating philosophical with scientific is apparent when we realize that men like Joseph Hooker and others sometimes applied the term "philosophical" to people whose practices and speculations departed sharply from what were ostensibly the correct methodological standards—while withholding it from others who seemed more deserving. This pattern of approbation suggests that certain people were philosophical *despite* their scientific ideas or practices; I will argue in this chapter that "philosophical" was often a description of individual behavior and character, especially of the courtesy and sociability that defined a gentlemen.

"We Are Gentlefolks Now"

The middle class, who felt their education entitled them to a share in the respect once accorded to hereditary landowners, were anxious to acquire the manners of the aristocracy, who in many cases were equally intent upon retaining long-standing distinctions. One result was constant shifts in the customs, manners, and etiquette of the "best people," with a concomitant growth in anxiety for those who wished to join them.[6]

As we would expect, over the course of Victoria's long reign, conceptions of gentility shifted repeatedly; for example, in the 1830s, breeding was still decisive, but by the 1860s, "character" was held to be the real test of a gentleman. A gentleman still needed money, of course, but he could actually exhibit his character by working to earn it, displaying determination and independence. Not only were such virtues admired, but they were often seen as incompatible with the inheritance of wealth and the leisured life that it allowed.[7]

Dinah Craik's hugely popular novel *John Halifax: Gentleman* (1856) provides a fascinating glimpse of these shifting values. Almost the first thing the reader learns about the eponymous hero is that his hands were "roughened and browned with labour" and that he refuses charity, wanting only work; his roughened hands are a badge of honor, envied by Phineas, the novel's sickly narrator, who yearns to be able to work as hard as John Halifax.[8]

The novel tells the story of Halifax's rise to wealth, thanks to hard work and his unshakable physical courage and moral probity. Craik ensures that her hero's life spans the industrial revolution (the book begins in 1794, during the French Revolution, and ends in 1834, just after the first Reform Act). And, just in case her readers might miss the allegory of the rise of the middle classes, it is steam power and industrialization that give Halifax the power (literally) to break the traditional aristocracy's (represented by the unsubtly named Lord Luxmore) hold over the land and its people.[9] Had Halifax been a real person, his life would undoubtedly have been chronicled by Samuel Smiles as yet another example of the virtues of honest toil. By contrast with Halifax, the aristocratic characters—especially Luxmore—are despicable, morally corrupt, and rendered worthless by luxury and idleness; Luxmore's eldest son, William Ravenel, is allowed to marry Halifax's daughter only after he has given away the family fortune to pay off his father's debts and spent many years laboring in obscurity in America.

Despite this strongly drawn political and social framework, Craik seems unsure of the virtues she supposedly advocates; when the dirty, impoverished, orphaned John Halifax first appears, his sole possession is a Greek Testament that had belonged to his father and bears the inscription "Guy Halifax, gentleman," on the title page.[10] Phineas, the narrator, is not surprised to find that such an honest lad is the descendant of gentlefolk, and he explains that his own father, Abel Fletcher, despite being only a tanner by trade, "held strongly the common-sense doctrine of the advantages of good descent": "For since it is a law of nature, admitting only rare exceptions, that the qualities of the ancestors should be transmitted to the race—the fact seems patent enough, that even allowing equal advantages, a gentleman's son has more chances of growing up a gentleman than the son of a working man. And though he himself, and his father before him, had both been working men, still, I think, Abel Fletcher never forgot that we originally came of a good stock."[11] Craik seems determined to duck the issue of whether gentlemen were born or made. If the gentry's social advantages were due to "qualities of the ancestors" that had been "transmitted to the race," why were the Luxmores so degenerate? But if true gentlemen were those who behaved as such, regardless of birth, why was it necessary to establish John Halifax's genteel birth? As one contemporary reviewer noted, "It is impossible to answer the question, 'What does the author mean by gentleman?' since this shadowy word in the book is a loophole through which she escapes from the charge of holding a very democratic view that a gentleman is a man of noble nature who leads an unselfish life." The *British Quarterly Review* lamented "the characteristic irresolution of this writer," which prevented her work from "conveying one of the highest moral truths," namely, that "men, in the sight of God, are equal, and that therefore all good men must be equals upon earth." As the reviewer noted, "We can't all of us find little Greek Testaments with the inscription 'Gentleman' after the names of our ancestors." [12]

Although seldom read today, Craik's novel was hugely successful; it was second only to *Uncle Tom's Cabin* in popularity in the early 1860s. It was pirated in the United States and ran through more than forty-five editions by the end of the century, by which time it had come to be regarded not so much a work of literature as a social force, a personal gospel to those who believed in its message of honest toil and upward mobility.[13]

Many readers of *John Halifax* shared the view, widely held by the early 1860s, that wealth alone (almost regardless of source) led to genuinely higher social status, a view that provides clear evidence of the changing nature of both gentlemen and gentility. In one of the novel's later scenes, Halifax refers to his humble origins in front of his family, and his son, Guy, "flushing scarlet," comments that "we may as well pass over that fact. We are gentlefolks now." To which his father responds, "We always were, my son." [14] But Halifax does not explain why he believes that; was it birth or behavior? The author's lingering doubts over whether good breeding could be entirely discounted also reflected common attitudes; the emergence of formalized professional codes of conduct and ethics (another form of etiquette book) suggests that members of the professions, especially the newer ones, felt they had to acquire the traditional values of gentility in order to raise their status.[15] The men of science were part of this larger debate, and so gentlemanly values—such as courtesy, disinterestedness, independence, and respectability—became part of the definition of "philosophical." [16] One way to think of naturalists like Hooker is as scientific climbers, aping the etiquette of the physical scientists in an effort to be accepted among them—to associate with them and join their clubs. In the highly sociable world of Victorian science, the concern the naturalists evince about the respectability of their science is invariably also a concern about their individual social status.

Botanists in a Broil

The connections between being a gentleman and being philosophical are apparent in a bitter dispute that arose in British botany during the mid-nineteenth century. In 1846, Hooker wrote to Darwin asking him to help keep "almost the only 2 Philosophical Brit. Botanists, out of a broil."[17] The two in question were Edward Forbes (fig. 9.2) and Hewett Cottrell Watson (fig. 9.3), and their "broil" resulted from Forbes's use (in his seminal 1846 "Connexion" essay; see chapter 8) of Watson's data, which led to Watson's accusing him of plagiarism.[18] Hooker was strongly averse to public brawling, later urging Darwin not to respond to his critics in the newspapers: "I cannot abide this lugging of Science before the public in Times & Athenaeum, & implore you my dear fellow not to do so again." Hooker added: "Science will be much more perfected if it keeps its discussions

9.2 Edward Forbes. By permission of James A. Secord.

9.3 Hewett Cottrell Watson. By kind permission of the Trustees of the Royal Botanic Gardens, Kew.

within its own Circle."[19] Hooker's reaction to Forbes's and Watson's dispute also reveals that "philosophical" described people's character and conduct as much as their ideas; indeed, ideas were often a secondary consideration in determining who was philosophical.

Hooker characterized Forbes and Watson as "almost the only 2 men who have looked on British Flora with the eyes of philosophers": "Watson in particular ranks in my opinion at the very head of English Botanists, whether for knowledge of species or of their distribution; he first wrote philosophically upon them & his works are of the highest order."[20] Watson's main work, the *Cybele Britannica* (1847–60), was an immense compilation of data on that most philosophical of topics, plant distribution. During the 1840s, Forbes was one of the few British naturalists whose distributional studies brought him prominence, and Hooker was clearly concerned that their argument might damage the scientific reputation of botanical geography.

Watson's and Hooker's work had much in common: both focused on distribution; both relied on correspondence networks to supply specimens;

both entailed analyzing collections gathered by others; and both used Brown's and Humboldt's botanical arithmetic to analyze data. Watson was also happy to describe himself as a "lumper of species."[21] He was as opposed as Hooker was to those splitters who "can discover nothing for themselves" and so seek to "acquire a sort of spurious claim" by conferring names, which resulted in "the rapid increase of false species." However, unlike Hooker, Watson chose to attack the authors of several British floras by name—including John Lindley, William Hooker, and Charles Cardale Babington—and he publicly imputed unworthy motives to those he felt were to blame for the proliferation of spurious species.[22]

While they had sharply different senses of etiquette, Hooker and Watson shared a commitment to correctly formulated inductive generalizations as the route to natural laws. Watson urged "scientific men" to distinguish "between those general remarks which are merely vague, and those which are truly generalisations of facts"; the former were all too "easily attained, and at small cost of time and thought," in contrast to "True generalisations," which "require much time and thought."[23] His position echoed Hooker's claim that colonial botanists were "prone to rely for information on . . . speculative subjects (which they seek with avidity) upon a class of works . . . by authors who have no *practical* acquaintance with the sciences they write about."[24]

Watson concluded his comments on true generalizations by observing that, since they took lengthy effort, "combined with a scrupulous regard to accuracy," they "are in consequence extremely rare."[25] When the *Gardeners' Chronicle* (whose editor, Lindley, was one of those Watson had attacked) reviewed the *Cybele,* it concurred, albeit noting that such generalizations were especially rare in Watson's own work: "It is . . . remarkable that whilst extolling the importance of generalisation on this subject, Mr. Watson should neither treat the great questions it involves himself, nor allow others to touch them." "We are induced to offer this remark in consequence of . . . an attack on Professor Edward Forbes," whom Watson had accused of "having made free with his labours, and of having taken credit to himself for results and generalisations, which he, Mr. W. had alone originated."[26] The attack in question appeared in an appendix to the first volume of the *Cybele,* which was when the Forbes/Watson "broil" went public. Watson referred to the "so-called 'floras,' announced by Mr. Forbes," which he claimed were not merely plagiarized but had been altered by Forbes in a way that revealed his lack of botanical expertise.[27]

Hooker admitted to Darwin that "most unfortunately" their friend Forbes "does not seem to know the Geographic Distrib. of the English

Plants." Hooker had "taken his modification of Watson's types of vegetation as correct" but on closer examination noted several errors. He commented that "I now see the cause for Watson's being so peculiarly savage & offering me proof that all that is correct is mere plagiarism."[28] And yet, as we shall see, despite Forbes's ignorance and his mistakes, Hooker continued to regard him as philosophical, ultimately indeed as more philosophical than Watson.

In a letter to Hooker, Forbes claimed: "Nobody is more ready than I am to proclaim Watson's [?excellent] & indefatigable work. I grant the excellence of his types & their truth," but since they did not directly address distribution issues, "surely I did more than enough in quoting them." What Forbes wanted was not Watson's static British floras but "those of *migration*" or "of *foreign relation* at any rate." Forbes also maintained that in using Watson's data, "I did not put them forth as mine—since they were simple statements of facts known to every British botanist":

> What I claimed & claim was the attempt to explain the causes of these things & through what series of events they were brought to pass (as essential a chapter in Botanical Geography & therefore in Botany as any other in flower-books). This Watson's want of a sufficient variety of natural history & Geological knowledge prevented his doing or being likely to do. I fancied—how vainly!—that he would have been pleased at my coming into the field with the collateral sciences of Zoology & Geology to aid Botany.[29]

As Roderick Murchison had said in praising Forbes's essay to the BAAS, it was "the first attempt to explain the *causes* of the zoological and botanical features of any region anciently in connexion."[30] Forbes knew that this was the basis of his claim to philosophical status and was one Watson could not make because of his lack of expertise in geology and zoology.[31] Watson's hostile reaction was partly prompted by resentment of Forbes for making a more philosophical use of his data than he had been able to do himself.[32]

In his initial attack on Forbes, Watson admitted that, while his discussion gave the matter "an aspect unpleasant and unfavourable to Mr. Forbes," the latter "must remember that his own neglect of . . . courtesy and justice" had prompted the attack.[33] However, in the eyes of Hooker, Lindley, and others it was ultimately Watson's lack of courtesy that decided the outcome of the dispute, since it contrasted so markedly with Forbes's gentlemanly restraint. Darwin suggested that Forbes's "noble indifference to fame" might explain his "not in some instances making proper acknowledgment," implying that someone who was above self-aggrandizement might not realize how jealous another man might be of receiving his proper due.[34] Hooker agreed with his friend's assessment of Forbes's character; after noting the latter's errors, Hooker asserted that "I still however quite acquit Forbes of any intentional piracy."[35] By contrast Watson was, in Hooker's words, "touchy & very *severe* when first offended, though he never holds a grudge long."[36]

This turned out to be a spectacularly inaccurate reading of Watson's character—he neither forgave nor forgot. In the fourth volume of the *Cybele,* published in 1859, he was still calling Forbes a "plagiarist" and describing the "Connexion" essay as "blundering and false"—despite the fact that Forbes had by then been dead for five years.[37] This violation of etiquette incensed Lindley, who took it upon himself to anonymously review Watson's book, commenting that "we must draw attention to the extraordinary language which its author employs when speaking of others": "Poor Forbes he calls blundering and false, a plagiarist, a man of reckless hardihood of assertion."[38] The review also ridiculed Watson's claim that "technical" botanists, such as classifiers, were "seldom profound reasoners": "Philosophical botanists on the other hand, to which class Mr. Hewitt [*sic*] Watson evidently claims to belong, are profound reasoners, and equal to any object in science. It is painful to undeceive him. But surely the man who asserts that a correct knowledge of species is the foundation of all his 'philosophy,' and at the same time announces that there is no correct knowledge of species, should have seen that *therefore* there is no philosophy." Lindley suggested that "so great a philosopher" as Watson should have "set to work to acquire a correct knowledge of species" by devoting his time to detailed systematic work, but instead he had "abused others for not furnishing him with the object of his desires."[39] Lindley's vehemence was probably prompted by Watson's unconcealed contempt for systematic botany and its practitioners, but it is worth noting that Watson's claim—that distribution studies must rest on "correct knowledge of species," which was currently lacking—was one Hooker and his fellow philosophical botanists were keenly aware of.

Watson's criticisms of his fellow botanists were not new; as early as 1845 he had rebuked British botanists for their "limited ideas regarding the scope and objects of the science": naming plants was "doubtless an agreeable kind of study," but one that "so lightly taxes the mind, as to be within the grasp of moderate capacities; for even children can learn Botany thus far." Systematic botany did not engage "those higher intellectual attributes of man, which are concerned in all trains of reasoning, and which lead to the knowledge of causation and dependence between the phenomena of creation."[40] Given the concern Lindley and Hooker shared over the status of their discipline, Watson's comments must have seemed a sort of treachery—a fellow botanist supplying ammunition to those who sneered at botany's scientific claims.

Watson had never been tactful; he believed intellectual debate should be impersonal and therefore that people should not take offense at his remarks.[41] Yet, while he was editor of the *Phrenological Journal* (from 1837 to 1840), he was so critical, tactless, and outspoken in his attacks on others that one of those he offended, Thomas Prideaux, published a pamphlet attacking Watson, in which he wrote, "It is not to be expected that individuals will quietly submit to be visited with the petulances and impertinences of Mr. Watson without retaliating," adding that "I would hint to Mr. Watson that it is very unwise policy in an individual, whose egregious vanity and conceit render him so peculiarly sensitive, to commence a mode of attack, from which he shrinks when retorted upon himself."[42]

Watson continued his aggressive attacks after he shifted his main scientific interest from phrenology to botany. When William Hooker gave Watson the opportunity to review for the *London Journal of Botany* in 1850, Watson used it to attack his fellow British botanists: "English botanists can repeat, or copy, or imitate very successfully, but they do not often originate new views and improved modes; they do not discover and invent in botany. They can detect and observe, can depict and describe, can catalogue and group individual facts and objects; but they do not connect their data, thus acquired, by the theoretic relations of cause and effect, nor do they often generalize their details, so far as to elevate these into the category of scientific principles."[43] These were, of course, the familiar complaints leveled at botany and at natural history more generally by some practitioners of the physical sciences, and once again Watson's criticisms would not have endeared him to his fellow botanists.

Watson's tactlessness extended to his personal correspondence, leading to breaches with both Babington and John Balfour, who had once been friendly correspondents. As Babington noted, "I do not think that [Watson] wants to quarrel with either of us but he has a manner which makes it very difficult to keep on good terms with him. He seems to think that he may say and print whatever he likes of others & that they must not even hint at anything on the other side."[44] In a letter to John Gilbert Baker, the Kew fern expert, Watson savaged Bentham as "a man who has never shown any originality of thought, discovered nothing, invented nothing, but has always been an imitative plodder."[45] And in a letter to Alphonse de

Candolle, he wrote: "Technical botanists are usually bad reasoners. Take Sir William Hooker in example. His talent in botany, as you are well aware, is depictive and descriptive. He represents and describes species or genera. You will admit that he has done good service to botany. And (making some allowance for haste and quantity) that he has been a good depicter and describer. What more is he? Nothing. Utterly unable to reason on any subject, whether botanical or not."[46] Even after de Candolle had published a warm tribute to Hooker's life and work in 1866, and sent Watson a copy, the latter still wrote back in equally rude terms, dismissing Hooker's intellect as "utterly incapable of ratiocination" and "incapable of connecting ideas in the way of cause and effect." He closed the letter by commenting that Joseph Hooker could not accept criticism with good humor.[47]

Watson's endless rudeness would have been enough to damage his relationships with other men of science, but it also violated accepted codes of scientific conduct. In 1847, the *Chronicle*'s review of the first volume of the *Cybele* had commented on his "querulous tone," which the reviewer thought "unworthy of a man of science," and his posthumous attack on Forbes provided Lindley with what had probably been a long-sought-for opportunity to savage Watson while defending systematic botany from his disparaging comments.[48] Nonetheless, the tone of the *Chronicle*'s review seemed excessive to Darwin, who wrote to Hooker that while Watson "does deserve punishment, . . . surely the review is too severe" and asked whether Lindley had written it.[49] Hooker replied: "Of course Lindley wrote the G.C. article on H.C.W. & upon my honor I do think it richly deserved. The sneering contempt with which he treats his enemies the virulence of his dishonest attacks on those he knows little of, & his patronizing air to those he approves, are beyond all whipping powers of reviewers."[50] Watson's "sneering contempt," "dishonest attacks," and "patronizing air" had cost him the title of gentleman and thus his place at the "very head of English Botanists."

"Forbes Is Often Rather Fanciful"

If ideas made a man philosophical, Watson—the cautious inductive plant geographer—should surely have qualified. Yet, by 1860, neither Hooker nor Darwin was describing Watson as a philosophical botanist, while Lindley was ridiculing his claim to the title, despite the fact that Watson's botanical *ideas* had not changed significantly (indeed, the lack of development in his thinking was explicitly attacked by Lindley). Meanwhile, Forbes retained the title despite having espoused ideas—most notably his theory of "polarity"—that were unacceptable. Understanding this apparent contradiction reveals gentlemanliness to be a decisive ingredient in being philosophical.

Forbes presented his theory of polarity during his presidential address to the Geological Society in 1854. He thought it was a major new principle of natural history, but his colleagues' reaction can be gauged from the fact that, after his premature death later in the same year, most of them decided his theory was a symptom of his illness.[51] Being a determined antitransmutationist, Forbes used polarity to explain away apparently progressive patterns of development among organisms, arguing that species were created in greater numbers in the earliest and in the most recent parts of life's history.[52] Darwin noted in the margin of one of Forbes's papers that he found this idea "absolutely unintelligible."[53] He commented to Hooker, "I think Forbes is often rather fanciful; his 'Polarity' makes me sick—it is like 'magnetism' turning a table."[54] Yet such "fanciful" ideas did not diminish their overall respect for Forbes (although had he lived and continued to promote them, they might have caused greater strain).

Even if Hooker and Darwin chose to politely overlook Forbes's polarity, they could not ignore the fact that other aspects of his thinking—and, even more importantly, of his practice—sat rather uneasily with their declared canons of methodology. Although Hooker was more receptive than Darwin to Forbes's Atlantis theory, both agreed he was rather too speculative. Darwin commented that "without any proof to speculate on Ireland & Portugal having been once connected: it staggers me."[55] Hooker shared some of his doubts, observing that Forbes's suggestion that the seaweed beds of the Sargasso Sea indicate "a previously existing line of coast is surely preposterous & untenable" and "I fear will tend to throw discredit on the rest of Forbes work."[56] Darwin concurred: "It is really a pity that Forbes is quite so speculative: he will injure his reputation."[57] Premature speculation upon inadequate evidence ought to have been a grave sin against inductivist precepts—especially when based on other people's inadequately acknowledged data—yet both Hooker and Darwin were mainly concerned that Forbes did not damage his scientific reputation as a result of his boldness.

Forbes's philosophical reputation survived despite the plagiarism dispute, his speculative theorizing, and his advocacy of polarity. To understand why, we need to examine those factors that went beyond one's published statements and that made a person philosophical. These included courtesy;

social, political, and religious respectability; sociability; open-mindedness; calmness; and sensibility. Naturalists had to be gentlemen before they could be considered philosophical.

A Gentlemanly Manner

Courtesy was widely regarded as the mark of a gentleman and of a man of science in particular.[58] In 1833, the *Westminster* felt obliged to take the scientific writer John Murray to task for being "a good deal given to the disputatious and controversial; a fault not at all uncommon among his brethren of modern science." Perhaps he had, as he claimed, been "unjustly attacked" and even "been robbed of discoveries," but the *Westminster* was convinced that "these are accidents which the philosopher should bear meekly."[59] Almost thirty years later, Darwin commented to Hooker, with evident amusement, on a review of the *Origin,* "Did you see in Literary Gazette that Prof. Clarke of Cambridge says the chief characteristic of such Books as mine is 'their consummate impudence'—mild & gentleman-like language!"[60]

In arguing that "the philosopher" should not respond to attacks, the *Westminster* was applying a standard traditionally associated with the gentleman. The nineteenth edition of *Hints on Etiquette* noted: "A man should never permit himself to lose his temper in society—not show that he has taken offence at any supposed slight—it places him at a disadvantageous position—betraying an absence of self-respect,—or at the least of self-possession." Not only should arguments be avoided, but "even slight inaccuracy in statement of facts or opinions should rarely be remarked on in conversation."[61] The anonymous author of *The English Gentleman* (1849) urged his readers to reflect on "the revulsion and disgust" invoked by those who are "petulant, irritable, and self-willed" or even "morose, and sulky"; such qualities "would be viewed in precisely the same light if seen by others in yourself. *A Gentleman should be incapable of them.*"[62]

Good manners eased one's path through society, and an early schooling in etiquette was essential, since it enabled one to participate in society, which further improved one's manners and conversation. The author of *Advice to a Young Gentleman, on Entering Society* (1839) argued: "In cultivating company, the advantages are as attractive as the duty is compulsive. The drawing-room will furnish the minds of some; it will refresh the spirits of all."[63]

While aboard the *Erebus,* Hooker wrote to his father that "if ever on my return I am enabled to follow up botany ashore, I shall live the life of a hermit as far as society is concerned; like [Robert] Brown perhaps."[64] William Hooker was concerned to ensure that Joseph did not turn his back on society; he told his son that Brown was no hermit: "he is really fond of Society and calculated to shine in it: and to my certain knowledge, never so happy as when he is in it."[65] Hooker seems to have taken his father's advice to heart, since contemporary accounts comment on his good manners and social skills; when Hooker first met Charles Bunbury in 1846, Bunbury wrote to tell Mrs. Charles Lyell that he liked Hooker "extremely, he is so modest, and at the same time so communicative and full of knowledge and talent." This positive impression was no doubt reinforced by the fact that Hooker went out of his way to comment on the "great assistance" he had derived from Bunbury's published notes on the flora of the Cape.[66]

A man who shunned company would miss the opportunity "for the correction of numerous indefinite evils, and the bestowal of some positive profits."[67] Not the least of the benefits of such socializing was that "conversation must be regarded as a most important means of acquiring knowledge" as well as of meeting the right people.[68] An inability to mingle in society characterized one form of that social impostor, the Gent, who was mocked in the *Natural History of the Gent* for being antisocial: "the dreary Gent" was one who "affects a drawling tone of voice, which he considers cool and fashionable," and refused to converse or dance.[69]

Above all, regular mixing in society refined a man's manners and made them easy and natural, a quality routinely praised in etiquette manuals. The author of *The English Gentlemen* praised natural good manners above "all the rules which have been laid down in books for our country cousins," telling them "how they are to enter a room" or "what subjects to speak on, and what to avoid." These vital matters "can only be learned by a continued intercourse with those that use them. It is utterly impossible that they can be taught by mere writing." Genteel manners were as much a form of tacit knowledge as botanical drawing or writing, and readers of *The English Gentleman* were warned that, if one attempted to learn the forms without practicing them in appropriate company, "the chances are, that . . . you give such an unnatural stiffness to your manner as completely to defeat your object."[70]

The importance of socializing for scientific men was also recognized by the BAAS president the Reverend Thomas Romney Robinson, who argued that one of the association's achievements had been to improve the tone of scientific discussion by checking "some evil elements of our nature": "the bitter disputations and petty hostility, which have too often disgraced the records of science, and made its followers contemptible. The most irritable man must feel less disposed to apply violent language, or attribute unworthy motives to one whom he has met in kindly intercourse, and whose character he has appreciated, than when he encounters a perfect stranger in the arena of the press." Thanks to their regular meetings, "the fault has nearly disappeared," and the association had also played a role in eradicating "a greater danger . . . that of self-esteem. The true philosopher does not incur it: he knows too well the proportion between his ignorance and knowledge."[71] Modesty was another characteristic of the gentleman of science.

Such views were widespread. For example, the *Westminster* claimed that courtesy ensured that "the great congregation of botanists is at peace with itself and friendly with its philosophical neighbours." There might be a few examples of "moroseness, or viciousness, or indigestion, or envy, (for naturalists being mortals are afflicted occasionally with these original sins,)" but they were very much the exception. Unfortunately, these exceptions included "Watson, indefatigable and deservedly illustrious in statistics, but grown misanthropic from working overmuch when in ill-humour"; such men "find a melancholy pleasure in attributing evil motives to his fellow-labourers."[72] As Hooker and Darwin probably knew, the anonymous author was Forbes, who, six years after the plagiarism dispute, was gentleman enough to put his adversary's bad temper down to the strain of overwork rather than attributing genuinely hostile motives—in sharp contrast to Watson's posthumous persecutions.

The naturalist's standards of behavior and sense of community were forged in contexts like the BAAS meetings, which were among the many social gatherings where the men of science mixed, became friends, and learned to disagree courteously. Forbes was a noticeably affable and sociable character. As an undergraduate, he belonged to a secret society, the Oineromathic (or "Friends of Truth") Brotherhood, devoted to the advancement of "wine, love and learning."[73] His botanical field trips always ended up at a pub "with lots of punch (in moderation) and good songs," some of which he composed himself.[74] The culture of fieldwork, which botanists and geologists shared, encouraged healthy outdoor exercise and manly sociability.[75] Like Arthur Henfrey and Hooker, Forbes was young and unmarried when they worked together at the Geological Survey and they often spent their evenings and weekends together. De la Beche even discouraged the younger men from marrying for fear that they would spend less time within the Survey's "family" if they did.[76] Forbes also founded the Red Lions dining club (named after a tavern), where he spent much time carousing with his fellow young naturalists, who spent many of their evenings together, drinking, playing billiards, attending *tableaux vivantes* (where scantily clad young ladies posed suggestively), picking up prostitutes, and attending ballet, theater, opera, and exhibitions.[77]

One important function of clubs like the Red Lions was to define insiders and outsiders by creating a forum in which the pretensions of the powerful were satirized and the claims of the young could be advanced. Forbes, the quintessential insider, shone in such situations, where he sang "some of his merriest songs and told some of his raciest stories, amid the vociferous growls and applauding tails of his leonine brethren."[78] By contrast, Watson was solitary and irascible. He told a friend: "I am by nature reserved almost to timidity, & so little accustomed to society, that I am scarcely fit for it; & hence have pretty constantly the idea that others must think 'what a dull & stupid person H.W. is, I wish he was gone.'"[79] Not long after Watson and William Hooker started corresponding in 1834, Hooker expressed the hope that they might meet at a Naturalist's Association meeting, but Watson responded that his "love of solitude makes me shun such scenes."[80] Watson avoided most of the London scientific societies, losing the chance to make friends and allies and to convince his peers of his moral character.[81]

Unsociability was only one factor that contributed to Watson's outsider status: he was a phrenologist, which was no longer widely regarded as a plausible science by this period; and he held radical political opinions, having been a supporter of Lamarckian transmutation since the 1830s.[82] By contrast, Forbes was a deeply religious and rather reactionary Tory whose respectability made it possible for him to be active in the more staid and conservative Philosophical Club of the Royal Society.[83]

The Philosophical Club was a typical, convivial place for distinguished men of science to meet; Forbes told Hooker that "Herschel, Beaufort, Rennie, Brown & Sedgwick were among the dinner party last week. We mustered 32 and had a lot of discussion, which did capitally. I am pressing you for first vacancy."[84] The club's members were expected to be gentlemen, and the sources of their income occasionally came under scrutiny.[85] In 1855, Darwin told Hooker he was sorry to hear that the chemist and astronomer Warren De la Rue was not among those newly elected. Darwin acknowledged that "he does not appear like a gentleman" but felt that "all that he says at the [Royal Society's] Council seems very gentlemanlike & nice."[86] Darwin gave no hint of why De la Rue did not "appear like a gentleman," but he was probably blackballed because he was the active and hard-working head of his family's stationery manufacturing business.[87]

For those who avoided the taint of trade—men like Hooker, Forbes, and

Owen—clubs allowed patrons to be cultivated and alliances to be made, gossip exchanged, and friendships cemented, and Watson's self-imposed exclusion from this world further damaged his reputation.[88] In the informal setting of early-nineteenth-century British science, dining clubs, friendships, and networking had not yet been supplanted by peer review and academic institutions; in such a context, a naturalist's character—including his manners—was as important as any practical or formal credentials in establishing him as philosophical.

The English had once been perceived as an antisocial nation, but—partly as a result of all this club-going—sociability became essential to the British conception of the man of science. In the 1830s, Edward Bulwer Lytton argued that the clubs which "form a main feature of the social system of the richer classes of the Metropolis" had "assumed a more intellectual character; every calling has its peculiar club." He believed that "the effect which this multiplicity of clubs has produced is salutary in the extreme; it has begun already to counteract the solitary disposition of the natives."[89] And James D. Forbes had told the BAAS in 1834 that "there has been . . . a general but most erroneous impression abroad, that philosophers are incapable of enjoying, and stoically superior to, the ordinary sociabilities of life,—that scientific ardour dwells only in the mind of the solitary." He dismissed this by noting the cooperative and collaborative nature of much of the work before the association.[90]

The male-only sociable ethos encompassed both the purely social clubs and the scientific societies.[91] The traditional ethos of the gentleman's club was to avoid controversial topics of conversation, especially political and religious matters. Similar habits were the norm for professional bodies, such as those that represented the relatively new profession of engineering; the need to find and keep clients in order to earn a living encouraged engineers to be discreet about their views. An air of religious and political conventionality, with no strong protestations of faith or belief, was the most appropriate demeanor for a gentlemanly engineer.[92] The same discretion was often to be found within scientific bodies, such as the Geological Society.[93]

Hooker seems to have found it natural to conform to these conventions. In a private letter to Asa Gray, he described himself as "a whig myself (if anything)."[94] Elsewhere he referred to himself as "a philosophic conservative, a strong Unionist, but not a Tory" (although Darwin teased him for having become a "jolly old Tory" when Hooker argued that natural selection explained the dominance of aristocrats in society).[95] In general, Hooker never expressed much interest in party politics and was publicly discreet about his religious views. In a diary entry, written after they had known each other for more than twenty years, Bunbury described Hooker as "in general a warm advocate of what are called 'liberal' and 'progressive' doctrines, though not violent or extravagant; not a *subversionist* like Huxley."[96] The contrast with Bunbury's perception of Huxley is telling, since in private, Hooker seems to have shared Huxley's agnosticism (in a letter to his friend, the clergyman and amateur naturalist James Digues de la Touche, Hooker expressed praise for Huxley's concept of "a religion of pure reason"). Unlike Huxley, however, Hooker was not "a subversionist" but observed a conventional, broad church Anglicanism. Nevertheless, he defended those who argued for liberalization in the church, and when agreeing to stand as godfather to Huxley's son, he told Huxley that, although "I hate and despise the spiritual element of the ceremony," he felt that "the pleasure of being in any recognised relationship to your child will sweeten any pill of doctrine that may be offered."[97]

The convention of avoiding controversy and not holding opinions too dogmatically helped prevent scientific disagreements from becoming bitter and personalized. Evenhanded, disinterested consideration of rival positions was a long-established aspect of the philosopher's character; the *Oxford English Dictionary* gives one sense of *philosophical* as "befitting or characteristic of a philosopher; wise; calm." Darwin used the word in a light-hearted version of this sense when he congratulated Hooker on the birth of his daughter, Harriet, in 1854: "You seem to have taken it very philosophically."[98] Dispassionate courtesy certainly characterized Darwin's and Hooker's disagreement over the value of Forbes's land-bridge theory and they managed to disagree without falling out. After Darwin showed that plant seeds could survive saltwater long enough to have migrated across oceans, he claimed that Forbes's ideas were now redundant and wrote to Hooker to announce that "we are come to a complete split on the philosophy of extension of continents: have not I been audacious in attacking poor dear Forbes's theory?"[99] Hooker refused to abandon Forbes, and Darwin wrote again a few days later to say, "It shocks my philosophy to create land, without some other & independent evidence."[100] This was a strong challenge to Hooker's commitments, yet the latter replied mildly: "I wish I could give you a rational account of my faith. I am far from wedded to the *continental* or Forbesian doctrine; the difficulties on all sides are so numerous & perplexing, that I really do not know where to turn for a satisfactory theory of distribution."[101] Despite jokingly referring to his "faith," Hooker acknowledged that no properly philosophical gentleman could be "wedded" to a "doctrine"; that would imply too much dogmatic passion, but he would not surrender his views without some better evidence. Their shared commitment to both a Lyellian-influenced methodology and conventions of gentlemanly politeness helps explain how Hooker and Darwin were able to correspond and collaborate during a period when Hooker was not persuaded by Darwin's views about migration.[102]

The moral character of the man of science included a capacity for sincere friendship, which was evidence of his sociability and courtesy.[103] The claim that botanists generally formed a particularly friendly group was an aspect of the perception that natural history—being in part an education in careful seeing, smelling, and touching—was also capable of educating naturalists' sensibilities; William Whewell made precisely this point in justifying its inclusion among Cambridge University's studies, long before there were science degrees on offer.[104] The perception that naturalists were especially sensitive was also common in novels of the period: Elizabeth Gaskell included a naturalist, Roger Hamley, in *Wives and Daughters*. In sharp contrast with the other men in his family, Roger was a sensitive soul: "so great a lover of nature that, without any thought, but habitually, he always avoided treading unnecessarily on any plant." This care prompts him to be similarly sensitive to the feelings of the heroine, Molly Gibson, and he tries to distract her from the unwelcome news of her father's impending remarriage by showing her "the treasures he had collected in his morning's ramble" under the microscope.[105] Gaskell helped create a positive image of the sensitive naturalist in other novels too; the *Westminster Review* compared the working-class naturalists who attended the BAAS Natural History section to one of her characters, asking rhetorically, "Who that has read the story of 'Mary Barton,' does not recollect the admirable picture of the quaint old artisan-naturalist, Job Leigh?"[106] Similarly, one aspect of the almost-feminine sensibility displayed by the eponymous lonely weaver of George Eliot's *Silas Marner* (1861), who surprises his neighbors by adopting and caring for a foundling child, is his knowledge of medicinal herbs, learned from his mother, which helps rescue him from misanthropic isolation.[107] To be a lover of nature was considered improving to one's moral character even in the latter part of the century, when the direct link between natural history and natural theology began to fade.

Of course, this emphasis on moral character extended beyond the sciences and the novel. This was reflected in a shift in the etiquette literature during the middle decades of the century; the etiquette manuals of the 1830s codified rules for avoiding embarrassment; by contrast, those of the mid-1840s were becoming more moral in tone, merging the rule-based etiquette with an earlier tradition, that of the conduct book, which argued that good manners were merely a side effect of good moral character. John Butcher's *Instructions in Etiquette* (1847) was a literal marriage of the two genres: five letters to parents on moral guidance for children were added to a plagiarized etiquette manual of 1828, because, as Butcher wrote in his preface, "children should be distinguished by the possession of moral and religious qualities, but also that those qualities should be accompanied by polished manners, and the charms of a graceful behaviour."[108] This renewed moral tone was a response to increasing concern about false advertising, imposture, and adulteration of commercial goods, especially foods and medicines, all of which paralleled the ease with which etiquette books allowed fake gentlemen to pass themselves off as the genuine article.[109]

In similar vein, *The English Gentleman*'s emphasis on naturalness and avoiding mere book-learned manners was partly motivated by a concern with detecting the fraudulent "Gent" and similar parvenus. A man "who has been at all accustomed to good company" cannot be confused with one who merely pretends to such connections: "No mere mechanical practice will ever confer a true grace on the manner and deportment."[110] As the *Hints on Etiquette* expressed it: "Gentility is neither in birth, manner, nor fashion—but in *the* MIND. A high sense of honour—a determination never to take a mean advantage of another—an adherence to truth, delicacy, and politeness towards those with whom you may have dealings—are the essential and distinguishing characteristics of a GENTLEMAN."[111] In addition to his manners, a naturalist's writings also revealed his character. When the *Athenaeum* reviewed Hooker's *Flora Novae-Zelandiae,* it congratulated him "on various points of great interest, not only to the botanist, but to the general naturalist." As the *Westminster* had done earlier, the reviewer also congratulated Hooker on the Baconian virtues, "indefatigable energy and activity of mind and body," which were presented as part of his moral character, his "devotion of spirit," and "simple truthfulness of purpose." The language in which he expressed his opinions was particularly praised: "his opinions,—formed, as they are, with a soberness and considerateness which constitutes their great safeguard,—with an authority and a claim to our respectful consideration which are rarely deserved by the propounders of scientific theories."[112] By avoiding impassioned invective, Hooker had managed to make the potentially dangerous topic of the origin of species a safe one (once again, the contrast with Watson is obvious). In this case, however, the reviewer's opinion had no doubt been shaped by Hooker's apparent decision not to explicitly endorse any of the controversial speculations he discussed, a point whose significance I will return to.

Species-Mongers

The various philosophical virtues were, as I have shown, also those that helped define a gentleman in the fluid social world of mid-nineteenth-century Britain; they helped the would-be scientific gentry stake a claim to cultural leadership on the basis of their moral character as well as their expertise.[113] Given the persistent equation of philosopher with disinterested gentleman, a man of science should be above caring about money; when the Linnean Society asked each of its members to contribute twenty pounds to the cost of moving to new premises, Darwin wryly commented to Hooker, "I have just sent in my name for £20 to Linn. Soc; but I must confess I have done it with heavy groans, whereas I daresay you gave your £20 like a light-hearted gentleman." [114]

I have argued that the careers of men like Hooker and Forbes should not be described in terms of professionalization, because they employed the term "philosophical" to contrast their approach to science with the merely commercial interests of those who pursued it for pay. In 1843, Forbes had closed his inaugural address at King's College with a quote from Bacon that reminded the student that science was not a "couch whereon to rest a searching and restless spirit," nor was it for pleasant relaxation, a tower of pride, a fortress "for strife and contention," nor—most importantly of all—"a shop for profit or sale." [115] Yet anticommercial ideals made life complicated for such men, whose efforts to be philosophical were always partly about their need for an income. In this context, it is interesting to reexamine Hooker's concern to head off the impending Watson/Forbes "broil" because if it were allowed to continue "all the dirty species-mongers will chuckle." [116]

"Species-monger" was a phrase Hooker and others used regularly, usually to describe those who multiplied species by splitting.[117] However, a "monger" was literally a trader or dealer (as in "fishmonger"); applying it to hairsplitters implied that they were out to make money by multiplying species. The Victorian culture of collecting included many commercial aspects: alongside those who traded in books, magazines, and various kinds of natural history equipment were those who sold specimens, which made it possible for men like Augustus Oldfield and James Drummond to think of making money from collecting.[118] Specimen dealers were common in both the metropolis and the colonies: Colenso complained that he had "seen several collections of Ferns" that were "prettily and fancifully got up for sale" in New Zealand, yet "I have never yet seen one such manufactured collection correctly named throughout; even the very names of the

Ferns are often mis-spelt on the printed labels!"[119] However, a specimen hawker's "mistakes" may not have been accidental; the opportunity to sell more specimens may have motivated those who multiplied species.[120] For example, the popularity of natural history specimens such as ferns created a mania for bigger collections full of rare or unusual forms; from the dealers' perspective, the more species the splitters christened, the more specimens they had to sell.

Hooker's use of the term "species-monger" linked all splitters with the slightly shady world of the specimen hawkers; yet those of his collectors who worked for money, like Oldfield and Drummond, were no more likely to be splitters than those who worked for love (indeed, Oldfield explicitly condemned the "lax notions" of naturalists who named new species on spurious grounds).[121] Colenso was the most vigorous of splitters, but he refused payment and sneered at the "professed fern-collectors" whose misnamed collections had, he thought, messed up the nomenclature of the ferns, "the chief botanical glory of New Zealand!" He claimed that "the other great natural Cryptogamic Orders . . . have hitherto escaped," but only because they were "far too difficult a study, and TOO UNFRUITFUL OF PAY!"[122]

Yet even if Hooker did not intend "species-monger" to make a literal equation between selling specimens and multiplying names, it nevertheless expressed anxiety about his own social position. As James Secord has argued, during the early nineteenth century most of the clearly professional aspects of science were perceived as low-status ones, such as specimen selling or instrument making.[123] Hooker was in the awkward position of wanting to mix socially with gentlemen of science like Darwin and Murchison while needing to earn a living from science (which they did not). His social position was not that different from Gunn's, who also wanted to be a gentleman while needing an income. The term "species-monger" linked splitting with commercialism but thus implies that properly philosophical, broad notions of species were the product of a properly gentlemanly and disinterested approach to natural history; whatever else a philosophical naturalist was, he could not be a species-monger.

The resistance to the term "professional" that I have identified gradually faded in the later decades of the century, in part because of a shift in the identity of the professions themselves. A great many trades transformed themselves into professions during the nineteenth century by such actions as forming professional associations. These were intended to protect their members from competition with the unqualified and were justified on grounds of public interest, since they supposedly protected the public from quacks. Claims of disinterested service to the public became a common feature of professional rhetoric.[124] As a result, we find a medical man in 1868 arguing: "The commercial spirit . . . is wholly foreign to, and at variance with . . . the forgetfulness of self . . . which [is] the habitual frame of mind of the philosopher and man of science, and which [is] essentially necessary to the due discharge of the medical practitioner's duties towards the public."[125] In very similar terms, the author of *The English Gentleman* urged his readers to set aside "your feelings of personal interest." In considering any public matter, "you should strive to divest it of every thing that relates immediately to yourself, and take it solely upon the rule of right and justice. You may individually suffer loss in one way; but you will gain considerably in another;—in the consciousness that you have done right, and the assurance that you have contributed by it to the general good." This principle "will save you from many actions which are inconsistent with a Gentleman's mind."[126] This identification of profession with disinterestedness and public duty represents perhaps the most important of the values that were transferred from the eighteenth-century idea of the gentleman, via the mid-nineteenth-century term "philosophical," to become part of the professional ethics of the late-nineteenth/early-twentieth-century professional scientist.

"Open and Expand Your Views"

The importance of gentlemanliness and its attendant virtues for nineteenth-century men of science is well known, but the connection with the term "philosophical" has previously been overlooked; the two are linked by the conspicuous absence of discussions about professionalization during the period and by the reasons for that absence.[127] The Forbes/Watson dispute illustrates that ideas alone could not make a botanist philosophical, nor was good methodology sufficient: to be a philosophical botanist one needed to be a gentleman, to be civil and courteous with one's fellow botanists—to understand not just botanical Latin but the phrase *de mortuis nil nisi bonum.*

By the 1840s and 1850s, good manners were expected to reflect a good character, rather than being a veneer or the result of "mere mechanical practice." *The English Gentleman* repeatedly emphasizes that a superficial politeness is pleasant enough, but "to be lasting, it must be built upon Principle."[128] This shift from the earlier etiquette books may help explain the desire of the scientific parvenus, the naturalists, to ensure that their discipline's standing was not the result of merely aping the manners of mathematics or the physical sciences but was built on sound, philosophical principles. It is striking that Hooker repeatedly emphasized the importance not merely of broad species but of broad principles. His friend Charles Bunbury commented that Hooker was "not only one of the greatest botanists now living, but has a great variety of information and of pursuits; a generally well cultivated and remarkable active intellect."[129] In a similar vein, he and his contemporaries were praised for the breadth of their knowledge or for enlarging the public's understanding: the *Edinburgh* praised the "late lamented Professor Edward Forbes" for his views on distribution, which have "greatly enlarged our views of the subject."[130] Broad-mindedness was an important aspect of gentlemanliness. Gentlemen were urged to be open-minded and inquiring: "I would have you open and expand your views upon every subject which may be brought before you; and take it upon the broadest principles you are able."[131]

The English Gentleman explicitly linked the open-mindedness of the gentleman with that of the man of science:

> You will be open to *facts and evidence*, wherever you may meet them. It will be nothing to you, whether they are averse from the established ideas and current notions of men. Nothing, whether they attack prejudices the most deeply rooted. The idea that knowledge can be stationary is a perfect chimaera. The annals of a single year, in any branch of science whatever, prove the utter fallacy of such a notion; and it is therefore a great point to suffer your mind to be carried on with the progress of events, and to keep pace with the advance of real intelligence.

The author goes on to note: "Your acquaintance with history, even if slight, will enable you to recall very many examples of the evils of prejudice to the cause of sound Knowledge and Truth. The well-known tale of Galileo has, in its spirit, been repeated a thousand times in other affairs besides those of astronomy. There is scarcely a science which has not suffered by it."[132]

Given a good character—and one could only be "given" such by one's peers, after they had met and conversed with you—a proper, gentlemanly philosophical inquirer could engage with unacceptable ideas, such as polarity, without being condemned. For most naturalists of the period, transmutation would have been beyond the philosophic pale, but obviously not in Darwin's case nor in Hooker's. This is an important reminder not to view this period through post-*Origin* spectacles; as I will argue in the next chapter, recognizing this allows us to reinterpret some of the clashes between

Richard Owen and the members of the X Club, an informal dining club, that have traditionally been interpreted as being over Darwinism but that I will argue were mainly concerned with rivalry between institutions.

Hooker's reluctance to engage in public controversy—over science, religion, or politics—was also part of what made him philosophical. However, it was equally important that a man of science, like a gentleman, be motivated by a disinterested love of truth. As a result, conflict with one's colleagues might be unavoidable; as we have seen, one might even need to "spring a mine" under them from time to time, but such disagreements still had to handled appropriately. This was partly a matter of *where* one pursued a quarrel (more could be said in a privately printed essay than in the columns of a newspaper) but was much more a question of *how* one argued. Descending to "sneering contempt" or "dishonest attacks" was never acceptable.

Avoiding unnecessary conflict may have been one reason why Hooker avoided committing himself on the species question until the *Flora Tasmaniae* (I will discuss others below). Although Hooker was willing to discuss transmutation in private letters with Darwin, in print he commented that "nothing is easier than to explain away all obscure phenomena of dispersion by several speculations on the origin of species, so plausible that the superficial naturalist may accept any of them." However, testing such speculations required "a comprehensive knowledge of facts" and the botanist also needed to "know the value of the evidence they afford."[133] One had to be qualified to speculate and the qualifications included extensive fieldwork, possession of a large (and therefore metropolitan) herbarium, and being a gentleman. He argued that the "general acceptance which the doctrine of the mutability of species has met with amongst superficial naturalists" was a result of their having too narrow a conception of species.[134] In addition to its other uses, a broad species concept was a guarantee against species-mongers and their unlicensed speculation.

Hooker's wish was for botany to ascend from generalizations to laws and thus to ascend the status and funding hierarchies of the BAAS. In 1856, he urged his friend William Henry Harvey, who had just been nominated to the Linnean Society, to help him raise the quality of its *Transactions:* "what with popular lecturing & economic botany & horticulture, the absolute science of Botany is all but lost sight of, & except for some few of us who cultivate science for science's sake will join hands & support it, the prospects of Botany in this country are low indeed."[135] Despite the importance of botany's practical applications when it came to justifying government funding for Kew or Admiralty funding for his own publications,

Hooker was nonetheless concerned to be seen as one of those "who cultivate science for science's sake," just as Gunn, Archer, and Colenso were all at pains to stress their scientific vocation. Two months later, he wrote to Harvey again to contrast the efforts of the "Physical Science men" who work continuously to further their science, "whilst Botanists stand by & depreciate their own efforts and studies. I wish I could get you here for six weeks & join in a general effort to lift Botany up in the scale of appreciated sciences."[136] As we have seen, such an effort would entail lifting up its individual practitioners as well, by showing that naturalists could genuinely claim gentlemanly status. Given the pervasive uncertainty over the definition of "gentleman," there could be no absolute standard for judging; it could be defined only by a specific group, and one became a gentleman only when the group one wished to join accepted you as such.[137]

The first four decades of Joseph Hooker's career, from the mid-1830s until he succeeded his father in 1865, saw him travel from the Antarctic to the Himalaya and rise from being a lowly assistant surgeon to director of the Royal Botanic Gardens. One might therefore be tempted to assume that, once he held his father's post, his worries about his social and scientific standing would have been alleviated, but an analysis of Hooker's directorship of Kew suggests otherwise. Hooker remained deeply sensitive about his status and that of his science, concerns that were brought to a head in a series of public controversies over the nature and role of Kew itself.

CHAPTER TEN

Governing

I am no Market Gardener,
I, In an apron of violet blue.
I do not know any Botany,
Of Breeding I boast myself as free;
Yet I am the King of Kew.
—"Song of a Noble Savage," *Punch* (1872)

The widely read weekly newspaper the *Gardeners' Chronicle* offered its readership practical tips on gardening and landscape design, news of agricultural and horticultural advances, and a rich diet of advertisements for everything Victorian gardeners might need: greenhouses and the equipment to heat them, exotic plants to fill them, new varieties of fruit trees and the seeds of superior vegetables, as well as the latest interesting variants on that quintessential tool of suburban self-improvement, the lawnmower. Famous gardens were visited and described, which doubtless inspired its predominantly middle-class readers to further efforts (and further expenditure); in 1876 readers were offered a special supplement describing the Royal Botanic Gardens at Kew, complete with handsome woodcuts. The writer guided prospective visitors around the gardens, describing the sights and noting that, although it "was not originally intended as a pleasure-ground," Kew had "become one of the most popular of such resorts near London."[1] Thanks in part to the growing popularity of gardening and the efforts of the gardening press, the crowds at Kew had been growing steadily over the previous few years, from less that 500,000 in 1867 to almost 700,000 in 1875.[2] The Victorian middle classes' taste for high-minded, improving entertainment was evident in the *Chronicle*'s description of "the primary object of Kew" as "its educational purpose—education in its widest sense, in science, in art, in culture, in taste."[3] Education was a great Victorian passion: just a few years earlier, in 1870, Parliament had provided for compulsory elementary school education; new universities and colleges were being built; metropolitan museums and art galleries were finally admitting the public; and provincial museums were spreading across the country. Science, art, culture, and taste were to be the great civilizing influences which would ensure that the masses, newly enfranchised by the 1867 Reform Act, would behave as the responsible citizens they now were.[4]

In addition to education, the people were increasingly being prescribed gardens as a source of fresh air and healthy exercise, providing rational recreation and an escape from the smoke and dirt of the ever-expanding cities.[5] Kew offered all these too, but only after 1 p.m. in the afternoon—the mornings were reserved for those pursuing serious scientific work. The opening hours encapsulated the hybrid nature of the gardens: in the mornings they were a scientific research center, in the afternoon, a public park, but they were intended to educate and civilize all day long. This slightly awkward combination of purposes could not hope to please everyone, and in 1874 a meeting of the local Anglican church's members, the Richmond Select Vestry, petitioned the government to extend the garden's opening hours. Their motion argued: "The present hour was very inconvenient for the public generally. Large numbers of excursionists came from London to the gardens, and great was their disappointment at finding that the gardens were not open till the afternoon. Many of these excursionists were poor people, who only obtained a holiday once or twice a year."[6] These "poor people" had been granted their holidays by the 1871 Bank Holidays Act. And the term "excursionist" had been coined to describe those who took advantage of the new railway network to enjoy a day out. New holidays and spreading railways made more places accessible to "the public generally"; the London and South Western Railway reached Kew in 1869, forcing the garden's director to build a new gate to admit the increasing numbers of visitors, and in 1876 the Metropolitan and District Railways brought fresh crowds.[7]

Among those who were not entirely comfortable with the garden's hybrid character was its director, Joseph Hooker, who, despite (or perhaps because of) the increasing demand for access to Kew, stubbornly refused to extend the garden's opening hours, making himself increasingly unpopular in the process. When the issue was debated in Parliament, one member claimed that "the great obstacle to any concession in this matter arose, not from public, but from personal reasons," arguing that Sir Joseph Hooker (he had been knighted in 1877) "was determined that his neighbours should not get the concession which they wished."[8] The *Garden* magazine was even more outspoken: "Sir Joseph's private objections are intelligible. For years Kew Gardens have formed a snug little preserve—a sort of happy hunting ground for the scientifically inclined members of the Hooker family. It is, doubtless, a little hard that the privileges so long enjoyed by the latter should have to give way, as they must, to public rights and convenience."[9] This was another aspect of Kew's hybrid character—it was government owned and run, maintained by the public purse, but it was also private, having been built up by William Hooker, often using his own money, to be inherited by his son. In the early 1870s, the Hookers were routinely congratulated for their disinterested services to the nation; the *Globe* newspaper, echoing its rivals' sentiments (if exceeding them in hyperbole), described Joseph Hooker as a man "whose whole public and private life, and the life of his father before him, have been shining examples of all that is devoted, self-denying, chivalrous, and exquisitely pure and good."[10] Yet a few years later some were accusing Hooker and his family of selfishness—even hinting at corruption—and arguing that their "privileges" should give way to "public rights." To understand how Hooker had both earned and forfeited the public's love through his governance of Kew, we shall need to look in detail at some of the more controversial aspects of his directorship.

As director of Kew, Joseph Hooker governed a botanical domain that spanned the globe, from Sydney to Jamaica, from the Cape to Calcutta. This network had enriched the empire by transplanting important crops like rubber to colonies where they could be grown profitably, and a grateful nation had rewarded Hooker for his part in this work (fig. 10.1). It had been a long journey, as Hooker told the naturalist Henry Bates in 1870: "I was 16 years before I had an average income of £100 clear from my Science," since despite getting his first position on the *Erebus* in 1839, "it was not til 1855 that I was independent of my father!"[11] However, I want to argue in this final chapter that—because of Kew's hybrid character—Hooker's grasp on his empire was never entirely secure, his independence was less complete than it appeared, and as a result his authority was always potentially open to challenge.[12] Like most emperors, Hooker was generally uninterested in the public's rights or their convenience; his autocratic manner and sensitivity to the dignity of science were one cause of the controversies that beset him while governing Kew, but the issues involved also reveal how much Britain had changed in the first half of Victoria's reign. As Hooker told Bates, life for a man of science was easier than when he first "threw myself in Science in 1839 as an Asst Surgeon in the Navy," not least because "*positions* and means of scraping together a livelihood have

10.1 Sample of India rubber collected by J. D. Hooker. By kind permission of the Trustees of the Royal Botanic Gardens, Kew.

multiplied."[13] However, many of those new positions were now government funded, and, as one member of Parliament asked when Kew's opening hours were being debated, "the interests of the taxpayer ought to be considered in the matter. What was the money voted for?"[14] Having finally found the means to scrape together a livelihood, the men of science faced a new threat to their gentlemanly status—accountability.

The Directorship

By the time William Hooker died, in August 1865, Kew had grown from eleven acres to over three hundred under his stewardship. Ten years earlier, the government had finally accepted that the elder Hooker—then seventy years old—could not cope alone and appointed Joseph as assistant director. It might seem that Joseph's future was finally assured; he had secured government employment at Kew and his scientific reputation was rising steadily. In 1854, the Royal Society had given him their Royal Medal and their president, William Parsons, the third Earl of Rosse, used the occasion to note that Hooker had investigated critically "one of the most difficult questions of natural science . . . the question of the origin and distribution of species." Rosse singled out for particular praise "the cautious and philosophical manner in which the subject is treated." As a result of having tackled such a potentially explosive question with such admirable circumspection (in the *Flora Novae-Zelandiae*), combined with the "accuracy and completeness with which each subject is treated," Hooker had won himself "a high place as a philosophical botanist." [15]

Yet despite his apparently secure employment and high scientific standing, Hooker told Darwin in 1863 that "I regard succession to my fathers place with horror" despite the fact that no "better scientific place [i.e., position] exists in the world." He was contemplating taking up the Indian flora again, a project he had abandoned after the failure of the *Flora Indica* to get East India Company patronage. India had come under direct British government rule following the Indian mutiny, and the government's India office now seemed willing to subsidize the project. Hooker told Darwin, "Pay would tempt me, but only because it would hold out a prospect of early retirement from the struggle of scientific work for one's livelyhood." Clearly, he did not feel that struggle was over and relished the prospect of "shaking the dust off my feet at the Govt & Kew Gardens." Far from embracing the directorship as an opportunity for public service, Hooker wrote, "I am beginning too to hate the οιπολλοι [hoi polloi] of Science—Huxley, Lubbock & half a dozen others are enough for me, of the workers, outside my own immediate pale which includes only yourself Bentham Oliver & Thomson." [16] Hooker's almost aristocratic disdain for the growing numbers of scientific workers who filled the multiplying government-funded positions is a reminder of the comment he had made to his father twenty years earlier that it was only "*if* I cannot be a naturalist with a fortune" that he would "take honourable compensation for my trouble." [17]

The complex world of mid-Victorian science and Hooker's often-contradictory reaction to it were apparent in the same letter to Darwin when Hooker registered his disgust at what he called "Murchisonian science & all that sort of thing, like K.C B^{s}.," referring to the way the distinguished geologist Roderick Murchison had been so adept at using social influence to advance his career and for which he had just been made a Knight Commander of the British Empire (KCB). Yet, Murchison's conduct was typical of the careers that an earlier generation, men like Hooker's own father, had made for themselves by relying on aristocratic patronage for advancement. If one could not be a naturalist with a fortune, and despised traditional patronage, the only option was to take "honourable compensation" from the government, to join the hoi polloi of science, but it was not a prospect Hooker embraced eagerly.

Nevertheless, despite his reservations about the directorship, Hooker succeeded his father in 1865; within a year, he had received honorary degrees from both Oxford and Cambridge Universities, and in 1868, he presided over the BAAS annual meeting in Norwich.[18] And despite his apparent contempt for Murchisonian science, Hooker used his influential friends—particularly Darwin, "Huxley, Lubbock & half a dozen others"—to protect his position at Kew. And despite having sneered at Murchison's KCB, Hooker accepted the Companion of the Bath (CB) in 1869.[19]

Like many of his contemporaries, such as Richard Owen and Thomas Huxley, Hooker's career was built on supplementing or replacing traditional patronage with that of the government, relying on a salary rather than an independent fortune, but a salary that was understood to be *honorable* compensation, an apt reward for what was still understood to be disinterested service rather than paid employment.[20] In the process, the men of science became chimeras, their gentlemanly heads grafted onto the bodies of public servants, who trailed, in Hooker's words, "a lengthening chain of correspondence after them" as they struggled to keep up their official duties while participating in scientific society.[21]

Hooker's career as a scientific chimera found its echo in Kew's hybrid character. In 1872, he protested to the government that "Kew never was regarded as one of the Parks and never should be: its primary objects are scientific and utilitarian, not recreational."[22] Yet his 1868 Annual Report had referred to changes being made "for picturesque effect or for purposes of reference and instruction." Hooker's work addressed both scientific matters and the garden's appearance and accessibility, sometimes simultaneously; shortly after he became director, he had some of the trees in the arboretum felled to make new paths for visitors, taking the opportunity to separate genera and families from each other, thus creating a more systematic arrangement.[23]

Hooker worked to attract the public while keeping them out; to seek honors while spurning them; and to accept both a government salary and informal patronage while seeming hostile to both. His apparent hypocrisy was a product of the awkward hybrid natures of both Victorian scientific careers and botanic gardens.

The Ayrton Incident

When he became director of Kew, Hooker told Darwin that he was ready to make the gardens more attractive, but only if he could ensure that the "scientific character of the Establishment" would not "go down one iota." Just as in the earlier years of his career, Hooker was concerned both with botany's status (the "scientific character" of Kew) and his own standing; "my *elevation,*" he told Darwin, "brings me no increase of income but a higher scale of living; as I now feel it my duty to give up Examinerships &c that yielded upwards of £300." This was a slight exaggeration, since his salary had been increased from £400 per year to £800, and he had received a better house (no. 55 Kew Green); nevertheless, his sense of public duty—that he must now devote himself to Kew—ensured that his income increased only very slightly.[24]

The tensions that characterized Kew's purpose and Hooker's role in achieving it came to public notice in what has become known as the Ayrton incident (c. 1870–72). As Roy Macleod has demonstrated, the background to the clash between Hooker and his political master, First Commissioner of Works Acton Smee Ayrton, was the growth of an increasingly confident and vocal scientific lobby in the late 1860s.[25] In his presidential address to the BAAS meeting at Norwich (1868), Hooker had demonstrated the new confidence he and his fellow men of science felt, celebrating the increasingly widespread acceptance of Darwinism: "So far from Natural Selection being a thing of the past, it is an accepted doctrine with almost every philosophical naturalist." The Darwinian doctrine gave Hooker the confidence to argue that to "search out the whence and whither of his existence, is an unquenchable instinct of the human mind," one that fueled both science and religion, two routes to the truth that could "work in harmony," as long as their devotees chose not to "fence with that most dangerous of all two-edged weapons, Natural Theology; a science, falsely so called," which he dismissed as being "to the scientific man a delusion, and to the religious man, a snare, leading too often to disordered intellects and to atheism."[26] Such sentiments were in sharp contrast to the pious homilies that had characterized the first meeting Hooker had attended, thirty years earlier.

The year after the Norwich meeting, the scientific weekly newspaper *Nature* was founded and rapidly became the voice of this new generation of scientific men, who hoped for new careers as the principle of state support for science became more accepted. The year 1870 saw a Royal Commission on Scientific Instruction and the Advancement of Science set up, and

Parliament approved the building of the new Natural History Museum at South Kensington, which was to hold the natural history collections of the British Museum. As Nicolaas Rupke has shown, the decision was a triumph for Richard Owen, who had lobbied long and hard to induce government to rehouse the collections.[27] Soon afterward, in 1871, the government approved support for the *Challenger* expedition.[28]

However, the comfortable relations between William Gladstone's government and the scientific community were not to last. Ayrton, the Radical member of Parliament for Tower Hamlets, had supported Gladstone over the 1867 Reform Bill and was rewarded with a junior ministry within the Treasury before being moved to the Office of Works in 1869. Ayrton was well known for his attacks on government waste and excessive spending and he was no friend of science; he had regularly attacked the proposed expenditure for the new Natural History Museum.[29]

Ayrton and Hooker first clashed when Kew's director requested government funding to attend the International Congress on Botany in St. Petersburg in 1869.[30] Ayrton, then at the Treasury, refused and Hooker was affronted; he wrote to complain that the government saw fit to send a fully fledged royal commissioner to attend "a second rate and mere horticultural show for trade purposes in Hamburg," and reminded Ayrton that the Russians had sent official delegations to a British congress two years earlier. Clearly, his political superiors did not understand the importance of the Russian congress; their failure to send Hooker as an official representative "will be regarded as evidence of something more than mere indifference to the place which science holds in this country" or perhaps as "mere disregard of international courtesy in scientific matters."[31] These two issues—"the place which science holds" and "courtesy in scientific matters"—were, as we shall see, the key issues at stake in their later dispute.

One of Ayrton's first actions on taking office as commissioner of works was to send Hooker a reprimand over some detail in Kew's accounts, a reprimand that Ayrton soon realized was based on his own mistake. This was an inauspicious beginning, especially when contrasted with the informal, friendly relationships both Hookers had enjoyed with Ayrton's predecessors. Like his father, Joseph Hooker had been used to filling every position at Kew by judging the suitability and specialist expertise of the candidates for himself, whereas Ayrton insisted that competitive civil service examinations should be introduced, with the result that Hooker ended up with a clerk he thought entirely unsuitable for the position.[32] Matters worsened further in December 1870; Ayrton had approached John Smith, Kew's curator and Hooker's subordinate, to discuss the possibility of a position outside Kew, but did so without consulting Hooker and—Hooker would later claim—even asked Smith not to tell Hooker about their discussion.[33]

It was, however, Kew's heating apparatus that brought matters to boiling point. Ayrton could see no reason why a botanist, however distinguished, should be responsible for building works or for issuing government contracts for the boilers needed to keep Kew's greenhouses warm. In 1870, Ayrton had created a new post of director of public works to supervise all the building and engineering done by his department; an experienced engineer, Douglas Galton, was appointed to the post, which gave him oversight of the new plant houses being built at Kew.[34] Previously, Hooker had ordered his own contractors to build whatever he needed, but Galton insisted that future contracts be advertised and put out to tender. Hooker believed he alone had the expertise to commission the appropriate heating apparatus, since the welfare of Kew's incomparable plant collections was his responsibility.[35] What was even more serious in Hooker's view was that he had never been told of the new policy; the existing arrangements had simply been scrapped "without fault found, inquiry made, or even intimation given."[36] Ayrton's secretary responded by reiterating Ayrton's decision and telling Hooker curtly to "have the goodness to govern yourself accordingly."[37] An outraged Hooker replied that he could not "help recognising, in this arbitrary act, a further indication of that disregard of the Director's office, or want of confidence in himself, of which I have had such conspicuous proofs since, and only since, the accession of the present First Commissioner" and announced he was taking the matter straight to Gladstone.[38]

While he awaited Gladstone's response, Hooker told Sir Henry Barkly, then the governor of Cape Colony and a keen botanist, that "the whole affair has become Club gossip, & I have great difficulty in keeping it from being brought before the House, which would [just?] prevent Mr. G. from doing any thing at all." For the moment, he wanted "to keep the matter quiet." However, Hooker could not restrain himself from adding, "My official letters accuse Mr Ayrton of telling him [i.e., Gladstone] *falsehoods*."[39]

Hooker began recruiting various influential friends to his cause. In October 1871, he persuaded Darwin to make his grievances known to Sir Henry Holland, the queen's physician and president of the Royal Institution (who was related to Darwin via his father-in-law, Josiah Wedgwood), since "Sir H is said to be a very intimate private friend of Mr Gladstone." Hooker told Darwin, "I have no intention of resigning & if I am not much mistaken Mr Gladstone would rather see Mr Ayrton turned out than he would me—not from the merits of the case, because he thinks that I have friends who would be troublesome, & that Ayrton has none!" "I should lose caste altogether if I did not stand up & fight. I am putting all this in plain language to Mr Gladstone."[40] Darwin urged Hooker not to quit, telling him "no one could possibly fill your place so well." With the interests of science in mind, "I think your duty is to hold on, & bear, as far as a gentleman can, his rule," comparing Hooker's situation to that of a ship's officer "under an odious Captain."[41]

Hooker soon told Darwin he had "at last driven Mr Gladstone into a corner, & obliged him to take up my grievances." Gladstone's secretary, Algernon West, had shown Hooker Ayrton's explanation of the various matters that had been complained of. Hooker told Darwin that, in response, "I have unhesitatingly pronounced Mr Ayrtons's [*sic*] 'explanations' to be 'a tissue of evasions, misstatements & misrepresentations' & I further charge him with telling the Prime Minister a *direct falsehood.*"[42] Whether or not Hooker was justified in this accusation, it was the kind of language that gentlemen did not use of one another and was to create considerable trouble for Hooker.

Gladstone decided that the whole matter was an internal departmental one and declined to get involved other than by sending Hooker various ambiguous reassurances through his secretary. After another letter to Ayrton failed to get a response, Hooker wrote to West repeating the charge he had made in private, once again accusing Ayrton "of evasion, misrepresentation, and misstatements."[43]

In March 1872, Hooker formally asked the Cabinet to restore his previous powers and clarify the division of responsibilities between himself and the Office of Works. Gladstone tried to calm him with a private reassurance that he should "be treated as head of the local establishment at Kew," but Hooker would not relinquish the fight. He got his fellow X Club member the physicist John Tyndall to recruit Lord Derby to his cause. When Derby's support failed to get results, Tyndall convinced Hooker that he must take his case to Parliament, and with the help of other X Club members (Thomas Huxley, the mathematician and physicist William Spottiswoode, and the member of Parliament and banker John Lubbock) Tyndall drafted a memorial to Parliament. This was signed by eleven of the leading figures in British science and was published in *Nature* on 11 July.[44]

The letter in *Nature* gave full details of Ayrton's behavior and Hooker's various attempts to resolve the problem. Hooker was, not surprisingly, depicted by his friends as blameless, a dedicated man of science who had undergone considerable hardship—in the Antarctic and the Himalaya—in his attempts to serve the nation. They mentioned his remodeling of Kew, the colonial floras, the transplantation of crops, and the garden's massive popularity among the public. Having worked in such a disinterested manner for the public good, Hooker had—quite naturally in the memorialists' view—become "accustomed to respect," which Ayrton had denied him, treating him instead with "personal contumely," insolent and contemptuous treatment that dishonored and humiliated him. The letter writers stressed Hooker's self-restraint, quoting a letter he had written to Ayrton in which he made it clear that he did "not for a moment question the First Commissioner's power to exercise arbitrary authority over the Director of Kew," but that he had given Ayrton no cause to do so.[45] The clear implication was that, as a gentleman should, Hooker knew his place and had acted accordingly—it was Ayrton who had overstepped the bounds.

"The Ordinary Practice of Gentlemen"

The letter in *Nature* brought the Ayrton matter to the public's attention, and over the next few months numerous scientific societies from around the world passed resolutions in support of Hooker.[46] There was considerable press coverage of the affair, mostly sympathetic to Hooker.[47] When it printed the letter to Gladstone, *Nature* had added the editorial comment that "there can be but one opinion [regarding] the systematic discourtesy and want of consideration with which Dr. Hooker has been treated."[48] This accusation of discourtesy became the central issue when the matter was taken up by the press. When the Ayrton affair was finally discussed in Parliament, on 8 August 1872, the *Spectator* reported the debate and accused Ayrton of using "language a King would not use to a footman, or a Judge to a convict at his bar," which in their view proved he "is incapacitated, by arrogance rising to a moral offence, for the service of the Crown."[49] The *Saturday Review* commented that Hooker "had been treated with systematic and persistent disrespect," while Ayrton had violated official procedure and "the most elementary rules of intercourse between gentlemen."[50] And the *Daily News* observed that Hooker had "found himself a butt for contumely, insult, arbitrary insolence, and ignorance."[51] The *Globe*, ever eager to outdo its competitors' language, described Ayrton as "a man, whom the thick breath of a turbulent suburban democracy has blown for a moment into patronage and power," thus allowing him to acquire the ability to "supplant the authority, and destroy the labours of a public servant whom all nations envy us, and whose loss to his country and to the interests of universal science would be absolutely irreparable."[52]

Ayrton had established a reputation for rudeness well before his dispute with Hooker, and the comic weekly *Punch* seldom missed an opportunity

10.2 A *Punch* cartoon of the Gladstone cabinet in which Ayrton (third from left) is saying, "And mind you're polite and gentlemanly—d'you hear?" ("The Ministerial Odd Man," *Punch* 63 [17 August 1872]: 69). Reproduced by kind permission of the Syndics of Cambridge University Library.

to chide him for his ill-mannered behavior (fig. 10.2). At the close of the 1871 parliamentary session, the magazine published a poem, "Nice Little Holiday Tasks (for Little Ministers)," in which the Cabinet were treated as naughty schoolboys and set appropriate holiday homework. Ayrton was to be paraded as a warning: "That no man in office / May follow his leading, / And suppose he may doff his / Respect for good breeding."[53]

While the Kew gardens affair was developing, Ayrton also fell foul of the Society of Antiquaries, rebuffing their hope that he would continue his predecessor's policy and bring in a bill to protect ancient monuments. *Punch* rose to a sarcastic defense of Ayrton, "the amene, the urbane, the amiable, the genial," rechristening him "the Noble Savage" and suggesting that the Antiquaries should understand his "pardonable eagerness to avail himself of the opportunity of giving a slap in the face to those who come up to him with the unwarrantable expectation of a shake of the hand," to "say nothing of the natural pleasure of making oneself disagreeable, for which a Minister who has little enough amusement, Heaven knows, can hardly be expected to pass by so fair an occasion."[54]

Punch redoubled its attacks when Hooker and Ayrton clashed, publishing a song in which Ayrton boasted, "I do not know any Botany, Of Breeding I boast myself as free; Yet I am the King of Kew," and describing Hooker as a gardener who thinks himself "A scientific gentleman" and "expects to have his way," but *Punch* had Ayrton sing that "I am no respecter of gentlemen, / Nor of scientific swells."[55] The comic weekly never explained the issues at stake; for them the matter was simple: one combatant was a gentleman, the other was not; "it is customary, between gentlemen, for a wrong-doer to offer an apology, and we have as yet heard nothing of the kind from Mr. AYRTON. Shall *Punch* dictate the terms in which the Chief Commissioner should express his regret?"[56]

The *Gardeners' Chronicle,* now under the editorship of Maxwell Tylden Masters, proved itself to be Hooker's staunchest supporter throughout the Ayrton incident. Masters had succeeded Hooker as chairman of the Royal Horticultural Society's scientific committee and had worked in the Kew Herbarium.[57] He demonstrated his support for Hooker by publishing a satirical "allegory" in the *Gardeners' Chronicle,* in which a "Chef" complains that his steward, a "M. Vandal," is constantly interfering with the kitchen arrangements: "he would try to make intrigue with my friends. He takes my *garçons* and sends them away—he sends me others that I will not have" (references to the staffing issues at Kew). However, M. Vandal's most grievous offense was that he insisted the chef use "filthy coal instead of charcoal" and that "he arranges my fire."[58] For the *Gardeners' Chronicle,* the control of Kew's heating apparatus was the central issue. Mindful perhaps that many of its readers were more concerned with practical agriculture and horticulture than with issues of scientific independence, the newspaper returned regularly to "the point which we have endeavoured to keep prominently before our readers," that Hooker must have "full control over the heating apparatus." Who ran the boilers was "the keystone of the present controversy."[59] Its editorials referred to Ayrton's "unfortunate . . . reputation for rudeness and discourtesy" and his "insolence, discourtesy, and meanness."[60]

The horticultural world had, apparently, accepted the *Gardeners' Chronicle*'s account of the matter, while the nonspecialist press had "not unnaturally, contrasted the established reputation of the Director with the obvious ignorance and incompetence of the Minister." And the editor agreed with them that, even if Ayrton had, for example, been justified in his decision to redeploy Kew's curator, "the underhand shuffling way in which it was done" was not "in accordance with the ordinary practice of gentlemen."[61] The constant reiteration of this point, from the gardening press to the national dailies, is unmistakable evidence of how important the issue of gentility was during this period.

However, while the press were almost unanimous in both their respect for Hooker and their contempt for Ayrton, their attacks achieved nothing. The debate in Parliament on 8 August was precipitated rather unexpectedly by Henry Fawcett, the member for Brighton, very much against Lubbock's wishes, and it caught the pro-science lobby unprepared. An acrimonious debate took place and the *Gardeners' Chronicle* reprinted selections from the speeches in the House of Commons. Ayrton sarcastically referred to those who had petitioned Gladstone as "a number of philosophers or scientific gentlemen" who claimed "great weight and great authority." "No doubt," he observed in the same vein, "they were gentlemen who were eminent for their knowledge of organic and inorganic matter," but while he claimed no superiority for himself, he felt constrained to point out that his philosophical accusers were not in "possession of all the facts."

Ayrton added, "Dr. Hooker had performed important duties as a botanist—he distributed thousands of interesting plants to persons who made botany their study, but he had done so in the discharge of a public duty." In other words, he was a paid public servant and not entitled to any special treatment or consideration. The view (which had been widely expressed in the press) that Hooker was irreplaceable was mistaken in Ayrton's view; "it would be a grave reflection upon the botanists of this country if it were to go forth that there was only one person competent to undertake the office now held by Dr. Hooker."[62]

In preparation for the Commons debate, all the papers relating to the issue had been collected and officially published. They form a substantial volume (almost 180 pages), snappily entitled *Copies of Papers Relating to Changes Introduced into the Administration of the Office of Works Affecting the Direction and Management of the Gardens at Kew.* In among the voluminous correspondence about the unsatisfactory clerk's poor handwriting and numerous assessments of the efficiency of Kew's boilers, Ayrton's defense of his actions was included. He noted that "there does not appear to have been at any time any organised code of instructions for the management of Kew Gardens," decisions being largely made after informal chats between the Hookers, father and son, and Ayrton's predecessors; this was something Ayrton felt needed formalizing. He replied to some of the specific charges Hooker had made, noting, for example, that Galton's appointment and role had been public knowledge, and so he "did not think it necessary to make any special communication to Dr. Hooker." The supposedly clandestine approach to Smith was a mere misunderstanding. Ayrton had simply told him that, if he were to accept the offered appointment, he would be reporting to Ayrton; in which case, Ayrton would inform Hooker, and hence there was no need for Smith to do so. Since Smith declined the appointment, Ayrton had never bothered to tell Hooker about it.[63]

As Macleod argues, Ayrton managed to win considerable support in the House with the conclusion to his speech: "It would be a principle fatal to the administration of the public service if you were to allow it to be proclaimed that there is any one person who occupies such a position that he is entitled to dictate to his official superior, who is invested with the discharge of public duties, or to the Government, the course which they are to pursue."[64] This was a passage the *Gardeners' Chronicle* did not see fit to reprint, and one can, perhaps, see why. Hooker had acted as if Kew were his by right, a point his supporters tended to emphasize by constantly recounting the story of his—and his father's—selfless sacrifices. The letter to Gladstone published in *Nature* gave extensive details of William Hooker's career, not least the fact that his tenure as director was undertaken "at a sacrifice of more than half his income." The letter writers suggested: "The liberality of his father and his own self-denying life in the public service have, we think, been sufficiently illustrated."[65] This was evidence that the directorship of Kew was a gentlemanly position, whose holder was no mere hired hand, and it reinforced the sense that Hooker had inherited his position, together with the gardens and herbarium, from his father. Yet, as Ayrton stressed, the directorship was a government appointment, overseeing government property; he could see no reason why a botanist, however eminent, should appoint staff without reference to normal civil service procedures or commission engineering works without their being subject to public tender, nor why one of Ayrton's subordinates should expect special consideration for work he had done "in the discharge of a public duty."

Gladstone's speech in the House tried to smooth things over, assuring Hooker's supporters in the House that there "has never been on the part of the First Commissioner of Works the slightest intention to wound the feelings or disparage the character or position of Dr. Hooker." However, an amicable settlement to the dispute depended on Hooker apologizing to Ayrton for the accusation of "evasions and misrepresentations," which Gladstone saw as "a fact of the gravest character," especially as Hooker's words had now been published in *Nature*.

Gladstone disingenuously suggested that it was "totally impossible it could have been done by Dr. Hooker's agency or permission," but since

Ayrton had been publicly charged with lying, "the whole House will see that it is absolutely necessary, if it cannot be sustained, that it should be distinctly and unconditionally withdrawn, and that regret should be expressed for its ever having been made." Gladstone concluded by observing that "scientific men, as they are called by the exclusive appropriation of a title which I must protest against, have a great susceptibility." And as they were "not accustomed to enter into our sturdy conflicts," they might be inclined to "take reproof in a much more serious manner than we who are hardened by long use are accustomed to do."[66] Those who could not take the heat of parliamentary debate should stay in their greenhouses.

The press reacted to the news with astonishment, the *Saturday Review* commenting on the "startling transformation scene" in which Ayrton "presents himself to our astonished gaze as the weak and helpless victim of a scientific tyrant," since this "most meek and pacific of men" has supposedly "for several years been subjected to cruel persecution by the director of Kew." The Commons, the *Review* opined, was probably too flabbergasted by this "monstrous perversion of facts and insolent justification of official misconduct" to express an opinion, allowing Gladstone to close the debate without a vote, but it was unfortunate that Gladstone had thus countenanced Ayrton's view "that good manners are not only superfluous, but out of place in the public service."[67]

Ayrton told the Commons that the letter to *Nature* was "a scurrilous and calumnious libel upon him" and that "the charge made against him by Dr. Hooker was that he had been guilty of evasion and misrepresentation, and of all those errors which were used by a slave to escape from the anger of his master."[68] (This was the language which the *Spectator* had condemned as that which "a King would not use to a footman.") However, Hooker's accusations, which he had so unwisely allowed to be made public, were—as the *Spectator* realized—to be his undoing. Gladstone's demand that Hooker apologize ensured that "the general question is to be forgotten in a side detail" since Hooker was to be "smashed for a hot-tempered indiscretion," while Ayrton would get away with "a cold, deliberate out-pouring of vitriolic contempt." Yet it was Hooker, "an excitable savan," who "must retract." There was, the editorial concluded, "no justice in such a decision, and little regard for the welfare of a service into which, under such treatment, gentlemen will not enter."[69] Once again, the key issue was gentility; Hooker in his "excitement" (even Darwin described him, privately, as "impulsive and somewhat peppery in temper") had overstepped the bounds of politeness, to the extent that even the boorish Ayrton could justifiably claim an apology.[70]

The *Spectator* attempted to defend Hooker's conduct by arguing that his was the "much smaller insult" since "he evidently intended to question the First Commissioner's method of argument, not his personal truthfulness."[71] The letter writers in *Nature* had also tried to soften Hooker's statement, claiming he had been "driven to it by the necessities of the case."[72] However, as we have seen from Hooker's letters to Darwin and others, he had every intention of questioning Ayrton's "personal truthfulness." It was, perhaps, just as well for Hooker that the following comment, which he had made in a private letter to his friend James Hector in New Zealand, never appeared in print: "Ayrton was a Bombay attorney, & his mother a native woman which sufficiently accounts for his vulgarity & intrigue."[73]

As we have seen in earlier chapters, as the conception of a gentleman shifted, good manners became an increasingly vital test of who was entitled to the label; it was a test Hooker almost failed. To call someone a liar without very firm evidence was a grave breach of decorum and so, largely as a result of Hooker's discourtesy, Ayrton escaped without even a reprimand. A few weeks after the parliamentary debate, *Nature* told its readers that "we may hope that the public have now heard the last of the unfortunate Hooker and Ayrton dispute. We learn that Mr. Ayrton has expressed himself satisfied with Dr. Hooker's explanation of the 'offensive' matter in his letter to Mr. West, and here the matter will probably rest."[74] However, Hooker told Barkly that "I had no hesitation however in withdrawing the personal implications conveyed by the charges but the charges themselves I could not withdraw; after which, after an explosion of impotent rage on Mr Ayrton's part, he has ceased his demands."[75] And there the matter was allowed to drop. Gladstone had stuck by his minister, despite Ayrton's unpopularity, but—perhaps to avoid any repetition of the incident—Ayrton was moved to the position of judge advocate general in August 1873.[76] Hooker commented waspishly to Darwin that "rehabilitating Ayrton with the Judge Advocacy, a sinecure suppressed by this Ministry, is the jolliest job I have heard of for many a long day."[77] However, Ayrton's job was not to last: Gladstone's government struggled on until the general election of January 1874, when Disraeli's conservatives took power; Ayrton lost his seat and never gained a new one.[78] *Nature* commented that the most important issue raised by the clash with Ayrton was whether "the Government recognises the principle that a servant selected to control a great scientific establishment must necessarily be entrusted with all the details of its management," but this was a question that had carefully been left unanswered.[79]

There were several moments during the Ayrton conflict when it could have been amicably resolved, especially as it became increasingly clear that Ayrton was perceived as a liability within a government that had little chance of surviving the next election, yet Hooker refused all compromise. Moreover, his intemperate language cost him a clear victory, even though much of the press awarded him the fight on points. There was clearly more at stake in the struggle than official discourtesy or greenhouse boilers; according to the press, behind Ayrton's rudeness there was a plot afoot to destroy Kew by giving its scientific collections to the new Natural History Museum at Kensington, or—according to whose paranoia one believed—there was a plot to further enlarge Kew's collections by stripping the British Museum of its botanical treasures before they could be moved to the new museum. Behind both conspiracy theories lay the unresolved question of what kind of control the government should expect to exercise over the science it funded, which resulted in an intense struggle over the nation's scientific collections.

Kew Is My Washpot

In its report on the parliamentary debates, the *Saturday Review* suggested that the only useful thing to come out of the Ayrton controversy was "that it has brought to light the project which is hatching for the flinging of the South Kensington shoe over Kew Gardens, and making the herbarium a washpot for COLE."[80] This reference to the triumphal verse of the sixtieth psalm ("Moab is my washpot; over Edom will I cast out my shoe") and to Sir Henry Cole (the civil servant responsible for the South Kensington site where the new Natural History Museum was to be built) implied that Kew's herbarium would be moved to the new institution. *Punch* made the same inference when it wondered whether, despite Ayrton's faults, "is there not a job somewhere? Kensington, eh? Is Kew to be made a Cole-cellar?"[81] Hooker shared these fears, telling Hector that "the Brute Ayrton has set his heart on disestablishing Kew as a Scientific Institution."[82]

The cause of these speculations was a question Ayrton had posed in his memorandum to Parliament: "Whether it is desirable on the grounds of science, public utility, efficiency or economy, that two museums [i.e., the herbariums of the British Museum and Kew] should be kept up, with their libraries, and staff of public servants to prosecute the science of botany." Might it not be better to appoint a single director for the two herbaria, especially as the "Kensington Museum will be close to one station, and Kew Gardens close to another, on a short line of railway, with telegraphic communication between one institution and the other." The new chief botanist could then shuttle between the two, and so "the sum now spent on the collections, library, and establishment for botany at Kew, might be expended in completing and improving the establishment at Kensington, or be saved." Ayrton felt he was not in a position to resolve these questions until "the circumstances which may exist at the time of the completion of the new museum are fully considered"; nevertheless, he felt it prudent to ensure that in the interim "no new expense is incurred at Kew."[83]

To assist him in reaching a decision, Ayrton had consulted Richard Owen, director of the zoological collections at the British Museum (and later director of the new Natural History Museum), on the matter. Owen's views appeared as an appendix to the parliamentary papers, in which he compared Kew to "the menagerie at Regent's Park" (the Zoological Gardens), whose "chief application is the instructive pleasure of the public." The zoo was not supported by the public purse and kept no stuffed or bottled collections of specimens, preferring to consult the British Museum as necessary. The same should be done by Kew, which had no need for its own herbarium, especially as those vital Victorian technologies, the railway and telegraph, would make the new Natural History Museum so accessible.[84]

Owen pointed out that some of the country's most important botanical collections, notably those that had belonged to Sir Joseph Banks, were housed at the British Museum, yet botanists who wished to compare them with more recent collections were "compelled to go from the Botanical Department at the British Museum to the Competing Department developed by Dr. Hooker at Kew." This necessity "has been created by the will and in my view, the misapplication of opportunities and influence of the present Director of the Royal Gardens at Kew." As a result, instead of "amity and co-operation to a common end of public utility," the two institutions "have been dragged into competition." Hooker's empire building had resulted in a needless duplication of effort, while "the British Museum has had no part in bringing about this unwise, unthrifty, and uncalled-for condition."[85]

Not content with attacking Hooker's motives, Owen attacked his science: "The scientific work of which a herbarium is the instrument has been defined by a great wit and original thinker as the 'attaching of barbarous binomials to dried foreign weeds.' This roughly expresses the net result of the application of a museum of dried plants." However, any "botanist who is desirous of wiping off this aspersion, should be by no means content with a list of names, he should study plants philosophically," by investigating their distribution and by bringing the study of plant physiology to bear on their medical properties and cultivation. "To raise the 'weed' to the condition of a plant, useful to man's estate," Owen concluded, "is the work of a Director of a national collection of living plants."[86]

One might think that was exactly what was going on at Kew (apart from physiological study, for which there were no facilities), but Owen proceeded to attack the garden's record on precisely these points, arguing that they had given inadequate support to the introduction of important potential crops. Kew's failures, Owen implied, resulted from the pointless competition with the British Museum; since the new museum would have the space to house both herbaria, it was surely time to combine them, leaving Kew free to entertain the public and do horticultural experiments, such as "the application of manures, [or] demonstrations of the fittest species of grasses for particular soil."[87]

To some who read it, Owen's statement appeared to have been hijacked by Ayrton as part of his assault on Kew's independence and scientific standing. *Nature* expressed regret at seeing Owen's name "dragged into the discussion on a point which has nothing whatever to do with the question . . . which is simply Mr. Ayrton's monstrous behaviour to a man of science."[88] In a similar vein, the *Gardeners' Chronicle* expressed their sympathy at that way Owen's "honoured name should have been introduced into the controversy at such a time and in such a manner," describing the publication of his statement as "the most characteristic bit of 'Ayrtonism' in the whole affair."[89] However, attentive readers of Ayrton's evidence might perhaps have noticed that his question about the costs and benefits of maintaining two herbaria so close to each other had appeared in almost exactly the same form in *Nature* in June 1870, but as part of an argument for doing precisely the opposite of what Ayrton and Owen were now suggesting; *Nature* appeared to propose giving the British Museum's herbarium to Kew.

Nature's editors had "been favoured with a copy of a memorial, drawn up as long ago as 1858," which had been submitted to the government of that day and concerned the reorganization of the British Museum's natural history collections. Since in 1870 the government had just approved the building of a new Natural History Museum, the editors felt it was appropriate to revisit the issue.[90]

Given *Nature*'s friendly relationship with Huxley, Hooker, and their fellow members of the X Club, Owen might reasonably have interpreted the journal's decision to reopen the matter of the British Museum's botanical collections as evidence that a pro-Kew conspiracy was being hatched in mid-1870. Similarly, his intervention in the midst of the Ayrton controversy would have struck the X Club as sinister, since Owen—unlike most of them—was on friendly terms with Gladstone.[91]

Whatever *Nature*'s editors' intentions in reprinting the memorial, Owen would undoubtedly have been alarmed by its resurrection, especially by the question it raised of whether there was "an adequate advantage in maintaining, at the cost of the State, two Herbaria or Scientific Botanic Museums so near together as those of London and Kew" (virtually the same question Ayrton was to ask in Parliament in 1872). The memorialists felt not and recommended that both herbarium and fossil plants be transferred to Kew "and that a permanent building be provided for the accommodation at Kew of the Scientific Museum of Botany so formed," while the British Museum would retain a small "Typical or Popular Museum of Botany."[92] Ironically, when this plan had first been mooted in 1858, Owen—who had then been on friendly terms with both William and Joseph Hooker and had no interest in botany—had supported it.[93] However, by the time the proposal resurfaced twelve years later, relations between Owen and Joseph Hooker had soured, partly because of Owen's opposition to Darwinism but also because the X Club were regularly involved in attempts to block Owen from gaining more institutional power.[94]

The recent reopening of this long-running dispute about who should control the national scientific collections was the background to Owen's statement to Ayrton, but its immediate cause was Hooker's evidence to the Royal Commission on Scientific Instruction and the Advancement of Science (the Devonshire Commission) in March 1871. Owen's statement quoted—somewhat selectively—from Hooker's evidence to support his claim that Hooker had called for "the abolition of the Botanical Department in the British Museum" and "that the Botanical Department to be transferred from London to Kew should be under one Head, that is to say, the Director of [Kew]."[95]

Hooker's response to Owen also appeared in the parliamentary papers but received a wider readership in *Nature* on 24 October 1872. He politely corrected Owen on several factual matters and noted that matters such as the proposed manuring experiments "would involve an enormous expenditure, and occupy many acres of ground now devoted to the legitimate purposes of a botanic garden." He added that "Prof. Owen is in error in asserting that the main end or drift 'of Dr. Hooker's evidence before the Scientific Commissioners is to impress upon them the necessity of the transfer of the collection of dead plants' from the British Museum to Kew. My evidence," Hooker asserted, "is unequivocally opposed to such a transfer."[96]

A close reading of Hooker's evidence and Owen's interpretation of it suggests that both men were partly correct. Hooker had argued that the new museum and Kew should keep their herbaria, each of which would remain under the control of the heads of the respective institutions. When one of the commissioners, the mathematician Henry John Stephen Smith, asked Hooker directly whether he proposed "to have an *imperium in imperio* [a sovereignty within a sovereignty] put upon the Superintendent of the Natural History collections," Hooker replied, "Distinctly not."[97] However, Hooker explained that he saw it as "part of the duty of the first herbarium in the country," namely Kew, "to supply the British Museum with as complete and well named a set of herbarium specimens from the several geographical areas as possible."[98] As a result, the independence of the British Museum's herbarium would be illusory, since the vital work of naming would take place at Kew. As Hooker explained, once the British Museum's collections had been rearranged according to Kew's system, adding new specimens from Kew would be "as simple as putting books on a shelf," adding "a subordinate could intercalate the additions."[99] So, while Hooker was strictly correct in telling *Nature*'s readers that his evidence had been "unequivocally opposed" to removing the British Museum's herbarium, it was clear that the new museum's herbarium would be entirely subordinate to Kew.

While the public debate went on, Darwin privately congratulated Hooker on his letter to *Nature,* calling Owen an "utter blackguard" and wondering "how you could have written so quietly."[100] Hooker was "greatly gratified by your hatred of Owen *for me*" but claimed, "I so despise him, that I feel I could afford to converse with him across a neighbour's table tomorrow—& yet I should be confoundedly angry if any friend of mine did so at the same table!" Interestingly, Hooker endorsed comments in the press that implied Ayrton had taken advantage of Owen, but Hooker believed Owen had been hoist by his own petard. He told Darwin that "the fun of the thing is, that Owen *never intended his letter to be made use of.*" It had apparently been "intended to be used by Ayrton, *when the* question of removal of the Brit. Mus: *contents* to S. Kensington was to be discussed, that is some years hence." Hooker believed (on what evidence, he does not say) that Owen had intended his statement should do "its work, of poisoning Ayrton's mind," in private, convincing him of the need to transfer Kew's herbarium in due course, and "It's [*sic*] premature appearance must have horrified Owen."[101]

Whatever the precise chain of events, Owen was clearly furious by the time he responded to Hooker's critique of the original statement. Owen wrote a rather intemperate letter to the editor of *Nature* (7 November 1872) in which he claimed that "Dr. Hooker's attack" had created "the emergency threatening a Department of Natural History in the British Museum," which he referred to in his letter as "the National Herbarium in London" or "the Metropolitan Herbarium"—an interesting rhetorical move that implicitly transformed Hooker's "first herbarium in the country" into a provincial collection. Owen again accused Kew of having "intercepted" collections of new plants and gave further details of Kew's failure to help introduce ipecacuanha (a small South American shrub that possesses emetic and purgative properties) into India, selectively quoting the official report of the Calcutta Botanic Garden's superintendent, George King, to bolster his claim. Owen concluded by saying that he looked forward to the day when the country would have "a botanic garden whose officers are not diverted from experimental work, not trammelled and obstructed by that wasteful weed—an overgrown herbarium."[102]

The columns of *Nature* were briefly afire with what Owen had christened "the Kew Question," with Hooker, Bentham, and William Carruthers (curator of the British Museum's herbarium) all writing in to dispute various points or correct one another. George King wrote to correct Owen's statements about ipecacuanha cultivation, asserting that Hooker had sent the first plant to Calcutta in 1866, "long before Government had begun to show any official interest in the matter," and it had been propagated successfully; hence "to Dr. Hooker is India indebted for the first beginning of this valuable cultivation."[103] It gradually became clear that Owen's statements contained numerous errors of fact, not surprisingly since at the time he was keeper of the zoological collections and—as Carruthers felt obliged to point out—"has no official relation to the Botanical Department."[104]

The dispute fizzled out early in 1873, after Gladstone received a petition concerning "the National Herbaria" signed by several eminent naturalists—most of whom were friends and allies of Hooker's.[105] In response to Ayrton's having raised the possibility of amalgamating Kew's herbarium with the British Museum's, a move the petitioners felt would be "highly detrimental to the progress of science," they urged the government to commit itself to maintaining the two separate herbaria, especially as herbaria were cheap to maintain.[106] If there had been any plan among Hooker's allies to annex the British Museum's herbarium, it had clearly been dropped, not least for fear that the new Natural History Museum should prove the winner. Two weeks later, a brief letter on Gladstone's behalf assured the writers that "Her Majesty's Government have not formed the intention of removing the collection to South Kensington," and no such move would be contemplated without consultation.[107]

However, the ending of the dispute was more than just a political compromise. The issue of *Nature* that contained the official announcement that no merger was imminent also carried a letter from William Thiselton-Dyer, then professor of botany at the Royal Horticultural Society in London, in which he explained exactly what a herbarium was, with the aim of helping "those who are not botanists to form an opinion upon the matter."[108] Hooker's patronage had already helped Thiselton-Dyer obtain the position at the society, and he was to be appointed assistant director of Kew in 1875, so he was hardly an unbiased witness in the debate.[109] However, his rationale for maintaining two herbaria is interesting. He emphasized that "it is absolutely necessary in a public herbarium that the specimens should be securely stuck down upon sheets of paper" to withstand "frequent handling," but many of those in the British Museum were "unfastened on paper" and thus could not be used in the way Kew's were. Moreover, "the British Museum specimens are mounted on paper of a very different size" to those at Kew; this, as I noted earlier (chapter 5), was a long-standing difference between Kew's practice and the museum's, as were their different approaches to preserving specimens from insect attack (using either poison or mothballs). Whatever the original reasons for these divergences in practice, they had the effect of making it extremely difficult to merge the two herbaria. Thiselton-Dyer therefore concluded that it would be best to allow the Kew Herbarium to remain, "as at present," as a resource for those who were "engaged in important works, as original memoirs and colonial or forest floras executed for the Government." Such people would be willing to travel to Kew, where they would enjoy its "tranquillity from the incursion of visitors less permanently occupied with botanical pursuits." Meanwhile, the British Museum collections would stay in London, to benefit those "who would use them rather for reference than for continuous study."[110]

Strikingly similar arguments had been put forward by William Carruthers of the British Museum; moving the Museum's herbarium to Kew would inconvenience "the active professional man, and the man of business, who devote their spare hours to botany," since only botany's full-time devotees could spare the time to visit Kew. However, Carruthers offered another practical reason why the herbaria could not be merged. He observed that "the most varied views are entertained by botanists as to the limits of a species, and consequently as to what constitutes a duplicate." His carefully chosen examples were Bentham, who recorded the British flora as 1,274 species, Hooker, who listed 1,473 in his *Student's Flora*, and Babington, who had 1,648. By contrast, the French naturalist Alexis Jordan would no doubt make three or four times as many species as even

Babington.[111] Carruthers commented, not unreasonably, that "it is quite obvious that these different botanists have each very different notions as to 'duplicates,' and that a distribution undertaken by Mr. Bentham would certainly result in the loss to the herbarium of plants which Dr. Hooker would consider good species," while if either Bentham or Hooker did the job, thousands of Jordan's species would go. The continuing disagreements about the limits of species and the instability of classificatory practice were cited by Carruthers as an excellent reason for maintaining the two separate herbaria, since their differing classificatory practices would serve to counteract each other and preserve specimens of potentially valid species from destruction.[112]

"My Future Estates"

Hooker's conflict with Ayrton was clearly deeply entangled with the dispute over the control of the national herbaria. There were, as several historians have noted, issues of institutional power and rival patronage networks in play, as well as the ideological clash over Darwinism, which often became a weapon in the institutional maneuverings. In analyzing these conflicts, it has sometimes been argued that Hooker, Huxley, and their allies represent a professionalizing or modernizing force, determined to free British science from the control of the Anglican establishment and the clients they patronized, such as Owen, replacing jobbery with meritocracy. However, as I have shown, this interpretation can be challenged on various grounds, and the dispute over Kew's herbarium gives us further cause to rethink the standard story of when and how the men of science became such an important part of Victorian society.

As we have seen, Joseph Hooker thought of his father's herbarium "as much my future estates to be cared for by me, as if they were landed property."[113] As Richard Drayton has shown, William Hooker had long campaigned to ensure that his son's inheritance would include Kew gardens. As early as 1846, he had written to the commissioners of woods and forests (who were responsible for Kew at the time), offering to leave his herbarium to the nation if his son were to succeed him; Sir Joseph Banks had made just such an arrangement when he had left his collections to the British Museum on condition that his employee, Robert Brown, be their superintendent. However, Hooker received no commitment from the government in the 1840s and concentrated on building up Kew's herbarium, by encouraging such people as Dawson Turner and George Bentham to leave their collections to Kew, in order to create a government-owned herbarium that would eventually require housing and someone—like Joseph Hooker—to look after it.[114] So, shortly before his death, William Hooker wrote a memorandum explaining that while he planned to leave his collections to his son, he realized that Joseph could not afford to maintain them. William therefore urged the government to consider "the propriety of purchasing the herbarium at a fair valuation and depositing it at Kew, as part of the Crown property attached to the Royal Gardens."[115] William Hooker got his wish; Kew got the herbarium, which the government purchased, and Joseph got the directorship; he received the letter offering him the appointment on 13 September 1865, exactly one month after his father's death.[116] By the 1860s the government could not have openly contemplated the kind of deal that Banks had been able to make forty years earlier; but nor could the state have acquired the largest privately owned herbarium in Europe if they had not offered Joseph Hooker the directorship. No doubt his publications and scientific reputation alone would have won him the position, had it been advertised, but it was not.[117]

During the Ayrton dispute, William Hooker's noble efforts to build up the national herbarium had been repeatedly stressed; the *Gardeners' Chronicle,* for example, described how he "came to Kew at a clerk's salary" and then "expended largely from his private fortune in the establishment and maintenance of the museum and herbaria." Hooker undertook all the expense and effort required until, "after some years the herbarium, which was always open for public use, became too costly and unwieldy for any private purse to maintain in efficiency, and it was sold to the government for a sum probably not equal to a sixth of what it had cost."[118] In similar vein, Hooker's friends had stressed the financial burden on William Hooker that herbarium building had imposed in their 1872 letter to Gladstone. William Hooker, "knowing that his son could not afford to be as regardless of pecuniary considerations as he had been," had nevertheless ensured that the collection would be "offered to the Government at the lowest valuation."[119] That deliberate undervaluation, putting the needs of science before profit, was typical of Hooker's disinterested behavior.

And so, instead of inheriting the collections, Joseph Hooker inherited the directorship and—by extension—Kew gardens itself, which thereby came to be regarded almost as the Hooker family's personal property. This point was not lost on the press. When the Ayrton affair was being debated, the *Pall Mall Gazette* suggested that the British Museum's desire to acquire Kew's herbarium lay behind the controversy, but "seeing that the botanical collections at Kew are wholly the result of the industry, and in great part of the munificence, of Dr. Hooker and his father, Sir William Hooker, it would have been difficult to have effected the reform during the present Director's tenure of office." Clearly, Joseph Hooker could not have been asked to simply retire, not least because his pension would have absorbed all the savings the move was supposedly intended to achieve, "but there was an alternative,—viz. to worry him into resignation." The writer was careful not to claim that "such a course of action was deliberately adopted by Mr. Ayrton" but observed that "it is not easy to find another explanation of what has come to be known as the case of Kew Gardens." [120]

Regardless of whether or not Ayrton and Owen were really trying to get rid of Hooker (he certainly believed they were), there is something frankly bizarre about the claim that the herbaria could not have been amalgamated while Hooker remained in office. As his friends had written to Gladstone, the result of William Hooker's action was that "collections which had previously been devoted to the nation's use became the property of the nation itself." [121] The fact that Sir Joseph Banks's "munificence" had originally brought the British Museum's herbarium into existence did not deter Hooker and others from suggesting it might be moved to Kew, despite having been left explicitly to the British Museum. Both herbaria were, after all, government property; it was surely a matter for an elected government to decide where they were to be housed and how they should be managed. Similarly, while Joseph Hooker's energetic and often dangerous travels had undoubtedly enriched his father's collections, and his publications had enriched botany, both had been done—as Ayrton had told the House of Commons—"in the discharge of a public duty"; Hooker had been paid by the Admiralty both during his travels and while writing up their results (albeit with some financial help from his father).

Despite the fact that the nation clearly did not owe Joseph Hooker a living, a herbarium, or a botanic garden, the press generally reported the Ayrton controversy as if it did; the endless attacks on Ayrton's conduct and manners reinforced the point that he was not a gentleman, but Hooker was; as the *Spectator* put it, if Ayrton were simply allowed to insult Hooker, the scientific service of the country would become one "into which . . . gentlemen will not enter." [122] Despite the fact that Hooker had worked for over forty years to get a government salary and a government house, he was still portrayed by the press as someone who, like Banks, served the nation altruistically. This vividly demonstrates the abiding importance of gentility as the defining characteristic of the man of science; in the eyes of his contemporaries at least, Hooker—far from being a modernizing professionalizer—was one of the last great gentlemanly naturalists.[123] However, in sharp contrast to someone like Banks, Hooker had no vast private fortune to support his enthusiasm and had to rely on the government to support his work. The clashes with Ayrton and Owen illustrate how fragile that support could be; Hooker could have been stripped of his empire at any time, which is why he was so ferocious in defending it, whether from economizing government ministers, from rival institutions, or from "mere pleasure or recreations seekers."

"Rude Romping and Games"

The growth of railways was, as I mentioned previously, one factor behind the increasing numbers of visitors to the gardens, while another was the introduction of bank holidays, introduced in 1871 (rather ironically, the member of Parliament who drafted the Bank Holidays Bill was none other than John Lubbock, one of Hooker's staunchest supporters during the friction with Ayrton; the first purely secular holidays in British history were popularly referred to as "St. Lubbock's days" in his honor).[124] In addition to the impacts of the railways and "St. Lubbock," the crowds were growing because gardening was becoming more popular—and because both Joseph and William Hooker had worked hard to make Kew gardens attractive. Their success had gradually become a mixed blessing, as Tyndall had noted during the Ayrton dispute when he had written to Lord Derby to say that it was precisely because Kew "has been made so beautiful and so attractive to the public" that there was a danger that "its immense scientific importance is likely to be overlooked."[125] This was the basis of Owen's scathing comparison between Kew and the zoo; entertaining the public was all very well, but it was not to be confused with genuine scientific research.

Ever the taxonomist, Hooker used his official report on the state of the gardens for 1871 to classify the visitors. Like the BAAS sections, visitors to Kew were divided into a hierarchy defined by their social class but also by their studies—the uses to which they wished to put Kew. Visitors were, in Hooker's view, dominated by "persons who thoroughly enjoy the garden as an instructive horticultural establishment," often "of the professional and upper classes" who tended to visit on Saturdays, when "the approaches are crowded with carriages." By contrast, on Sundays, "the trading class is the most numerous, and private carriages are, to a great extent, replaced by omnibuses," since the upper classes gave way to "persons of the middle and lower grades of life especially, who throng the plant houses and museums in search of general or special information." These in turn disappeared on Mondays, when "the artisans come in great numbers" and "the omnibus on this day gives place to the van."[126]

The conduct of these diverse visitors was "uniformly good" in Hooker's view, and "there is annually more consideration and attention paid to the rules." Nevertheless, there were a few improprieties, including "indecorous conduct"; he noted that "offences of this last class, from indiscretions downwards, are of all the most difficult to deal with." Such behavior was most common among those whom Hooker rather contemptuously described as "mere pleasure or recreation seekers . . . whose motives are rude romping and games," but fortunately "this section, with that of their invariable followers, the 'roughs,' becomes smaller each year."[127]

Hooker argued that the gradual improvement in the visitors' behavior was "in no respect attributable to the means of repression that are at hand" but "to the natural beauties of the place, and the evidences of system and order with which visitors are confronted on entering the gates, and the proofs these afford of the solicitude of Parliament and the public that everything should be well-ordered and attractive, as well as instructional and scientific."[128] The suggestion that Kew's carefully contrived "natural beauties" were a key part of its civilizing mission was part of its claim to public support.[129] Like his father before him, Joseph Hooker knew Kew's popularity was vital to its future. Public recreation and edification helped to justify the use of public money to fund Kew's other functions, from supplying plants to distant colonies to attaching "barbaric binomials to dried foreign weeds," activities whose benefits to Britain's taxpayers were not always obvious. The *Gardeners' Chronicle* commented in 1875 that, while botanists and horticulturalists "might feel disposed to grudge the time, labour, and money expended on . . . fashionable garden decoration," the editors had to acknowledge "the wisdom of maintaining and increasing the popularity of Kew. Kew is maintained by the public purse, and those who help to fill the public purse not unnaturally like to see something for their money." Though most visitors were "incapable of appreciating the higher branches of horticulture," much less of "botanical science," and could not be expected to take more than "a languid and vague interest in the supply of plants to the colonies," there was no doubt that, as long as they "see something they can understand and enjoy," they would "willingly contribute their quota to the carrying out of other and more important matters" (fig. 10.3).[130] The supposed moral improvement that flowed from the contemplation of natural beauties helped bridge the divide between recreation and science by turning landscape gardening into a moral education; as early as 1849, a contributor to the *Florist* had proposed to tempt working people "away from the gin-palace, the public house, and the beer-shop" by having "these delightful Gardens opened at nine o'clock; ay, and we would have in

10.3 Beautiful exotics were also vital products of empire. Illustration of *Hodgsonia heteroclita* by Walter Hood Fitch (named after Brian Houghton Hodgson, one of Hooker's friends in India). From J. D. Hooker, *Illustrations of Himalayan Plants* (1855). Reproduced by kind permission of the Syndics of Cambridge University Library.

the beautifully kept ladies' cloak-room a building where they would have an opportunity of partaking of any refreshments they might bring. As to misconduct, there are plenty of ways to prevent that." [131]

Striking the balance between science and recreation became increasingly difficult as successive commissioners of works ruled that Kew must cater to those whom Hooker dismissed as "pleasure seekers." He began to protest about the rising cost of providing public gates and public conveniences, salaries for extra gatekeepers and police patrols, and garden ornaments and elaborate ornamental displays of seasonal bedding plants, asserting that they were absorbing almost three-quarters of Kew's annual income.[132] The *Gardeners' Chronicle* felt that a few purely ornamental displays were "desirable for the pleasure of the throngs of taxpayers who visit the garden," as long as they did not waste "resources which might be more profitably expended on other objects."[133] By contrast, the popular garden journalist William Robinson used his magazine the *Garden* to criticize Kew's bedding, decrying the lack of "picturesque beauty."[134] The outspoken Robinson remained one of Hooker's most vociferous critics in the debates over public access to Kew; Hooker had earlier complained to Darwin that "Owen, Ayrton & Robinson, the Editor of a paper called 'the Garden,' are all at work against Kew."[135]

The *Times* reported that the 1877 August bank holiday had brought 58,000 visitors to Kew, the highest ever total for a single day.[136] Inevitably, the "mere pleasure seekers" engaged in a little "rude romping," but the *Richmond and Twickenham Times* claimed that Hooker's obstinacy over opening hours was exacerbating the very problems he wished to avoid: "a large number of people came down in the morning [of the last bank holiday] expecting admission to the Garden, but finding the gates shut betook themselves in true British fashion, to drinking and dancing, and then some 2 hours later sought to refresh exhausted nature by falling asleep in the grass."[137] Hooker seems to have decided that, by comparison with the sight of visitors sleeping it off amid the flowers, early admission was the lesser of two evils; in October 1877, he agreed to open the gardens at 10 a.m., but only on Bank Holidays. He explained that "if opened the whole day the Gardens will be regarded as a Park. Park-licence will insinuate itself & demands for luncheons, pic-nics & bands of music will follow."[138]

The year after early opening had been introduced, Hooker raised the possibility that—far from extending the opening hours—the sheer numbers might force him to restrict access; "it may eventually prove impossible to admit the public on exceptionally crowded days, at any rate except in restricted numbers, to visit certain of the collections, which must be eventually exterminated in the process."[139]

In response to every demand for greater access, Hooker responded that the scientific purposes of the garden would be damaged if "swarms of nursery maids and children" were allowed in before lunchtime.[140] In response to his claims, Robinson's *Garden* commented sarcastically that science "requires that Kew Gardens be closed to the public for half the day," and that "if we could look over the walls we should no doubt see a multitude of students and philosophers walking to and fro in the grounds during the forenoon." It was only when these savants had taken themselves "to their cells to ruminate upon their morning's experiences, and in order also that they may not be contaminated by contact with the ignorant public," that the latter were allowed "to invade their sacred haunts." The *Garden* concluded that "we do not suppose that anybody believes for a moment the plea of 'science' put forward by the Kew authorities."[141] It was a few months later that the magazine complained that Kew was "a snug little preserve . . . for the scientifically inclined members of the Hooker family."

Thanks to the campaigns of the *Garden* and others, Kew's opening hours became the subject of a parliamentary debate in 1879, during which Sir Trevor Lawrence rose to defend the British public from "the rather serious charge" Hooker had made "against the people that they resorted to the woods for immoral purposes." Having visited the gardens himself and consulted the police, he had found no evidence of such behavior, but admitted that he had been "himself some years ago making protestations of an honourable character to a lady in the gardens (laughter); he knew no fitter place for such a purpose, and he would only say, *Honi soit qui mal y pense.*"[142] However, the new commissioner of works, Gerard Noel, repeated Hooker's claim that earlier opening of the gardens "would deprive them of the scientific character which it was originally intended they should maintain," quoting Hooker's own words that "the original objects of the institution were purely scientific, utilitarian, and instructional, and that to these has been superadded the recreation of the general public." The claims of science and imperial service ensured that the motion was lost.[143]

Hooker's reluctance to make concessions to Kew's recreational functions was a product of his long struggles over the status of botany and its practitioners. Together with the Ayrton incident and the clash with Owen over the national herbarium, the opening-hours controversy is an illustration of how fragile Hooker felt his grip on power to be.

Imperial Kew

Despite the fact that Hooker faced some serious challenges to his status and that of his institution, there were many signs that both were increasing steadily during the 1870s. As we have seen, he was elected to the presidency of the Royal Society in 1873, an event which the *Gardeners' Chronicle* noted was "peculiarly gratifying to botanists and horticulturalists," since he was the first naturalist to hold the positions since Banks's day. Hooker had received a just reward for his "scientific work, carried on with equal steadfastness and simplicity of purpose in antarctic seas and Himalayan slopes as in the geological and botanical establishments of the metropolis," but the scientific community had also recognized Hooker's courage in protecting "the honour of science" from the "onslaughts of those who were incapable of understanding its procedures, and unable to appreciate its objects."[144] Ayrton's name was not mentioned, but the *Gardeners' Chronicle* was in no doubt that he had, albeit inadvertently, helped Hooker attain this latest honor.

Hooker's service to the nation also resulted in his knighthood in 1877, and the choice of Knight Commander of the Star of India was a recognition of his services to India in particular. The introduction of new crops to the subcontinent was the most important aspect of Kew's work, with cinchona being the most important example of the economic role that Kew had gradually carved out for itself under William Hooker's direction. The "transfer" (or theft, as it undoubtedly appeared to the rival governments who had previously held a monopoly) received great publicity. As we have seen, the cinchona transplantation was not entirely successful, but that did not prevent Joseph Hooker and his staff from using the example of cinchona to assert increasing control over Kew's colonial satellites in the 1880s and 1890s. Kew's role in the introduction of another South American plant, ipecacuanha (also something of a failure), although disputed during the debates between Hooker and Owen, was another example of Kew's value. The relative failure of some of the early transplants contributed to the fragility of Hooker's reign, but later, more successful transfers—especially those of rubber and sisal—would eventually allow him to exercise considerable patronage, especially in appointments to colonial gardens.[145]

The continuing growth of Kew's herbarium received much less publicity than the crop transplantations but was even more significant in shoring up Hooker's scientific reputation and authority. In his evidence to the Devonshire Commission in 1872, in response to a question about whether Kew was "much resorted to by foreign botanists," Hooker had proudly replied: "No botanical monograph is considered complete which has not been worked up with the materials at Kew."[146] Kew's role as de facto gatekeeper to botanical publication was a combination of several factors; the sheer size of its collections made it an attractive place to work, but this carrot was supplemented by the stick of the Kew Rule (see chapter 7), which was designed to prevent provincial and colonial naturalists from evading the supervisory role that Kew had claimed for itself.

As we have seen, Kew's authority eventually reached the point that it could prevent names from appearing, even when they had been sent elsewhere to be published. By the end of Hooker's tenure as director, Kew's influence was global, subordinating both metropolitan and colonial naturalists to its rule. This power is a final, graphic illustration of why Hooker battled so hard to enlarge his herbarium, to try and wrest the right to name species from the British Museum, and—above all—to resist any move to strip Kew of its herbarium. Its contents might be no more than dried foreign weeds, but they were the foundation of Hooker's empire.

Seen from the colonies, Hooker may have looked like an all-powerful emperor, but in the metropolis he was often depicted as a selfish autocrat who might at any time be led to the guillotine by an angry mob of mere pleasure seekers. The public awareness that he had inherited his position may have fostered the suspicion that he had not really earned it, especially among those who neither understood nor cared for his scientific work; to some he seemed like a survivor of the old scientific aristocracy rather than one of the new, self-made men. Yet in trying to understand the place of science in Victorian society, it is important to consider Hooker's resistance to greater public access alongside his landscaping efforts and the evident pride he took in them. We should be wary of assuming that the directors of the major Victorian scientific institutions, especially museums and gardens, were privately devoted to a narrow vision of science but reluctantly submitted to the demands of public recreation to get funding for their real interests. Hooker's commitment to the education and moral improvement of the public is evident in everything from his tree planting to the many hours he spent on government committees aimed at reforming and improving scientific education.[147] Kew's right to public support was founded as much on the provision of ornate flowerbeds as on service to the colonies, as was abundantly clear from the terms in which the press defended him against Ayrton's interference.[148]

The fierce debates over where the national herbarium should be housed also emphasized these issues of public education, with each institution's supporters trying to argue that their rival's public mission did not require a herbarium, while theirs did; however, neither side attempted to draw a distinction between an institution's scientific work and its duty to admit, edify, and educate the public. The power that the herbarium conferred on its possessor obviously motivated this clash, but its resolution, the decision to maintain two separate herbaria, was more than a political compromise; as we have seen, the practical aspects of herbarium practice also proved decisive—revealing scientific practice to be at the heart of even the most overtly political aspects of the emergence of Victorian scientific institutions.

Governing an institution was another practice that a Victorian man of science might need to master, just like collecting, classifying, or corresponding. The seemingly unscientific business of providing even the poorest members of the public with access to beautiful flowerbeds proves to be a crucial aspect of this practice. It was a central component of Hooker's claim to be a gentleman, engaged in disinterested public service. This persona meant that, despite the fact that he received a government salary, the press were largely disposed to treat him as a man of high scientific attainments who undertook landscaping work purely because of his selfless dedication to the public good.[149] Good manners were, as we have seen, another aspect of Hooker's gentlemanly status which also proved vital to his career—as is evident in the press coverage of his arguments with Ayrton, whose bad manners were the focus of repeated criticism. As Macleod has argued, the clash was a crucial stage in the emergence of a "professional 'scientific civil servant,' subject as any other civil servant to the rules of central departmental authority," but it was clearly Ayrton, not Hooker, who was the professionalizer, determined to impose the discipline of modern government bureaucracy on a "scientific swell" who treated Kew as his private domain.[150] Hostility to Gladstone's failing government made it convenient for many in the press to attack the ill-mannered Ayrton, ridiculing the "turbulent suburban democracy" which had produced him, but within a few years the suburban masses were at Kew's gates, demanding admission. The *Times* reported Hooker as claiming that the "the agitation for the earlier opening of the gardens . . . is so far entirely local in character," the agitators being primarily "the inhabitants of houses recently erected in the neighbourhood." In Hooker's view, those who demanded easier access were forgetting that Kew was "an imperial and not [a] local institution."[151] That contrast, which was central to so many aspects of Hooker's career, helped him claim that higher imperial purposes—especially Kew's service to the colonies—outweighed the demands of local pleasure seekers. However, it was not a resistance that could last forever: Hooker allowed daily opening to be moved from 1 p.m. to midday in 1882. In 1898 his successor, Thiselton-Dyer (who had married Hooker's daughter Harriet in 1877, ensuring that Kew was kept in the Hooker family), allowed the gardens to open all day during the summer, a concession that was finally extended to the winter months only in 1921—long after Hooker's death and his successor's retirement.[152] The family's "snug little preserve" was finally public.

CONCLUSION

In September 1901, just a few months after Queen Victoria's death, Sir Joseph Hooker had the honor of opening the new botanical laboratory at Glasgow University. He gave his audience a sense of some of the changes that had happened in the eighty years since his father had been professor of botany at Glasgow, describing, for example, the hazards of William Hooker's early voyage to Iceland—on the way home, his wooden sailing ship caught fire, destroying all his notes and collections. Since then, as Joseph Hooker noted, "great changes have been introduced in the method of botanical teaching," not the least of which were improved microscopes and the use of chemistry and physics to elucidate "the elementary problems of plant life."[1]

Perhaps the most significant change, however, was that Hooker was opening a purpose-built botanical laboratory, something neither he nor his father had ever felt they needed. Joseph Hooker had seen the importance of experimental work and had helped ensure Kew got its first laboratory, the Jodrell, in 1876, but he never worked in it himself, recognizing that it was for younger men to take up the new experimental approaches. As the twentieth century began, the young men who would use these new laboratories at Kew and Glasgow were unmistakably scientists, and proud of the title. Dressed in their white coats and looking forward to careers in government- and industry-funded labs like the one they were to be trained in, they must have regarded the eighty-four-year-old Hooker as a creature from another time, almost another world.

The transformation of Britain's men of science into modern scientists had been a slow, uneven process that had taken most of the nineteenth century, and whose course had varied according to discipline and institutional setting. The change had hardly begun when Hooker attended his first meeting of the BAAS in Newcastle in 1838, but it was largely complete by 1901. Although Hooker was one of those who brought about that change, I hope that my reevaluation of his career and its significance has demonstrated that some existing accounts of the professionalization of the sciences fail to do justice to the richness of the wider social and cultural changes that the men of science shaped and were shaped by. Scientific practitioners fashioned new types of careers against a background of steam and industry, of growing cities and growing democracy, and, above all, of Britain's expanding empire, yet throughout all these changes, they had always to remain gentlemen. The demands of gentility created a conflict between the need for an income and the ideal of disinterestedness—a conflict that was common not only to many other men of science but to other groups, such as men of letters and engineers.

One reason this book is called *Imperial Nature* is that Britain's political and economic empire provided crucial opportunities for men like Hooker to build careers; without the empire, men like Hooker, Huxley, Darwin, and Wallace would have had very limited access to the natural world beyond Britain's shores. Nevertheless, I have tried to emphasize how uncertain and varied these careers were—as the contrasts in the careers of the four men I have just named demonstrate.[2] Key aspects of Hooker's later career only make sense in the light of his continuing worries about financial security and its links to concerns about personal and disciplinary status. To understand the way his imperious nature affected the governance of Kew, we need to remember that Hooker was an ambitious young man in search of a job long before he was a botanical emperor.

The existing history-of-science literature on Hooker usually situates him within wider nineteenth-century debates, particularly those over plant distribution and the origin of species. As a result, historians emphasize the content and significance of his ideas, but I think this inverts both the structure and the priorities of his career; he was a traveler and field collector in search of "honourable compensation" for his work long before he was an essayist, and the content of his essays cannot be understood without first considering his material practices and the objects and people they involved (fig. C.1). The practices of collecting and classifying were prior to those of charting distribution and philosophizing in two related senses: chronologically, in that they were what he did first; and physically, in that they provided the specimens about which the books were written. To begin an analysis of Hooker with his ideas is, I argue, to stand him on his head; focusing on practice is my attempt to set him back on his feet.

C.1 Joseph Hooker at his desk in 1906, still hard at work on classification. By kind permission of the Trustees of the Royal Botanic Gardens, Kew.

Moreover, a focus on practice serves to overcome a long-standing historiographical tendency to divide the factors and influences that shape science into those which are internal to science (such as objectivity and careful experimentation) and those which lie outside (e.g., political, religious, and economic factors). Most recent historians of science acknowledge the need to deal equally with all these factors and avoid privileging one or other set.[3] Yet I want to argue that even these sophisticated approaches tend to divide individual lives and careers (and the ideas, publications, and institutions that emerged from them) into scientific and nonscientific categories, inadvertently sustaining divisions that invariably prove unhelpful. I have chosen instead to examine a broad range of practices—from gardening to classifying, from philosophical essay writing to collecting and drying flowers—that characterized the communities Hooker moved in.

A detailed study of scientific practice helps relate apparently esoteric matters, like theories of geographical distribution, to mundane matters like the practicalities of earning a living. However, the relationship between these is always complex. To take just one example, the relationship between the outdoor work of daily collecting and the work needed to build and maintain a herbarium reveals the way that something as seemingly abstract as Hooker's species concept was rooted in a wide range of issues: the way specimens were made; the practicalities of reducing both the numbers of specimens arriving and of existing herbarium sheets; Hooker's need to create an appropriate distance between himself and his collectors—by making the colonies peripheral and the metropolis central; and Hooker's concern to distinguish himself from the "species-mongers" (in the sense of both "splitter" and "specimen seller") in the eyes of his metropolitan colleagues. However, I have also shown that creating and stabilizing a species concept was not something that could happen in the herbarium alone: it had to be exported to the colonies, by being encapsulated in the classifications Hooker wrote and sent to his correspondents; and then it had to return to the metropolis, embodied in specimens collected according to the standards he was attempting to inculcate. The physical movement of specimens and books, accompanied by letters and gifts, embodies various processes of negotiation, over compliance with metropolitan standards, the value of local knowledge, the social and scientific hierarchies within which Hooker and his collectors were maneuvering, and ultimately the location of the center and periphery.

Analyzing the minutiae of daily practice reveals a second important sense in which nature was imperial. Hooker's work—in the field and herbarium, through words and pictures—remade nature in empire's image. Despite visiting each country only briefly, he defined the floras of New Zealand, Tasmania, and, in later years, of British India and other countries; he persuaded each country's inhabitants to accept that he alone knew how many species of plants their land held and what each was called. The definitions of genera that he and George Bentham produced in their *Genera Plantarum* remain a foundation of modern botanical taxonomy.[4] And in 1881, Charles Darwin wrote a codicil to his will informing his executors that he had "promised to Sir J. Hooker to pay about £250 annually for 4 or 5 years, for the formation of a perfect M. S. catalogue of all known plants."[5] This "perfect catalogue," the *Index Kewensis*, began publication in 1893 and is still being added to at Kew; it instructs the botanical faithful to believe only in the species it lists, effectively pronouncing anathema on all other plant names. Hooker's "philosophy of system," a philosophy shaped by the need to establish and manage his own botanical empire, is still very much with us, embedded in the names of plants from almost every country.

Yet, as we have seen, Hooker's empire was never wholly secure, nor was it unchallenged. Despite his influence, he never had the power to simply impose his views even on his fellow workers in Kew's herbarium, much less on the world's botanists. The need to negotiate compliance with his standards shaped his practices and the conclusions he drew from them. Recognizing how complex these relationships were has led me, like many of my colleagues, to become critical of those approaches to the history of imperial science that tend to collapse the complex worlds and cultures of imperial and colonial science into a simple process in which Western scientific ideas simply "diffuse," as if by osmosis, from a concentration in Europe out into a vacant and passive periphery.[6] Yet while recent work has helped overcome this inadequate conception, analyses of imperial centers and peripheries still tend to assume that these are static categories—they were not. The location and definition of the empire's center and its periphery were being continuously redefined and negotiated through the practices of collecting and classifying—and cannot therefore be used to explain how those practices worked. Instead, I have tried to show how the colonists' responses to the demands of the metropolis can more usefully be thought of in terms of compliance (or the lack of it), in that collectors could choose to adopt or reject metropolitan standards by conforming or refusing to conform with explicit or tacit expectations. These standards include uniform methods for collecting, drawing, labeling, and classifying, but they can be extended to encompass social conventions, such as gentlemanly behavior and courtesy, both of which were essential to making friends and joining networks. Indeed, scientific and social status interlock in such complex ways in the mid-nineteenth century that the two categories can scarcely be considered distinct.

It might at first seem that what was most significant about the relevant standards was that colonists could only choose whether or not to comply; they could play no part in actually setting the standards—that was done almost entirely in the metropolis. Yet, as illustrated by my analysis of classification, not everyone in the metropolis agreed; the standards themselves were still being worked out. It is crucial to remember that the specimens which collectors sent were vital to these metropolitan debates; Darwin's theory of evolution by natural selection is only the most famous of the Victorian theories about the laws governing the natural world that were shaped by these interactions with now-forgotten colonial contributors to science. Understanding the classification debates, in which Darwin was as much a participant as Hooker, depends on recognizing that classification itself began in the field, because it was an essential part of naming specimens.

It might seem that, despite all that I said above about negotiation and barter, Hooker's metropolitan location and his father's influence (to say nothing of his library and herbarium) made him immensely powerful by comparison with the collectors, who were cut off from scientific societies, equipment, and books and had little access to powerful friends, but as we have seen Hooker was also isolated—from the plants he wished to study—while his collectors often had unique access to the plants. Hooker was tied to Kew because he needed his father's help and access to his collections; meanwhile, the plants that were necessary for his work were on the other side of the world, made inaccessible by distance and the endless difficulties of transporting them. By contrast, men like such Colenso and Gunn were close to the plants and could, had they wished to, have denied Hooker access to them (in the early stages of his career at least). By focusing on the plants and the material practices used to collect them, rather than on the people and their ideas, we can see a sense in which Hooker and Kew were "peripheral" while Colenso and his house at Paihia were "central." This point should not be pushed too far: if the full range of resources, assets, and opportunities at each man's disposal are compared, Hooker's dominance is clear, but we should not forget how he built that power.

Friendships were crucial to Hooker's career, and none more so than that with Darwin; they wrote to each other regularly for over forty years, until Darwin's death, and their surviving letters—over 1,300 of them—record one of the most important scientific friendships of the nineteenth century. However, I have deliberately chosen to keep Darwin in the background throughout this book, partly because his story is so well known and has been so well told but also because his enormous historical significance has tended to overshadow the lives of his contemporaries, such as Hooker. However, I want to conclude by bringing Darwin back to the center of my story and suggest that a history of everyday, mundane scientific practices forces us to reevaluate even this most eminent of Victorians.

On 29 December 1859, just one month after the appearance of the *Origin of Species*, Joseph Dalton Hooker published "On the Flora of Australia" (the introductory essay which completed the *Flora Tasmaniae*), in which he announced his support for "the ingenious and original reasonings and theories by Mr. Darwin and Mr. Wallace" (fig. C.2). In the *Flora,* he acknowledged that just a few years earlier he had been arguing for "the still prevalent doctrine" that species were "created as such and are immutable," but "in the present Essay I shall advance the opposite hypothesis, that species are derivative and mutable."[7] He thus became the first man of science anywhere to embrace Darwinism.

At first sight, this is pretty unsurprising. It seems to fit very nicely into conventional pictures of the "Darwinian revolution," in which the *Origin*

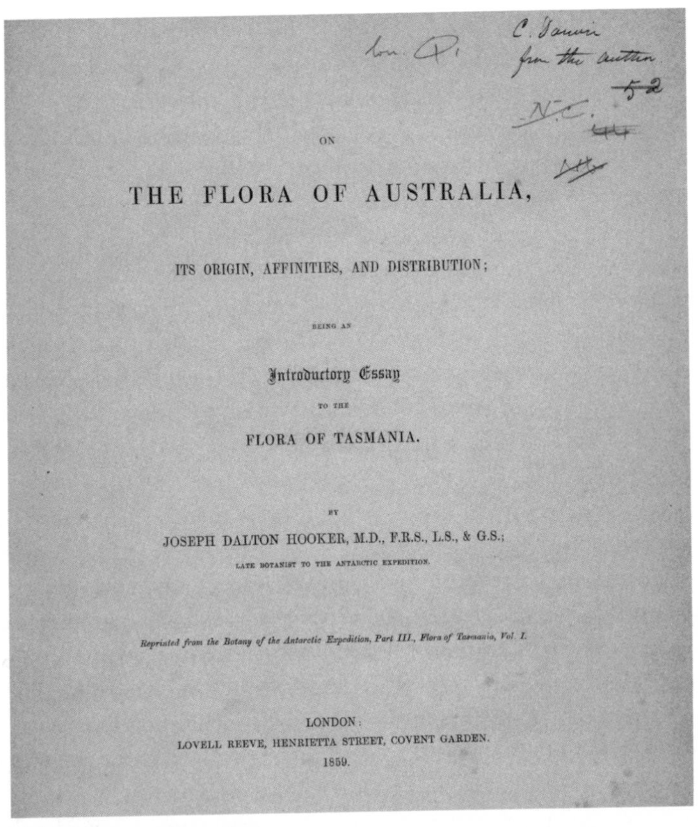

C. Darwin
from the author

ON

THE FLORA OF AUSTRALIA,

ITS ORIGIN, AFFINITIES, AND DISTRIBUTION;

BEING AN

Introductory Essay

TO THE

FLORA OF TASMANIA.

BY

JOSEPH DALTON HOOKER, M.D., F.R.S., L.S., & G.S.;

LATE BOTANIST TO THE ANTARCTIC EXPEDITION.

Reprinted from the Botany of the Antarctic Expedition, Part III., Flora of Tasmania, Vol. I.

LONDON:
LOVELL REEVE, HENRIETTA STREET, COVENT GARDEN.
1859.

C.2 "On the Flora of Australia" essay title page. Darwin's own copy; signed by Hooker. Reproduced by kind permission of the Syndics of Cambridge University Library.

of Species struck Victorian Britain like a thunderbolt and overnight the world looked very different as the scales fell from naturalists' eyes. But I want to argue that—when we look more closely—there is something rather curious about Hooker's "conversion" to Darwinism.

Hooker's essay often seems to be playing down the impact of his friend's theory. "I must repeat," he wrote, "what I have fully stated elsewhere, that [Darwin's] hypotheses should *not* influence our treatment of species, either as subjects of descriptive science, or as the means of investigating the phenomena of the succession of organic forms in time, or their dispersion and replacement in area." At one point he acknowledged that Darwin's theory would enable naturalists to form "more philosophical conceptions on these subjects"—which was, as we have seen, the highest possible praise—yet on the same page he stressed that "the believer in species being lineally related forms [the evolutionist] must *employ the same methods of investigation* and *follow the same principles* that guide the believer in their being actual creations."[8] This implies that Darwinism would change nothing about the way naturalists worked; and it is surely a strange revolution that does not change anything.

I am not the first person to have been puzzled by Hooker's comments; the apparent ambiguities in his writing have led historians to two almost diametrically opposed views. Some have concluded that he was a late, perhaps even a reluctant, convert to Darwinism.[9] While others have argued that—even back when Hooker was publicly claiming that species were immutable—he was in fact a "closet Darwinist," who had long been persuaded but felt he could not give any public hint of Darwin's ideas until Darwin himself chose to do so.[10]

So exactly what was Hooker's view of Darwinism? The question might seem unimportant, but I believe that answering it forces us to revise our understanding of the reception of Darwinism and of many other things about nineteenth-century science and its practices. Most accounts of the period are structured around the claim that the central problem for Victorian naturalists was "the species question"—and they are right. The trouble is that I think historians have rather misunderstood what that question was. For many nineteenth-century naturalists, "the species question" was not simply whether or not they evolved. That was one—admittedly important—part of a much bigger question: how can species be defined objectively, so that they can be counted and their distributions mapped, so the laws that governed their geography could be deduced, not least so as to discover and acclimatize valuable new plants?

The introductory essays to the *Botany of the Antarctic Voyage* and the *Flora Indica* were virtually Hooker's only extended discussions of the theoretical aspects of botany, so historians have tended to turn to them to discover when Hooker became a convert to natural selection.[11] The historians' curiosity is understandable, given Hooker's well-known role in the publication of Darwin's theory.[12] As we have seen, once Darwin had published, Hooker became a prominent, public defender of natural selection; he spoke at the famous Oxford debate of 1860, alongside the more combative (but, on this occasion at least, less effective) Thomas Huxley.[13] So, given Hooker's role as an advocate of natural selection and that he was the first person ever to hear of it, it seems reasonable to look for evidence as to when he became a convert.

As we have seen, the ambiguities of Hooker's first post-Darwinian essay are echoed in his earlier ones. In the introductory essay to the *Flora Novae-Zelandiae* (1853), Hooker had assumed that "species, however they originated or were created, have been handed down to us as such." This seems straightforward, but Hooker immediately dropped some mud into this limpid pool by adding, "I wish it to be distinctly understood that I do not put this forward intending it to be interpreted into an avowal of the adoption of a fixed and unalterable opinion on my part." [14] The waters were muddied further a few pages later, when he discussed naturalists' various theories about how species originated: many "consider species as permanently distinct," but "another large class do not consider species as permanent at all," a third group believe in "progressive development," while a fourth group "subscribe to various shades of these opinions, or blend them." Hooker then discussed the range of evidence that might be offered for or against each of these views, but did not identify himself with any of them—he simply warned his readers against their "hasty adoption." [15]

A contemporary reviewer in the *Athenaeum* was convinced that Hooker's essay had provided strong evidence against transmutation: "the arguments on which Dr. Hooker supports the first proposition [that species are "permanently distinct"] appear to us conclusive." [16] The *Athenaeum*'s reviewer was not the first to read Hooker's essay as opposing transmutation; Charles Bunbury was surprised to hear from Charles Lyell that "when Huxley, Hooker, and Wollaston were at Darwin's last week, they (all four of them) ran a tilt against species farther I believe than they are deliberately prepared to go." [17] Bunbury replied: "You surprise me by saying that Hooker was one of the party who `ran a tilt' against species. In all his writings, even in the most recent, his `Flora Indica,' he distinctly and explicitly maintains the reality of species, though he holds (and I have no doubt he is right) that a large proportion of the species admitted in our systematic works are *not* valid." [18]

As the comments by both Bunbury and the *Athenaeum*'s reviewer make clear, Hooker's essays were read as arguing for "the reality of species," and thus as opposing transmutation, but it is not clear that that was what he intended. Indeed, I would argue that far from clarifying his own view, Hooker went out of his way to obscure it. In the *Flora Novae-Zelandiae* essay, what he said was that, "if the expression of an opinion be insisted on" (on the subject of transmutation), he would give the one offered by a famous astronomer who, when pressed as to whether he thought the other planets were inhabited, "replied that the earth was so, and left his querist to argue from analogy."[19] So, Hooker simply asserted that, as far as species are concerned, first, they become extinct; second, we have not seen them "varying into other forms"; and finally, "from which established premiss the speculator may draw his own conclusions."[20] In other words, transmutation had probably not occurred, but Hooker's real message to his readers was that such speculations lay, for the time being at least, outside the proper realm of science.

The *Flora Tasmaniae* seems to substantiate the view that Hooker simply could not make up his mind, since its introductory essay seems as contradictory as the earlier ones. However, this "doubting Darwinist" view is hard to reconcile with, for example, the fact that Darwin certainly did not think Hooker's support had been half-hearted, telling him he was "the one living soul from whom I have constantly received sympathy."[21] After the presentation of the Darwin/Wallace joint paper to the Linnean Society, Hooker told Asa Gray he was "most thankful" "that I can now use Darwin's doctrines—hitherto they have been kept secrets I was bound in honor to know, to keep, to discuss with him in private—but never to allude to in public, & I had always in my writings to discuss the subjects of creation, variation &c &c as if I had never heard of Natural Selection—which I have all along known."[22] This expression of relief suggests that Hooker, far from being a reluctant supporter, had been eagerly awaiting the chance to make use of Darwin's ideas.

So, was Hooker eagerly waiting to make use of his friend's theories, or was he a late convert, only reluctantly pinning his colors to the Darwinian mast? I want to argue that the simple answer is "neither," because asking when Hooker became a Darwinian is asking the wrong question. "When was Hooker converted?" is a product of the popular myth of the Darwinian revolution, the idea that the world changed abruptly in 1859; it is like asking how Darwin was able—in a single book—to persuade his fellow Victorians to change their minds and accept evolution. Again, the simple answer is that he did not; a huge range of complex factors were at work in the first half of the nineteenth century that made Darwinism attractive to various Victorian communities, including some, but not all, naturalists. The more important question, I would suggest, is "what made natural selection useful to Hooker?"—not least because this question reminds us that the practices and debates that shaped Hooker also shaped Darwin.

Instead of dividing the world of nineteenth-century natural history into static or anachronistic groups such as professional or amateur, it can be more helpful to think about communities who shared common practices.[23] In one sense, Hooker and Darwin were members of the same community, one that practiced classification—plants for Hooker, barnacles for Darwin—and they both did so in the metropolis (at least, Down was no further from London than Kew was). Given the complex concerns and interests of this community, it comes as no surprise to discover that both Darwin and Hooker saw classification as vital to larger philosophical issues, especially distribution, nor that they both used the natural system, disliked splitters, and were concerned with the stability of classification and nomenclature.[24] However, there were equally important differences between them: classification occupied all of Hooker's adult life but only eight years of Darwin's (1846–54); Hooker worked in a large, government-funded institution, Darwin classified at home; Hooker built up and maintained a vast collection of specimens, Darwin borrowed most of his. However, this division is not exhaustive: Hooker and Darwin also belonged to different communities in that their systematic work was botanical/distributional and zoological/anatomical respectively (communities that do not map on to the first ones). Other contrasts could also be drawn.

The most important difference was that Hooker earned his living from classification, whereas Darwin did not need to. Darwin was wealthy enough to display a disinterested curiosity about barnacles, pigeons, orchids, primulas, earthworms, or indeed anything else that caught his fancy, including the origin of species; his speculations might hurt his reputation or his wife's feelings, but they could not hurt his income. Hooker, as we have seen, was in a very different position, dependent on government largesse, his father's patronage, and, more importantly, his network of colonial collectors: without them, he had no specimens; without specimens, he could not publish; without books, he could not build his reputation; and without a reputation, he had no chance of obtaining the kind of paid position he required.

Identifying these differences in Hooker and Darwin's communities of practice allows us to see both how Hooker used Darwin's ideas and why that usage created the apparent ambiguities in his essays.

One very important source of the apparent ambiguity in Hooker's writing was the conventions of gentlemanly behavior that governed such speculations; as we have seen, when the Earl of Rosse gave Hooker the Royal Society's medal, he praised "the cautious and philosophical manner" in which the subject of species was treated. As Hooker had written, "I am very sensible of my own inability to grapple with these great questions, of the extreme caution and judgment required in their treatment, and of the experience necessary to enable an observer to estimate the importance of characters whose value varies with every organ and in every order of plants."[25] Such caution and care were the hallmarks of the respectability befitting a gentleman.

For men from Hooker's community, those who practiced classification for a living, the species question was not primarily about the transmutation of species but about their definition. As he noted, to anyone who has studied botanical classification it "cannot but be evident that the word *species* must have a totally different signification in the opinion of different naturalists."[26] And as we have seen, in broaching this topic he constantly needed to deter his colonial audience from speculating. It is not therefore surprising that on the very page where he praised Darwin for providing "more philosophical conceptions," Hooker stressed the need for all botanists to employ "the same methods of investigation and follow the same principles." Whatever their "philosophical conceptions," each botanist must use the same method of classification; as Hooker put it, "the descriptive naturalist who believes all species to be derivative and mutable, only differs *in practice* from him who asserts the contrary, in expecting that the posterity of the organisms he describes as species may, at some indefinitely distant period of time, require redescription."[27] In other words, even to an evolutionist species were stable over such long periods that they could still be treated as "definite creations" for classificatory purposes. As Hooker had written before the *Origin* appeared, the difference between transmutationists and their opponents "is very wide perhaps, but not so wide as to allow of their employing different methods towards the advancement of Botany in any one of its departments."[28]

In his essays Hooker stressed the continuity of his practices, even though his view of the transmutation question was apparently changing. One strategy for emphasizing continuity was to refer to Darwin's theory as "the hypothesis that it is to variation that we must look as the means which Nature has adopted for peopling the globe with . . . species." Characterizing natural selection as the "variation" theory, as Hooker often did, was important—he had always maintained that species were variable; it was the key tenet of his broad species concept. Calling natural selection the variation theory emphasized its continuity with his earlier views; as he noted in the *Flora Tasmaniae:* "my own views on the subjects of the variability of existing species" remain "unaltered from those which I maintained in the 'Flora of New Zealand.'"[29] So successful was Hooker at stressing the continuity of his views that, after reading both the *Origin* and the *Flora Tasmaniae,* Bunbury observed that, while it was "a great triumph for Darwin" to have converted both "the greatest geologist of our time and the greatest botanist" (i.e., Charles Lyell and Joseph Hooker), he nevertheless concluded that, although Hooker "adopts the Darwinian theory," he "considers it not as proved, but as a hypothesis, quite *as* admissible as the opposite one of permanent species, and far more suggestive."[30]

If we read every Victorian discussion of species in the light of the post-Darwinian world, there is a strong tendency to confuse stable species with divinely created ones. Hooker told Darwin in 1845 that "those who have had most species pass under their hands as Bentham, Brown, Linnaeus, Decaisne & Miquel, all I believe argue for the validity of species in nature"; Richard Bellon cites this as evidence that, at this early stage in his career, Hooker believed species to be divinely created.[31] That is certainly a possible interpretation but not, I would argue, the most plausible one, since Hooker went on to write that these distinguished naturalists had all refused to pay attention to the unimportant features of organisms that were "taken advantage of by the narrow-minded studiers of overwrought local floras," that is to say, provincial splitters. Hooker made the comments in the course of their discussion of Frédéric Gérard's work on species, which was mainly concerned with deciding who was entitled to pronounce on the topic. Hooker, as we have seen, dismissed Gérard as typical of those "who have no idea what thousands of good species their [*sic*] are in the world," someone who "evidently is no Botanist." As early as 1845, Hooker's main preoccupation was with the stability of species and how to protect them from the splitters, "the narrow-minded studiers of overwrought local floras."[32] In an 1858 letter to Gray, Hooker described one of his chief goals in the introductory essay to the *Flora Tasmaniae* as being "to harmonize the facts drawn out with the old Creation doctrine & the new Natural Selection doctrine."[33] Like the earlier letter to Darwin, this strongly suggests stability was the main issue; assuming each species was specially created was one argument that achieved this goal, but Hooker was wedded to the goal, not the argument. Thanks to the stately pace of Darwin's slow, gradual version of evolution, natural selection did not undermine stable species, while its "more philosophical conceptions" could enhance the importance of Hooker's endeavors.

Conceptions that were more philosophical would help Hooker gain respect in the metropolis and compliance in the colonies, but they would also help Hooker earn a living as a philosophical botanist. Yet the need to earn a living meant Hooker had to minimize natural selection's potentially destabilizing effect on classification—hence his careful equivocation over natural selection's impact, which has been misinterpreted by some as a lack of enthusiasm.

As I have argued, Hooker's essays have relatively few parallels in their period; because there were no journals of philosophical botany to publish them in, he had to take advantage of his floras, hoping that the quarterlies would review and quote from them. Their reception was distinctly mixed; the difficulties of creating a new genre of philosophical botanical essay parallel the difficulties of creating a new career as a philosophical botanist.

In the hands of the "superficial naturalists" and the "thoughtless aspirants," natural selection might be used to multiply species, but had that been its only impact on Hooker's work, he is unlikely to have adopted it. In practice, Darwin's ideas had many positive implications and Hooker's assertion that he was "most thankful" to be able to "*use* Darwin's doctrines" refers to more than their tactical value for his battles over status; Darwinism provided intellectually satisfying answers to some of the major questions he confronted. As Hooker wrote, natural selection explained "many facts in variation," but more importantly he considered it "the most fatal argument against 'Special Creation' & for 'Derivation' being the rule of all species."[34] Six months after he published the *Flora Tasmaniae* essay, Hooker gave a paper at the Linnean Society entitled "Outlines of the Distribution of Arctic Plants," in which he argued that despite ill-defined species, such distribution studies were useful as long as the species "treated conjointly" were united by closer affinities than "those which exist between those treated separately." In Hooker's view, as long as these broader patterns were observed, it was "a matter of secondary importance to me whether the plants in question are to be considered species or varieties." He went on to note that if "we consider these closely allied varieties and species as derived by variation and natural selection from one parent formed at a comparatively modern epoch, we may with advantage, for certain purposes, regard the aggregate distribution of the very closely allied species as that of one plant."[35] Not only did natural selection allow the botanist to finally stop worrying about how to delimit species and varieties (as Darwin had noted, the latter were simply incipient species anyway), but it also allowed the distribution of numerous "very closely allied" species to be treated as "that of *one* plant." What more could even the most ardent lumper ask for? Not only did Darwinism not threaten Hooker's species concept, but it provided it with a firmer philosophical underpinning.

A similar argument applies to Hooker's systematic method of condensing geographical variation into the "phantom type"; according to Darwin, the type was no phantom at all but the common ancestor of all the species concerned. As he said in the *Origin,* if one assumed that "the natural system is founded on descent with modification," it became clear that the characters a good systematist used to show "true affinity between any two or more species" "are those which have been inherited from a common parent, and, in so far, all true classification is genealogical; that community of descent is the hidden bond which naturalists have been unconsciously seeking."[36] Some Darwin scholars might nevertheless want to argue that this last quotation shows Darwin "converting" Hooker to natural selection, but as I have argued throughout, that is a distinctly implausible interpretation. Hooker found natural selection useful because it meshed so well with his preexisting practices—practices that were not produced by abstract speculation about origins but were in many respects responses to the pragmatic needs of the diverse cultures of natural history. It is at least as plausible to note that Hooker had often discussed his classificatory approach in letters with Darwin; perhaps the latter was the one converted—his comments on classification in the *Origin* certainly appear to bear the stamp of Hooker's thinking.[37]

As a systematist, Hooker's immediate worry was Darwinism's impact on nomenclature, which led him to play down its impact. Because of the theory's longer-term benefits, however, which included an objective basis for his species concept and thus a securer position from which to overrule the splitters, he did not reject or oppose it. His distribution theories had similar concerns; he opposed multiple centers because they opened the floodgates to species-mongers; he adopted Edward Forbes's land-bridge theorizing in part because it allowed him to keep those floodgates tightly closed. Hooker had once feared that transmutation would reopen them, but Darwin was gradually able to persuade him that natural selection could account for both disconnected and representative species as effectively as Forbes's theories, and perhaps even better.[38] Indeed, among the many reasons Hooker adopted natural selection was that it was the best available tool with which to counter the species-mongers.

However, as we have seen, "species-monger" not only expressed Hooker's aversion to proliferating names but also hinted at his concern that earning a living from botany might compromise his hoped-for philosophical status. This anxiety is a vital part of understanding Hooker's approach to managing his collectors, to classification, to distribution, and even to Darwinism. The starting point in explaining all the varied forms his practices took must be the derisory £6 that natural history received from the BAAS at Newcastle in 1838, which vividly illustrated the equation between scientific status and hard cash. As Hooker's description of himself as a *rara avis* shows, he was conscious that there were almost no precedents for the type of career he was trying to construct. Hooker's networks, classifications, and philosophical essays were all improvised in the absence of guidelines, but always with one eye on his bank balance.

In analyzing what I believe are some of the more important Victorian scientific practices, I have tried to focus on the material aspects of each practice, hence, for example, my decision to begin my discussion of classification with herbarium sheets and the drawers they were stored in rather than with more abstract issues. My intention has been to avoid prejudging the categories into which practices might fall, and I hope this kind of approach helps achieve a richer understanding of how science works. The kinds of points I have made about the connections between Hooker's practices, his ideas, and his career could also be made about William Colenso, Alfred Russel Wallace, Asa Gray, and many other figures who are all currently languishing in Darwin's shadow. Analyzing their scientific practices could help us illuminate the details of the social, political, and economic circumstances that shaped their careers, and understanding the diversity of those practices and the resultant careers would—I believe—also give us a better sense of exactly what kind of impact Darwin's books had.

Surveying and analyzing the range of reactions to Darwinism in different communities of practice is more productive than asking questions such as "when did Hooker become a Darwinist?" The minor debate around that question is unimportant in itself but serves to illustrate a larger problem: his career appears to be a paradigm case of the professionalization of the sciences during the nineteenth century, which makes his eventual adherence to Darwinism entirely predictable. After all, Darwinism was supposedly the banner under which Britain's young professionalizers united in order to drive the Anglican gentry out of natural history once and for all. This brief summary of prevailing historiography is something of a caricature; nevertheless, it captures the sense of inevitability that often pervades accounts of nineteenth-century scientific careers: the major naturalists were born Darwinians (indeed, the "major naturalists" are often defined as the Darwinians). But as we have seen, there was nothing inevitable about Hooker's career. Once Hooker's essays are read from the perspective of his career and concerns, it becomes clear that he was not ambivalent about natural selection; the apparent ambiguity in his writing was motivated by his need to play down the importance of Darwin's ideas for one of his audiences while emphasizing them for another. Hooker was not hesitating but deliberately equivocating.[39]

Previous historians have left Hooker's life, his botanical practices, and his ideas buried under an avalanche of isms—professionalism, Darwinism, and colonialism. Once these are cleared away, we not only see Hooker more clearly but can also see how earning a living from the sciences, the reception of Darwin's work, and the way science was practiced in its colonial settings are all aspects of a richer picture of the diversity of nineteenth-century scientific practices and cultures.

A crucial attraction of natural selection for Hooker was that it was slow; on the time scale of individual lives, and botanical careers, species did not change at all. There really was a sense in which Darwinism changed nothing. In this crucial regard, the shift in Hooker's position from the *Flora Novae-Zelandiae* to the *Flora Tasmaniae* was insignificant—hence his emphasis on the continuities between his earlier and later positions. That emphasis is only a surprise if one assumes that the *Origin* turned the world upside down.

The question of when Hooker became a Darwinist turns out to be the wrong question because it presumes that nineteenth-century natural history revolves around evolution. It also privileges the "intellectual"—theories and concepts—over the practical—collecting and classifying. If instead we ask what use Hooker made of Darwin, considering the question from the perspective of Hooker's career and interests, it becomes clear that for him, as for many of his contemporaries, stabilizing species was the big question, not least because stable species were the key to transforming botanists from mere "species-mongers" into respected members of the scientific community. In one of Hooker's earliest letters to Darwin, he commented that "not being a good arranger of xtended [*sic*] views I rather fear the Geographical distribution, which I shall not attempt until I have worked out all the species."[40] In response, Darwin congratulated him for not indulging "in the loose speculations so easily started by every smatterer & wandering collector," assuring him he was more than capable of tackling questions of distribution.[41] However, Hooker continued to leave the big questions until he had "worked out all the species"; as early as 1860, he had considered writing "a Darwinian book on Botany" that would encompass "Classification, Distribution and the origin of species."[42] But even after his retirement, this project was deferred while Hooker spent the last decade of his life reclassifying the genus *Impatiens* (the balsams), on which he wrote thirteen papers, the last of which appeared just before he died. Working out "all the species" remained an unfinishable task.

NOTES

INTRODUCTION

1. Barton 1998; White 2003: 1–2.

2. Gilmour 1993: 111. For examples of the divisions within the scientific communities, see Winter 1998.

3. Some examples are given in Allen 2001a.

4. For Banks, see Miller et al. 1996.

5. For a recent overview of the complexity of these colonial and imperial relationships in nonscientific contexts, see Hall 2002.

6. See Endersby 2000 for a discussion of barter and its importance in such contexts.

7. Richard Bellon (2006) has come to similar conclusions based on his studies of Hooker.

8. I am grateful to James Secord for first drawing my attention to this discrepancy (personal communication). See also Jardine 2000b: 401.

9. On practice, see also Gooding et al. 1989; Pickering 1992; Larsen 1996; Camerini 1997.

10. Bellon 2001: 53.

11. MacLeod 1974.

12. Sinnema 2002: xix–xx.

13. Galton 1874: 12. Galton had also used the same pairing in the title of an address to the Royal Institution earlier that year (Gillham 2001: 191–92).

14. Anon. 1901: 552.

15. Allan 1967: 57–58, 71–72; Drayton 2000: 145–46; FitzGerald 2004; Fraser 2004.

16. The Linnean Society has always spelled its name with only one *a*, whereas the Linnaean system of classification is spelled with two.

17. Bellon 2005.

18. White 2003: 1–2.

19. Hall 2002: 27; Barton 2003: 79–80.

20. Similarly, Henry De la Beche lost the family fortune and became paid director of the Geological Survey, thus falling from genteel to professional status (R. Porter 1978: 825).

Steven Shapin (1984) notes the low status and uncertain incomes of Scottish university chairs at this period.

21. However, wealth from any source was already competing with landownership as a source of high social prestige during the late eighteenth century (Corfield 2000: 10–11).

22. William Hooker received a small salary from the university but relied on students' lecture attendance fees. Even Oxbridge professorships were much less well remunerated than most of the clerical livings in the gift of the colleges, so fellows tended to wait for the latter (Heyck 1982: 72–73).

23. Hoppen 1998: 40–45.

24. Apothecaries had hoped the passing of the 1815 act would raise their standing, but the Royal Colleges of Physicians and Surgeons, jealous of their traditional status and privileges, ensured that the apothecaries' lowly status was maintained (Loudon 1992: 234–35). Although the act applied only to England and Wales, it prompted both Edinburgh University and Glasgow University to expand their botanical teaching to meet its requirements (Comrie 1932: 519; Allen 2001b: 351–52). Botany remained part of an apothecary's training until 1886; see the Society of Apothecaries' *Court of Examiners Minute Book, 1890–97*, loose leaf at fol. 244, Archives of the Society of Apothecaries, London.

25. For the status problems besetting apothecaries, see Hamilton 1981: 162–64; Loudon 1992: 236; Morrell 1996: 980; Corfield 2000: 137–73; Allen 2001b: 335. In his old age, Hooker recalled that in his youth "Botany was not taught as now, as a branch of biology, but as an introduction to, and as handmaid of Materia Medica" (J. D. Hooker 1887: 3).

26. The "taint of trade" persisted well into the last quarter of the century. In Anthony Trollope's *The Vicar of Bullhampton* (1870), Miss Marable maintains "that when a man touched trade or commerce in any way he was doing that which was not the work of a gentleman" (quoted in Hoppen 1998: 43–44). For other examples, see Best 1979: 272–73.

27. Eliot 1872 (1994): 143.

28. Loudon 1992: 237, 41.

29. The Faculty of Physicians and Surgeons claimed the sole right to license surgeons in the area, which led to disputes with the university during the early nineteenth century, especially over the relative importance of formal academic study and practical, hands-on experience. See Dow et al. 1988: 229–30; Hamilton 1981: 160–61.

30. Geyer-Kordesch et al. 1999: 208–9, 193–94.

31. J. D. Hooker to Dawson Turner, 5 February 1845, in L. Huxley 1918a: 196.

32. [Lankester] 1857. When authors' names are given in brackets, the piece was originally published anonymously; for sources of attributions, see the bibliography.

33. Morrell 1997: 190–91; Barton 1998: 421.

34. For more on Scottish universities at the time, see Hamilton 1981: 154–58; C. Smith 1998: 119–22; Bellon 2005: 26–27.

35. Drayton 2000: 146–47. William Hooker's books on ferns included *Genera Filicum* (1838–42), *Species Filicum* (1846–64), and *Icones Filicum* (with R. K. Greville; 1827–32). His periodical publications appeared in *Botanical Miscellany* (1830–33), *Hooker's Journal of Botany and Kew Garden Miscellany* (1849–57), and *Icones Plantarum* (1836–64).

Lindley's *Ladies' Botany* (1834) is a conspicuous example of a work aimed at a popular, predominantly female audience. His books on orchids included *Genera and Species of Orchidaceous Plants* (1830–40), while his periodical publications appeared in *Botanical Register* (1829–47) and *Gardeners' Chronicle* (1841–65). For full details see R. Desmond 1994. For the Victorian fern craze, see Allen 1969.

36. J. D. Hooker to James Hector, [after 23 December 1868], in Yaldwyn et al. 1998: 107. Wherever dates are given in brackets, the original was undated and the date has been deduced from other evidence.

37. Barton 1998: 426–27.

38. Some of the meanings of "philosophical" are discussed in Rehbock 1983, but as Adrian Desmond (1989) has shown, different audiences had local and incompatible definitions of the term.

39. Sloan 1996.

40. J. D. Hooker to James Hector, [after 23 December 1868], in Yaldwyn et al. 1998: 107.

41. Bentham 1865; Babington 1843.

42. Lindley 1862.

43. J. D. Hooker 1870: vi–vii. Lamarck coined the term "key" to describe his analytical tables (Spary 2000: 80–81).

44. J. D. Hooker 1859: xxv.

45. J. D. Hooker 1870: v–viii.

46. "Macmillan and Co.'s New Books," *Publisher's Circular* 33, no. 785 (1 June 1870): 335.

47. My thinking about these topics has been deeply influenced by the work of Bruno Latour, Michel Callon, and some of those who have put what is known as actor-network theory to work on historical topics. See, e.g., Latour 1987, 1988; Callon 1985; Star et al. 1989; Stemerding 1991; Miller 1996.

48. E.g., "there is no science which demands more minute accuracy of observation, more patient research, or a more constant exercise of the reasoning faculties, than that of Botany" (Lindley 1830: xv). Arthur Henfrey (1849: 240) made similar comments.

49. J. D. Hooker and Thomson 1855: 20 (emphasis added).

50. J. D. Hooker to C. Darwin, 14 September 1845. See also J. D. Hooker to C. Darwin, [4–9 September 1845]; C. Darwin to J. D. Hooker, [10 September 1845]. This correspondence can be found in Burkhardt et al. 1987: 250, 53, 54.

51. Burkhardt et al. 1987: 253n4. Darwin would also have known that the gentlemen of the Geological Society stressed firsthand knowledge gained through travel and fieldwork as the only real geological education and boasted explicitly of their lack of formal, book-learned geological education (R. Porter 1978: 818–21). For Darwin's barnacles, see Browne 1995: 470–86; Stott 2003a.

52. J. D. Hooker to C. Darwin, [25 March 1846], in Burkhardt et al. 1987: 304.

53. C. G. Ehrenberg to C. Darwin, 11 March 1846, in Burkhardt et al. 1987: 298–99, 393–94; C. Darwin to J. D. Hooker, [24 March 1846], in Burkhardt et al. 1987: 302.

54. J. D. Hooker and Thomson 1855: 13.

55. For the economic value of plants at this time, see Lindley 1838; Brockway 1979; Crosby 1986; Grove 1995; R. Desmond 1995: 191; Cain 1999; Drayton 2000.

56. J. D. Hooker 1859: ii.

57. [Brewster] 1833: 40–41.

58. Hooker referred to including "notes on anatomical physiological etc points" in the letter previously quoted (J. D. Hooker to James Hector, [after 23 December 1868], in Yaldwyn et al. 1998: 107), but in the preface to the first edition of the *Student's Flora* he explained that these had been omitted because "the number of such observations was so great" and their value so often disputed (1870: viii).

59. J. D. Hooker 1853: xvi.

60. J. D. Hooker and Thomson 1855: 9.

61. C. Darwin to A. Gray, 24 August [1856], in Burkhardt et al. 1990: 208.

62. J. D. Hooker to W. J. Hooker, 18 May 1843, in L. Huxley 1918a: 165.

63. J. D. Hooker to W. J. Hooker, 18 May 1843, in L. Huxley 1918a: 165; Fielding 1775: 2.329.

64. W. J. Hooker to W. Baxter, 12 April 1830, RBGS (MRC).

65. J. D. Hooker to G. Bentham, [1853], KEW (JDH/2/3). Part of this letter is also quoted in L. Huxley 1918a: 473. Hooker's reference is to the phrase *rara avis in terra, nigroque similima cygno* ("A rare bird on earth, like a black swan"; Juvenal, *Satires*, 6).

66. J. D. Hooker to Asa Gray, 26 August 1858, in Stevens 1997b: 126.

67. J. D. Hooker to C. Darwin, [11 April 1857], in Burkhardt et al. 1990: 369.

68. Hooker mentioned the eight thousand species in his introduction (1859: xxx).

69. J. D. Hooker to C. Darwin, 8 June 1860, in Burkhardt et al. 1993: 249. Of course, Harvey, being Irish, was also in some senses a colonial botanist; his comments about Gunn sometimes suggest he was anxious to have Hooker acknowledge that Dublin was also a metropolis, or at least not as peripheral as Tasmania.

70. Ruth Barton (2003) argues persuasively that the categories of amateur and professional were of comparatively little importance to the Victorian scientific community, not least because of the need to foster unity among its diverse members.

71. J. D. Hooker to T. H. Huxley, [17 April 1861]; quoted in Barton 1998: 427–28.

72. For example, Frank Turner is one who sees Hooker as part of a self-conscious "young guard" of professionalizers, whose Darwinian ideology and professional goals brought them into conflict with "the religiously minded sector of the pre-professional scientific community" (1993: 183–84). However, I think this is an argument that confuses Hooker's concerns with those of Thomas Huxley. For a subtle analysis of Hooker's and Huxley's differences, see Barton 1998: 434–37; White 2003: 109–10.

73. As T. W. Heyck (1982: 14–15) argues, using actors' categories helps us realize that the early and mid-Victorian terminology of "men of letters" or "cultivators of science" was not synonymous with later ones. For more recent and comprehensive analyses of the range of terms that were in use, see Barton 2003; White 2003.

74. J. D. Hooker 1851–53: 1.

75. David Allen comments on George Bentham's use of the term "amateur" to describe the audience for his *Handbook of the British Flora* (1858) and notes that Bentham also used "professed" as the alternative, not "professional" (2003: 224, 34).

76. J. D. Hooker and Thomson 1855: 3.

77. Corfield 2000: 19; Eliot 1866 (1995): 150.

78. I am indebted to Dorinda Outram (1984: 8–9) for her analysis of the importance and contingent nature of the formation of a vocation in Cuvier's career.

79. Ann Shteir (2003: 238) has noted the fluid nature of the categories of professional and amateur in the late 1850s and early 1860s. See also Barton 2003: 80–81.

80. [Owen] 1860: 487–88; originally anonymous but attributed to Owen by David Hull (1973: 171–215). Other examples of the usage include Rees 1819: Botany; [Lindley] 1833b; [Leifchild] 1859: 537.

81. [Day] 1839: 134–35; Corfield 2000: 19–20.

82. Corfield 2000: 48–49, 57, 63.

83. John Tosh (1999) comments on the increasing middle-class use of "calling" to mean an occupation, profession, or trade, which allowed some of the status of a vocation to be applied to any means of earning a living.

84. Barton 2003: 90.

85. Forbes 1843: 3.

86. After serving as naturalist aboard HMS *Beacon* (1841–42), Forbes became professor of botany at King's College, London, in 1842 and custodian of the Geological Society's collections (thanks, in part, to Lyell's patronage); he was appointed to the Geological Survey in 1844 (Browne 1983: 114).

87. Huxley to J. D. Hooker, 19 July 1859; quoted in F. M. Turner 1993: 183–84.

88. F. M. Turner 1993: 184–85.

89. For some challenges to the teleological assumption, see Morrell 1996; Allen 2001a; A. Desmond 2001; Barton 2003: 74–75. A similar critique of professionalization is provided by Roy Porter's work on the Geological Survey, which stresses the ways the Survey reinforced rather than subverted gentlemanly geology, since it paid people to do what they wanted to do anyway. Porter (1978: 809) also notes that the dangers of the professionalization model becoming teleological had already been recognized at the time he was writing, notably by Nathan Reingold, Jack Morrell, and Susan Cannon. Roy MacLeod (1983: 55–56, 78–79) has offered similar criticisms, notably that earlier accounts have perceived institutional and related developments through the lens of an account of professionalization that is "at best oversimple, and at worst untenable." See also Barton 1998, 2003; A. Desmond 2001; Morus 2005; White 2003.

90. Allen 2001a: 8–9.

91. Bellon 2001: 51–56.

92. Carruthers 1872a: 451.

93. De la Beche had asked W. J. Hooker to recommend a botanist for the Geological Survey and Hooker had suggested his son. Apart from a single project in the 1880s, Hooker never worked on fossil botany again (L. Huxley 1918a: 207).

94. L. Huxley 1918a: 207–8, 12–14. It is not clear whether the position at the Survey would have been continued had Hooker not decided to go to India. De la Beche was apparently enthusiastic about retaining his services, but no formal application ever appears to have been made for the funds that this would have required.

95. Outram observes that Cuvier's Muséum d'Histoire Naturelle did not have the kind of stability and continuity that might now be thought of as defining an institution. Clear and continuing policies (of any kind) were thus also rare (Outram 1984: 170–71).

Emma Spary (2000: 3) has made similar points about the changing roles of the Jardin des Plantes.

96. Drayton 2000: 154–56.

97. Nevertheless, Paris's Muséum was a respectable institution because it had evolved directly from its ancien régime predecessor. In 1836, it was taken as an explicit model for the reform of the British Museum, and four years later, the British naturalist William Swainson described the French institution as the world's foremost museum (Limoges 1980: 211–13). See Drayton 2000: 130–69, for an excellent discussion of the political and economic complexities at work in the making of the new Kew.

98. Rupke 1994: 1–11.

99. McOuat 2001: 5.

100. Quoted in Briggs 1990: 64.

101. Dickens 1838: 61–62.

102. White 2003: 33, 48.

103. J. D. Hooker to C. Darwin, [13 May 1863], in Burkhardt et al. 1999: 413.

104. J. D. Hooker to H. W. Bates, 13 May 1863, APS (JDH).

105. For the ideal of the gentleman, see Smiles 1859 (2002); St. George 1993; Morgan 1994.

CHAPTER ONE

1. Hooker's reminiscences are quoted in F. Darwin 1888: 19. He does not give the date, but a letter to his father written in mid-July recorded that he had just returned from London (J. D. Hooker to W. J. Hooker, 13 July 1839, in L. Huxley 1918a: 44). Darwin himself remembered this first meeting as having been in Park Street, off Hyde Park, not Trafalgar Square. There is also evidence that they may have met earlier; see C. Darwin to J. D. Hooker, [29 July 1865], in Burkhardt et al. 2002: 210–11n22.

2. J. D. Hooker, quoted in F. Darwin 1888: 20.

3. For Humboldt, see Browne 1983; M. Nicolson 1987; Pratt 1992; Dettelbach 1996, 2005.

4. Cawood 1979: 494, 507–11.

5. Cawood 1979: 502, 505.

6. J. D. Hooker to C. Darwin, [26 February 1854], in Burkhardt et al. 1989: 178.

7. J. D. Hooker to W. J. Hooker, 27 April 1839, in L. Huxley 1918a: 41.

8. J. D. Hooker to W. J. Hooker, 27 April 1839, in L. Huxley 1918a: 42–43.

9. J. D. Hooker to W. J. Hooker, 3 February 1840, in L. Huxley 1918a: 163.

10. J. D. Hooker to W. J. Hooker, [May/June?] 1839, in L. Huxley 1918a: 44.

11. J. D. Hooker to W. J. Hooker, 3 February 1840, in L. Huxley 1918a: 163.

12. [Forbes] 1852: 397. For the importance of artisan naturalists, see A. Secord 1994a, 1994b, 1996.

13. I.e., the geologists William Buckland, Charles Lyell, and Adam Sedgwick and the philosopher and astronomer John Herschel. Richard Taylor (1781–1858) was a printer, naturalist, and undersecretary of the Linnean Society (Brock 2004), not to be confused with the Reverend Richard Taylor, the New Zealand missionary. The "Hutton" was almost certainly William Hutton (1797–1860), geologist and paleontologist. He was an honorary

curator of mineralogy and geology of the Natural History Society of Northumberland, Durham and Newcastle upon Tyne and vice president of the Newcastle Literary, Scientific and Mechanical Institution (A. Newman 2004).

14. J. D. Hooker to Dawson Turner, 8 October 1838, KEW (JDH/2/3/13); Yeo 1981 (2001): 73; MacLeod 1983: 76.

15. Cawood 1979: 507–8.

16. J. Herschel to Grey, 27 December 1841, ACCL (GL21).

17. James D. Forbes explicitly described the role of the BAAS annual reports as classifying existing discoveries (and, by implication, the disciplines that produced them). Quoted in Yeo 1981 (2001): 75; MacLeod 1971: 325.

18. J. D. Hooker to Dawson Turner, 8 October 1838, KEW (JDH/2/3/13).

19. Of the £24,848.86 distributed, natural history received £1,455.25 and botany received only £259.24.

20. I.e., £2,016.47 and £9,575.6 respectively. Figures are from the official reports of the BAAS for 1833–44. Expenditure on botany has been calculated by adding all funds given for research purely on plants together with half of any amounts given for work on plants and animals—a procedure which probably overestimates botanical expenditure. See also Morrell et al. 1981: 273. It might be argued that natural history was simply less expensive than some other sciences, but further evidence of relative statuses is given by Morrell and Thackray in a table of the various BAAS lobbies of government (1831–44). There were thirty in all: three were geological, two zoological, and none botanical; Leonard Jenyns's 1834 suggestion that the BAAS lobby for greater support for natural history voyages was one of those ignored (Morrell et al. 1981: 327, 42–43).

21. Cawood 1979: 517.

22. Morrell et al. 1981: 524; L. Huxley 1918a: 170, 46–48. John Tosh (1999: 12) estimates that an income in the range £100–£300 per year (such as Hooker's assistant surgeon's pay) placed one at the bottom edge of the middle class; over £300 was needed for a proper middle class existence, while £1,000 represented serious prosperity. By way of contrast with Hooker's income, Richard Owen's salary at Hunterian rose from £120 in 1827 to £300 in 1833 and to £500 by 1847 (Rupke 1994: 19).

23. Summaries are compiled from the official reports of the BAAS (1839–46). The brief reports of the association's specialized sections show a similar bias, with Section A receiving the most space and the reports of Section D being dominated by zoology. A similar pattern can be discerned in the Royal Society's priorities, which showed a strong bias toward the experimental sciences (MacLeod 1971: 329–30).

24. See Marchand 1971; Shattock 1989.

25. [Brewster] 1833.

26. [Brewster] 1833: 33. All emphases in quotations are as in the original sources unless otherwise stated.

27. Among the common motivations for natural history collecting noted by Anne Larsen (1996: 361) are "evangelical fervour"; the Romantic view of nature; healthy, rational exercise; and various kinds of acquisitiveness.

28. There is a great deal more that could be said on the gender aspects of botany during this period, but that topic is beyond the scope of this book. For a more detailed discussion, see Shteir 1996, 1997a, 1997b; Schiebinger 1996; Gates 1998.

29. Babbage's *Reflections on the Decline of Science in England, and on Some of Its Causes* (1830) claimed that British savants were falling behind their European competitors because there was no "profession" of science (and thus few rewards for its cultivators) (Heyck 1982: 50–51). For Brewster as a reviewer, see Brock 1984.

30. Morrell et al. 1981: 41–47; Morus 2005: 32–39.

31. [Lindley] 1834c: 402. Lindley's own position as professor of botany at University College London was also badly paid, and Edward Forbes also found the salary of the botanical chair at King's College inadequate; see Allen 1994: 75.

32. [Lindley] 1831: 262.

33. [Brewster] 1833: 69.

34. See Endersby 2000 for one example of a colonial garden and its relations with Kew.

35. Richard Yeo (1993: 78) notes that unsigned articles in review journals added to the intrigue of the articles (although the reviewers' identities would have been known to insiders) and to the impression that the journal itself was taking a position, especially at a time when journals were so clearly associated with particular political views.

36. Drayton 1993, 2000: 168–69.

37. An anonymous work, *Thoughts on the Degradation of Science in England* (1847), suspected to be by Babbage, renewed the declinist debate during the late 1840s, albeit in more muted tones (MacLeod 1983: 73–74, 1971: 326).

38. [Lankester] 1846b: 704. Adrian Desmond (1982: 19–55) also stresses the lack of material rewards for science in Britain right up to the 1850s.

39. [Lankester] 1857.

40. Cawood 1979; Morrell et al. 1981: 275–369.

41. Morrison-Low 2004.

42. [Brewster] 1833: 69.

43. [Brewster] 1833: 33.

44. Murchison 1846: xxxiii. Murchison's reference was to Forbes 1845. Given the approval of Murchison and others, it is not surprising that Forbes was the only naturalist to receive the association's financial support for one of his publications (Morrell et al. 1981: 342–43).

45. Whewell, BAAS *Report* (1841): xxxiii; quoted in Morrell et al. 1981: 273.

46. Murchison 1846: xxxv.

47. Stafford 1989 (2002): 19.

48. L. Huxley 1918a: 207; J. A. Secord 1986a.

49. Secord 1986b, 1997; Rudwick 1985; Stafford 1989 (2002).

50. Cawood 1979: 493.

51. Whewell, *History of the Inductive Sciences*; quoted in Morrell et al. 1981: 353–54. See also Heyck 1982: 60–61; Cawood 1979: 518.

52. J. D. Hooker to Dawson Turner, [September] 1839, in L. Huxley 1918a: 46.

53. L. Huxley 1918a: 46–48.

54. L. Huxley 1918a: 54–55.

55. L. Huxley 1918a: 6.

56. "Cryptogam" was Linnaeus's term for the flowerless plants (including fungi,

which are now classified outside the plant kingdom). Plants with flowers were known in the nineteenth century as phaenogams or phaenogamic plants.

57. L. Huxley 1918a: 76.

58. J. D. Hooker 1844–47: 209–10.

59. The closest continent to Kerguelen's Land is, in fact, Antarctica, but—as previously mentioned—the existence of the Antarctic continent was only suspected prior to the Ross expedition.

60. J. D. Hooker 1844–47: 239. The name "Pringlea" comes from that of Sir John Pringle (1707–82), author of a work on scurvy.

61. J. D. Hooker, *Antarctic Journal*; quoted in L. Huxley 1918a: 77.

62. C. Darwin to J. D. Hooker, [13 or 20 November 1843], in Burkhardt et al. 1986: 408.

63. J. D. Hooker to C. Darwin, 29 January 1844, in Burkhardt et al. 1987: 7.

64. C. Darwin to J. D. Hooker, [11 January 1844], in Burkhardt et al. 1987: 2.

65. J. D. Hooker to C. Darwin, 29 January 1844, in Burkhardt et al. 1987: 7.

66. J. D. Hooker to W. J. Hooker, 3 February 1840, in L. Huxley 1918a: 68–69.

67. L. Huxley 1918a: 154–55.

68. J. D. Hooker to W. J. Hooker, 16 August 1840, KEW (JDH/1/3).

69. L. Huxley 1918a: 44.

70. Gunn (1808–81) was born in the Cape Colony and immigrated to Van Diemen's Land in 1830 (Baulch 1961: xiii–xiv; R. Desmond 1994). See also Maiden 1909; Barker et al. 1990; A. M. Buchanan 1990.

71. Baulch 1961: xv.

72. Colenso (1811–99) was a Cornishman and first cousin to the more-celebrated John Colenso, bishop of Natal (Mackay 1990: 87–89). See also Bagnall et al. 1948; A. Buchanan 1990; Baulch 1961; Glenn 1950.

73. The surviving records of Anderson's College show Hooker attending just one class, but they are extremely patchy and it is not possible to be certain which other classes (if any) he attended. However, as he does not appear to have taken any university medical classes beyond Jeffray's anatomy class, he must have attended the extramural colleges in order to qualify for his degree; AUR (SR); UGR (JAS).

74. The Swiss botanist Alphonse de Candolle, whose work Joseph Hooker knew and admired (see chapter 8), was also a pioneer of statistics and his ideas may also have influenced Hooker's (T. M. Porter 1986: 28–29).

75. Dow et al. 1988: 249; Hamilton 1981: 156. For the university's resistance to such reforms, see C. Smith 1998: 121–22; Crowther et al. 1988.

76. Forbes 1843: 9–10.

77. The key figures were the following. James Adair Lawrie was professor of surgery at Anderson's College, Glasgow, from 1829, a member of the Royal Botanic Institution, and a pioneer of the use of ether as an anesthetic. Robert Perry was a surgeon and physician who worked at Glasgow Royal Infirmary (GRI) for thirty years and was a member of the faculty from 1812. After a study of four thousand cases, Perry was the first to distinguish typhus from typhoid fever; his work was published in the *Edinburgh Medical Journal* in 1836 and was first read at the Glasgow Medical Society. In 1844 he published *Facts and Observations on the Sanitary State of Glasgow*, which linked crime, poverty,

and disease. Richard Millar was lecturer in *materia medica* and professor from 1831. His *Clinical Lectures on the Contagious Typhus* was published in 1833 and included detailed mortality statistics from tabulated medical cases at the GRI in 1828. For Andrew Buchanan, Robert Watt, and Robert Freer, see Geyer-Kordesch et al. 1999: 315–18. Dr. James Cleland was superintendent of public works for the Glasgow Corporation from 1817 to 1826. He published a *Description of the City of Glasgow; Comprising an Account of Its Ancient and Modern History, Its Trade, Manufactures, Commerce, Health, and Other Concerns,* which used statistical methods to argue the case for sanitary reform (Hamilton 1981: 181–82, 84, 201).

78. J. D. Hooker to W. J. Hooker, 5–8 April 1842, KEW (JDH/1/3). George Walker-Arnott eventually succeeded to the Glasgow chair in 1845 (R. Desmond 1994). It may also be relevant that Glasgow in the 1830s was particularly socially divided, with new and old money and numerous other subgroupings vying for preeminence and mistrustful of each other's social pretensions (McCord 1991: 102).

79. J. D. Hooker to W. J. Hooker, 5–8 April 1842: KEW (JDH/1/3).

80. For a full list of Hooker's published works, see appendix B to the *Life and Letters* (L. Huxley 1918b: 486–506).

81. W. B. Turrill (1953: 47) notes that, despite the use of the plural pronoun throughout, both the internal evidence of style and Hooker's correspondence at the time of writing make it clear that he was almost solely responsible for the essay.

82. R. Desmond 1994.

83. J. D. Hooker to W. J. Hooker, 24 August 1841, in L. Huxley 1918a: 113–14. George Gardner (1812–49) was a former student of William Hooker's who collected plants for subscribers in Brazil in 1836–41 and sold his vast collection on his return home. He became superintendent of Peradeniya botanic gardens, Ceylon, in 1844, where he remained until his death. I discuss other specimen sellers, such as James Drummond and Augustus Oldfield, in later chapters.

84. J. D. Hooker to W. J. Hooker, 18 May 1843, in L. Huxley 1918a: 165–66. Gunn tried to tempt him back to the colony by telling him he could earn £200 a year, plus a house, for being secretary to the Horticultural and Botanical Society in Hobart; Gunn to J. D. Hooker, 20 August 1844, KEW (KDC218).

85. J. D. Hooker to Bentham, 3 September 1851, KEW (JDH/2/3). The "Gray" in question was J. E. Gray, who worked at the British Museum.

86. Dorinda Outram (1984: 7) argues that friends and family were also central to George Cuvier's patronage networks.

87. Hooker discussed his publication plans in letters to his father: J. D. Hooker to W. J. Hooker, 1–17 March 1840 and 25 November 1842, in L. Huxley 1918a: 82.

88. J. D. Hooker and Thomson 1855: 37.

89. For John Stevens Henslow's life and work, see Walters et al. 2001.

CHAPTER TWO

1. For collecting in general, see Pearce 1994; Pearce et al. 2000–2002. For natural history collecting, see Allen 1969, 1996; Keeney 1992; Secord 1994a; Larsen 1996; Camerini 1997.

2. The most obvious example of the "passive periphery" problem is Basalla 1967, but more recent works sometimes draw surprisingly similar conclusions, e.g., Moyal 1976: 4; Sheets-Pyenson 1988: 15. For an overview of the recent debates over imperialism and science, see Drayton 1999, 2000; Driver and Martins 2005.

3. This is touched on in Pearce 1994: 13.

4. Polanyi 1967; Kuhn 1970 (1975); Latour et al. 1979; Golinski 1998 (2005).

5. For Brown's life and work, see Mabberley 1985.

6. L. Huxley 1918a: 64.

7. William Hooker told a friend that "Joseph has quite won Brown's heart by bringing him some Van Diemen's Land plants which the boy had been studying with considerable attention" (W. J. Hooker to W. Burnett, 18 January 1839, in L. Huxley 1918a: 39–40). In old age, Joseph Hooker recalled that as a youth, still living in his father's home, "I knew Robert Brown, the companion of Flinders intimately" (J. D. Hooker 1887: 7).

8. J. D. Hooker to W. J. Hooker, 7 September 1840, KEW (JDH/1/3).

9. J. D. Hooker to Bentham, 27 April 1842, KEW (JDH/2/3).

10. Anne Larsen (1993: 131–35) discusses the need for practical firsthand experience.

11. James Britten, "Flowering Plants and Ferns," in Britten 1896: 117. James Britten (1846–1924), botanist and Catholic propagandist, worked at the British Museum and edited the *Journal of Botany, British and Foreign* from 1879 until his death in 1924 (Stearn 2004).

12. Larsen (1993: 294–96) notes the importance of such correspondence as a source of information about activities usually not recorded.

13. The Lyell and Hooker families were old friends, and the geologist had been one of Joseph's early heroes. See, e.g., L. Huxley 1918a: 24–25, 66, 136.

14. For gifts in natural history, see Barber 1980: 39; Secord 1994b. My analysis of gift-giving in natural history owes a good deal to the work of anthropologists who have studied the phenomenon in various non-European societies, notably Strathern 1988; Thomas 1991, 1994, 1997.

15. J. D. Hooker, KEW (JDH/1/1): 1. Also quoted in L. Huxley 1918a: 46–47. According to Leonard Huxley, all Hooker's additional equipment was provided by his father. "Vascula" is the plural of "vasculum," a metal collecting box in which freshly picked plants were kept until they could be pressed. For the history of the vasculum, see Allen 1959, 1965.

16. A quire is 25 sheets, or one-twentieth of a ream (500 sheets).

17. William Purdie, list of articles, [n.d.], KEW (KCL/11); William Milne, equipment list, 17 April 1852, KEW (KCL/9/1). For more details on Purdie and Milne, see R. Desmond 1994.

18. Ward was a London physician and examiner in botany to the Society of Apothecaries (R. Desmond 1994). He is usually described as the "inventor" of the cases, but similar devices were already in use when he made improvements to them; his main contribution was to promote their use (Underwood 2004).

19. For the role of the domestic Wardian case, see Allen 1969. For the wider impact of glass on Victorian gardens, see Carter 1984: 67–91; Elliott 1986: 13–14; Hoyles 1991.

20. Quoted in Ward 1852. See also Hoyles 1991: 176–78.

21. [Lankester] 1853.

22. Agreement between W. J. H. and Purdie, 14 April 1843, KEW (KCL/11).

23. Rowland 1970: 115–20.

24. J. D. Hooker to Grey, 31 May 1868, ACCL (GL21).

25. J. D. Hooker to Grey, 19 December 1865, ACCL (GL21). Similar problems and solutions were discussed with other New Zealand correspondents, such as the Reverend Richard Taylor; see, e.g., J. D. Hooker to Taylor, 20 August 1857, ACCL (TL).

26. J. D. Hooker to Hector, 30 May 1877, in Yaldwyn et al. 1998.

27. Larsen (1993: 226–87) also notes the importance of exchanges in botanical networks. I discuss the point more fully in Endersby 2000.

28. Ward 1852.

29. [Lankester] 1853. Larsen (1993: 49–51) notes the impact of glass taxes on zoological collecting; in the early decades of the century the cost of glass jars in Britain prohibited extensive collections of specimens preserved in alcohol. One result was that British naturalists were unable to use those French classifications that used the characteristics of animals' perishable internal organs. John Lindley successfully campaigned for the abolition of the glass taxes in the *Gardeners' Chronicle* (Stearn 1999: 59).

30. For Purdie and Milne, see n. 17. Hooker's equipment in India included "a small vasculum containing cutting tools"—presumably gardening ones. J. D. Hooker, equipment list, 18 November 1847, KEW (KCL/1/1).

31. J. D. Hooker to Hector, 8 November 1869, in Yaldwyn et al. 1998: 126.

32. J. D. Hooker 1859: ii–iii.

33. Allen 1965: 106.

34. [Forbes] 1852: 395. The term "*vasculiferi*" was coined by Edward Forbes, in a letter to J. H. Balfour, 1843; quoted in Wilson and Geikie 1861 and in Allen 1959: 147.

35. J. D. Hooker to Dawson Turner, 8 October 1838, KEW (JDH/2/3/13).

36. Allen 1994: 96; L. Huxley 1918a: 13. As early as 1818 George Graves, in his *Naturalist's Pocket-Book and Tourist's Companion,* mentioned that "herborizing boxes can be purchased from Messrs. G. & H. Knight, Forster-Lane, London." And by 1856 they were "procurable in any large town" according to S. Thompson's *Wanderings among the Wild Flowers* (1856) (Allen 1959: 145).

37. For more on Cunningham's career see McMinn 1970; Gilbert 1986: 17–74; Endersby 2000.

38. Cunningham to Colenso, 24 October 1838, ATL (CP4).

39. Cunningham to Colenso, 4–11 December 1838, ATL (CP4). The letters in the Turnbull Library are typescript copies and the drawing has been copied by hand; despite numerous inquiries, I have been unable to trace the originals of these letters and would be most grateful for any information about them.

40. Allen 1959: 140.

41. As Allen (1959: 144–45, 1965: 107) has shown, vascula were originally simply candle boxes or sandwich tins (and often reverted to the latter purpose on field trips).

42. J. D. Hooker to W. J. Hooker, [after 17 November 1839], in L. Huxley 1918a: 94. Gunn shared Hooker's concern with protecting specimens from the sun: "I pop the specimens between paper as I gather them—(I carry no vasculum) so that they are pressed as naturally as I believe it is possible"; Gunn to J. D. Hooker, May 1844, ML (RCG).

43. Bentham 1865: xlvi.

44. See also C. W. Short, "Instructions for the Gathering and Preservation of Plants for Herbaria," in his *Letter to a Young Botanist* (1833, Lexington, USA). See Allen 1959: 147, 1965: 106. For Landsborough, see R. Desmond 1994. The "India rubber seed bags" that Purdie took to Jamaica (see above at n. 17) could presumably have served this purpose too, and oiled silk bags, being waterproof, might presumably have doubled as seed carriers.

45. A late-nineteenth-century guide to pressing specimens lists "blotting paper" as especially useful for seaweeds and aquatic plants and recommends "botanists drying paper" as preferable, describing it as "a coarse, spongy, brown felt paper" with "a fine capacity for absorbing moisture." See A. B. Hervey's *A Collectors Guide and an Introduction to the Study of Marine Algae* (Boston, 1881); quoted in DiNito et al. 1999: 141–49.

46. J. D. Hooker took some to India with him (equipment list, 18 November 1847, KEW [KCL/1/1]), and it was recommended by name in William Hooker's "Botany" section in the Admiralty *Manual* (W. J. Hooker 1851: 418). See also Balfour 1849: 617; Allen 1994: 139. Bentall's Botanical Paper also illustrates how specialized papermaking had become. For more on the development of the nineteenth-century paper industry, see Coleman 1958: 206–11; Mandl 1985; Magee 1997.

47. Bentham 1865: xlvii.

48. Prior to 1860, British paper exports were modest, and although Australia (along with Canada and India) was one of the main markets, it was mostly writing and printing papers that were exported (Coleman 1958: 211–12). For examples of the shops built to serve the aquarium craze, see Stott 2003b; Brunner 2003.

49. Harvey to J. D. Hooker, 28 March 1845, KEW (JDH/2/1/11). For more on Drummond, see J. D. Hooker 1859: cxxvi; Barker et al. 1990; E. C. Nelson 1990, 2002; R. Desmond 1994: 217–81; Lines 1996; Home et al. 1998.

50. Drummond to W. J. Hooker, 24 November 1839, KEW (KDC76).

51. Bentham 1865: xlvii.

52. Hamlin 1971: 696.

53. Gunn to J. D. Hooker, 13 March 1844, KEW (KDC218).

54. J. D. Hooker to W. J. Hooker, 7 September 1840, KEW (JDH/1/3).

55. Gunn to J. D. Hooker, May 1844, ML (RCG).

56. Gunn to W. J. Hooker, 9 May 1844, in Burns and Skemp 1961: 98–99.

57. Gunn to J. D. Hooker, 20 August 1844, KEW (KDC218).

58. J. D. Hooker to Gunn, October 1844, ML (GC8). The work done in the herbarium is discussed in chapter 4.

59. J. D. Hooker to Harvey, 17 October 1844, KEW (JDH/2/3/5). Part of this letter is in L. Huxley 1918a: 191.

60. Harvey to J. D. Hooker, 7 December 1844, KEW (JDH/2/1/11).

61. J. D. Hooker to Gunn, October 1844, ML (GC8).

62. J. D. Hooker to Gunn, October 1844, ML (GC8).

63. For example, the outline appears in J. D. Hooker 1864–67 and 1872–97. Bentham's introduction was also used in most of the colonial floras written at Kew in the late nineteenth century (Stevens 2003: 195).

64. Bentham 1865: xlvi.

65. W. J. Hooker to Wilford, [1857], KEW (KCL/13/1). The notes are undated but were written specifically for Wilford, presumably shortly before he set off in 1857; see R. Desmond 1994.

66. J. D. Hooker, collecting instructions for Wilford, [1857], KEW (KCL/13/1).

67. Bentham 1865: xlvi. *Sedum* is a large genus of succulent plants whose fleshy, water-storing leaves are hard to dry; *Orchis* is a genus of the orchid family, which also retain considerable moisture in their leaves and stems.

68. Pyroligneous acid (also called wood vinegar) was made by the destructive distillation of wood; it was once a commercial source of acetic acid, which is its primary constituent.

69. Quoted in L. Huxley 1918a: 47–48.

70. Cunningham to Colenso, 24 October 1838, ATL (CP4).

71. Larsen (1993: 53–54) discusses the destruction of colors and the means zoologists used to overcome this problem.

72. The secondary literature on microscopy and its history is extensive; see G. L. Turner 1980; Ford 1985; Palmer et al. 1971. Gooday 1997 has a useful introductory bibliography.

73. Different microscopes and their uses were explained in Balfour 1849: 612–13.

74. Bentham 1865: xlviii.

75. J. Newman 1845.

76. Steward 1867.

77. Harvey to J. D. Hooker, 18 January 1845, KEW (JDH/2/1/11). Joseph Decaisne was professor of plant cultivation at the Muséum d'histoire naturelle, Paris. Thomas Fleming Bergin was a Dublin engineer who donated plants to Glasnevin Botanic Garden.

78. Bentham 1865: xlviii.

79. Anne Secord (1994a) discusses precisely this point, and I am indebted to her analysis of such botanists' pride in their skills, which informs my own account.

80. Knight to Grey, 28 September 1852, ACCL (GLNZ21). Knight had worked for Grey in South Australia and accompanied him to New Zealand when Grey became governor in 1845. For Knight's career, see Galloway 1990; R. Desmond 1994: 405.

81. Knight to Grey, 28 September 1852, ACCL (GLNZ21). Knight referred to the opticians Smith and Beck, 31 Cornhill, London (later Smith, Beck and Beck).

82. Cunningham to Colenso, 4–11 December 1838, ATL (CP4).

83. Colenso to Cunningham, 12 July 1839, ATL (CP4).

84. Lawrence, the son of a Tasmanian landowner, had initially been asked to collect for Hooker by Thomas Scott, a local merchant whom Hooker had approached (Burns and Skemp 1961: xiv, 18).

85. Gunn to W. J. Hooker, 18 August 1832, in Burns and Skemp 1961: 21–23.

86. Gunn to W. J. Hooker, 1 July 1833, in Burns and Skemp 1961: 30–31.

87. Gunn to W. J. Hooker, 30 September 1844, in Burns and Skemp 1961: 100–101. Even this seems not to have worked, since the following year, he told the widow of his friend Edmund Hobson that he would like to bid for his friend's microscope as long as the price did not go too high; Gunn to Mrs. E. Hobson, 1 May 1848, SLV (RCG).

88. Colenso to J. D. Hooker, 31 December 1884, KEW (JDH/2/1/11). Lady Franklin was the wife of Sir John Franklin (1786–1847), lieutenant governor of Tasmania, to whom

Gunn was briefly secretary. The Hepaticae are more familiarly known as the liverworts, small plants related to mosses.

89. Colenso to J. D. Hooker, 31 December 1884, KEW (JDH/2/1/11). In quotations from manuscript sources, underlining is shown as *italics*, while double-or triple-underlining is indicated in **bold**. John Browning, of 63 Strand, London, described himself as "Optical and Physical Instrument Maker To Her Majesty's Government." His catalog lists prices from £36.10 for an "Extra Large Model Microscope" to £50.0 for "The above Instrument, with Stephenson's binocular arrangement, for use with high powers," whereas "Houston's Botanical Dissecting Microscope, in case, complete," was just £0.6.6 (Browning 1881).

90. "Mitten" was William Mitten (1819–1906), who published extensively on mosses and was a member of the Botanical Society of London. "Wilson" was probably William Wilson (1799–1871), a solicitor who described the mosses for J. D. Hooker's *Flora Antarctica* (R. Desmond 1994).

91. Gooday 1997.

92. [Lankester] 1847.

93. J. D. Hooker to T. H. Huxley, 2 December 1869, in L. Huxley 1918a: 430. The naturalist under discussion has not been identified.

94. For "phaenogamic" and "cryptogamic," see n. 56 in chapter 1.

95. J. D. Hooker to Gunn, October 1844, ML (GC8). *Jungermanniae* are a genus of mosses, named after the German botanist Ludwig Jungermann; William Hooker was an acknowledged expert on their classification; see W. J. Hooker 1816.

96. Colenso to W. J. Hooker, 31 January 1853, KEW (KDC74). None of the letters from either J. D. or W. J. Hooker to Colenso have been found.

97. Yaldwyn et al. 1998: 70.

98. Colenso to J. D. Hooker, 24 August 1854, KEW (KDC174).

99. Colenso to W. J. Hooker, 31 January 1853, KEW (KDC74).

100. J. D. Hooker to Gunn, October 1844, ML (GC8).

101. Gunn to W. J. Hooker, 18 August 1832, in Burns and Skemp 1961: 22.

102. Gunn to J. D. Hooker, 13 March 1844, KEW (KDC218).

103. Colenso to W. J. Hooker, 31 January 1853, KEW (KDC74).

104. Colenso to J. D. Hooker, 23 February 1855, KEW (KDC174).

105. J. D. Hooker, *Erebus* Letters and Journal, [1839?]; quoted in L. Huxley 1918a: 46–47.

106. Colenso to J. D. Hooker, 1841, KEW (JDH/2/10). Kurt Sprengel was professor of medicine and botany at Halle, a pioneer of botanical physiology and distribution studies. J. P. Vaucher was the author of *Histoire physiologique des plantes d'Europe* (Paris, 1841) and many other works on cryptogamic botany. "Brown" was, of course, Robert Brown.

107. These were presumably gifts from William Hooker, since they included "some numbers of the Botanical Magazine & Companion," both of which he published; J. D. Hooker to Gunn, 19 August 1840, ML (GC8).

108. William Purdie, list of articles, [n.d.], KEW (KCL/11); William Milne, equipment list, 17 April 1852, KEW (KCL/9/1).

109. Hooker had observed that Don's *Dictionary of Gardening and Botany* was "useful, so far as it goes," but Lindley's works "ought to be in the hand of every student," espe-

cially the *Vegetable Kingdom.* He also recommended works by G. G. Walpers, Alphonse de Candolle's *Prodromus,* and Loudon's *Encyclopaedia of Plants* (W. J. Hooker 1851: 435–36). The Admiralty *Manual* was typical of published lists of collecting instructions, which, as Larsen notes (1993: 39–45, 146, 49), were produced in Britain as early as the seventeenth century.

110. Bentham 1865: xlviii.

111. J. D. Hooker to Wilford, [n.d.], KEW (KCL/13/1).

112. J. D. Hooker to Gunn, October 1844, ML (GC8).

113. Larsen (1993: 33) also notes the importance of standardization as a feature of natural history collecting practices.

114. Stevens 1997a.

CHAPTER THREE

1. For an example, see Endersby 2000.

2. In addition to Bruno Latour's concept of cycles of accumulation (see n. 47 in the introduction, above), E. P. Thompson's (1984, 1991) notion of a moral economy regulating each participant's expectations has also proved invaluable for my analysis.

3. The quantity of surviving correspondence (in terms of both numbers and length of letters and the period over which they corresponded) reinforces the impression given by Hooker's published acknowledgment of the importance of these two men's contributions. The surviving evidence also suggests that they contributed more specimens than any other collectors.

4. The agreement between Purdie and William Hooker specifies that Purdie was to receive a salary of £100 per year, and Kew would pay all his expenses other than food and clothing. In exchange, the gardens were to get all plants "living & dried"; W. J. Hooker to Purdie, 14 April 1843, KEW (KCL/11).

5. W. J. Hooker to unknown, 8 December 1857, KEW (KCL/9/1). A sum of £248.1.8 over approximately four years is recorded in Kew's accounts for 1856; however, the 1858 accounts show a final cost of £436.6.9½; Kew Accounts, 31 March 1856 and 13 December 1858, KEW (KCL/9/1).

6. Untitled accounts, 27 April 1850, KEW (KCL/11). A breakdown of this figure shows that it included fitting out and sundries, £19.6.3; passage to Jamaica, £26.13.4; carriage of plants, etc., £16.15.8; general expenses, £746.15.9½; and wages, £345.16.8; undated, p. 5, KEW (KCL/11).

7. Joseph Hooker's collections, drawings, and notes belonged to the Admiralty, but he collected duplicates of everything, and in any case, they not only gave him access to the collections but paid him to publish them (L. Huxley 1918a: 144–46, 70).

8. William Hooker to W&F, 4 December 1850, KEW (KCL/1/1). See also Drayton 2000: 139.

9. Colenso to W. J. Hooker, 12 February 1840, KEW (KDC73). Larsen (1993: 307–12) notes the importance of factors such as letters of introduction and exchanges of gifts in establishing trust.

10. Gunn to W. J. Hooker, 18 August 1832, in Burns and Skemp 1961.

11. Gunn told Hooker that "you must suppose me a perfect bibliomaniac, but having few other enjoyments but my Books to keep me up to the progress of matters in Europe I like to get as many as my limited means can purchase and the very great generosity of my British friends will furnish"; Gunn to W. J. Hooker, 10 May 1849, in Burns and Skemp 1961: 124.

12. Colenso complained that he had been overcharged and sent the wrong books by London dealers; Colenso to J. D. Hooker, 24 October 1863, KEW (KDC174).

13. J. D. Hooker to Gunn, October 1844, ML (GC8). There are numerous other letters about book buying in the archives, e.g., J. D. Hooker to Gunn, 12 August 1846, ML (GC8); Colenso to J. D. Hooker, 24 August 1854 and 23 February 1855, KEW (KDC174).

14. Gunn to W. J. Hooker, 16 November 1836, in Burns and Skemp 1961: 60.

15. J. D. Hooker to Gunn, 13 May 1844, ML (GC8).

16. Gunn to J. D. Hooker, 14 April 1845, in Burns and Skemp 1961: 111.

17. Colenso to J. D. Hooker, 24 August 1854, KEW (KDC174).

18. Colenso to J. D. Hooker, 24 August 1854, KEW (KDC174).

19. For example, Alfred Russel Wallace depended on local guides; see Camerini 1996: 50–51.

20. Colenso to J. D. Hooker, 6–10 August 1846, KEW (KDC73). For Dieffenbach, *Travels in New Zealand* (1843), see McLean 2006; R. Desmond 1994: 207.

21. Colenso to J. D. Hooker, 24 August 1854, KEW (KDC174).

22. Gunn to W. J. Hooker, 30 March 1835, in Burns and Skemp 1961: 41–43.

23. Bentham 1865: xlvi.

24. I have discussed the importance of seasonal collecting in Endersby 2000.

25. Drummond to W. J. Hooker, 24 November 1839, KEW (KDC76).

26. Lines 1996: 278, 97, 99, 301. For a discussion of colonists' attitudes to the Aborigines, see Clark 1957; Martin 1993; Macintyre 1999.

27. George Caley, who collected in New South Wales for Joseph Banks from 1800 to 1810, is one of a handful of exceptions (Webb 1995: 49, 125). My thanks to Anne Secord for drawing Caley to my attention.

28. Belich 1986.

29. Colenso to W. J. Hooker, 14 February 1840, KEW (KDC73).

30. Mackay 1990; McKenzie 2002.

31. Colenso to J. D. Hooker, 23 November 1869, KEW (KDC174).

32. Colenso told Hooker that he felt safe traveling in the interior because he was well known to the Maori "and I never had any of their Lands"; Colenso to J. D. Hooker, 3 June 1865, KEW (KDC174); Mackay 1990.

33. Colenso commented that even the mistakes in the Maori vocabulary were "not original" to Dieffenbach but were derived "from Polack's book on New Zealand, published a few years before" (1883: 10–11).

34. Taylor 1848: xiii. Taylor arrived in the Bay of Islands in 1839 and ran the missionary school in Waimate North. He purchased Maori land but only so as to allow its traditional owners to return to it. Like Colenso, he was present at the signing of the Treaty of Waitangi. See R. Desmond 1994; Owens 1990.

35. Taylor 1848: xii–xiii.

36. Colenso 1883: 11–12.

37. Colenso to J. D. Hooker, 24 August 1854, KEW (KDC174). Only Maori were referred to as "New Zealanders" at this time.

38. As Colenso proudly claimed, "No white man has tramped over more of N. Zealand ground than myself"; Colenso to J. D. Hooker, 3 February 1852, KEW (BNZ). Gunn made similar comments; see Gunn to W. J. Hooker, November 1845, in Burns and Skemp 1961: 115; Gunn to J. D. Hooker, 21 December 1846, KEW (KDC218).

39. J. D. Hooker to Gunn, October 1844, ML (GC8).

40. As was becoming common for surgeon-naturalists after voyages, Hooker was nominally appointed to one of the royal yachts, without duties, at £136.10 per year, which eventually rose to £200 (L. Huxley 1918a: 170).

41. Gunn to J. D. Hooker, 26 September 1844, KEW (KDC218).

42. Gunn had earlier complained of the visiting naturalist Thomas Short that he collected "by money alone." Buying specimens was evidence that Short was not a gentleman. See Gunn to W. J. Hooker, 31 March 1837, in Burns and Skemp 1961: 62.

43. Gunn to W. J. Hooker, 14 September 1834, in Burns and Skemp 1961: 39–40.

44. Gunn to J. D. Hooker, 13 March 1844, KEW (KDC218).

45. Gunn to J. D. Hooker, 28 May 1845, KEW (KDC218).

46. Gunn to J. D. Hooker, 28 May 1845, KEW (KDC218).

47. Gunn to J. D. Hooker, 13 March 1844, KEW (KDC218).

48. Marilyn Strathern (1988) and Nicholas Thomas (1991) both note the important distinction between gifts that are reciprocated and those that are not; for example, the latter create obligations and thus dependence.

49. Gunn to J. D. Hooker, 13 March 1844, KEW (KDC218).

50. Anne Secord quotes the zoologist Edward Turner Bennett, who said "a man of character is respectable whatever may be his rank in life, and one who collects with a view to Science and not to Profit I should esteem as an Entomologist" (1994b: 384, 93). Maintaining status by avoiding monetary exchange was a long-standing tradition; see Ogilvie 2006: 13–14.

51. The first mention of Gunn in Archer's journal records that in January 1848 he "Visited Penquite and inspected Mr Gunn's specimens of botany and works on the subject: Hooker's 'London Journal of Botany' and 'Icones Plantarum' seem desirable works." See UTA (Journal).

52. Archer had been born in Launceston, Tasmania, but trained in England as an architect, which is presumably how he acquired his artistic skills. See LSL (Orchids); R. Desmond 1994: 20.

53. J. D. Hooker to Ross, 7 September 1847, in L. Huxley 1918a: 217.

54. J. D. Hooker to Harvey, 17 October 1844, KEW (JDH/2/3/5). The same letter is partially transcribed in L. Huxley 1918a: 191. Gunn had sent Hooker £50 for a complete bound edition of *Curtis' Botanical Magazine*; when this was sent, Hooker told him: "The other books we send as presents, thinking we have sweated you enough in Curtis"; J. D. Hooker to Gunn, October 1844, ML (GC8). For other letters about book buying, see n. 13 above.

55. Colenso to J. D. Hooker, 13 September 1862, KEW (KDC174). Colenso became involved in local politics in 1858 and was elected to the General Assembly in 1861.

However, five years later his opposition to Maori land sales led to his being ousted by Donald McLean and the runholders (Mackay 1990; Ward 2001).

56. Archer to J. D. Hooker, 26 July 1854, KEW (KDC218). Gunn was elected to the Launceston seat in Tasmania's Legislative Council in 1855 but soon retired to win the Selby seat in the House of Assembly. Retired from Parliament in 1860, he became deputy commissioner for Crown lands in northern Tasmania (Burns and Skemp 2006: 492–93).

57. J. D. Hooker to Gunn, 29 January 1854, ML (JDH).

58. Cunningham to Colenso, 11 April 1839, ATL (CP4).

59. Colenso to J. D. Hooker, 17 May 1843, KEW (JDH/2/1/4).

60. Mackay 1990.

61. Colenso to J. D. Hooker, 24 August 1854, KEW (KDC174).

62. John Tosh (1999: 11–26) argues that failure to maintain a respectable, well-run, disciplined house (with an obedient wife and with children who went on to successful careers) was a major disgrace to a Victorian man. In which case, the scandal of Colenso's marital difficulties would have made it impossible for Hooker to mix socially with him had they both been in Britain; the colonial distance had its advantages, since Colenso seems to have successfully concealed many aspects of his situation from Hooker.

63. Colquhoun 2003. For example, with the help of Sir William Hooker, John Lindley, the son of a nurseryman, became assistant librarian to Sir Joseph Banks and then assistant secretary to the Royal Horticultural Society and was finally appointed as the first professor of botany at University College London. Kew's superintendent, the elder William Aiton, also rose from humble origins to become manager of the royal "physic garden" as a result of patronage, and his son, the younger William Aiton, then inherited the position. Drayton (1993: 195–96, 2000: 139–40) gives other examples. Further examples can also be found in Camerini 1997: 358, 64, 1996: 45–48; Keeney 1992: 1; Payne 1994;

64. Gunn eventually acquired an estate of 107 acres near Launceston, where, in 1856, he built Newstead House, a substantial home in which he lived for the rest of his life (Baulch 1961: xv).

65. Alex Buchanan (1990: 189–90) notes that he found no Gunn collections that date to after 1860, which is roughly when the letters to Hooker cease.

66. Barker et al. 1990: 56–59. Gunn also mentioned that his children helped him collect; Gunn to J. D. Hooker, 8 September 1842, KEW (JDH/2/10).

67. Drummond to W. J. Hooker, 28 May 1839, KEW (KDC76). William Baxter is another example of a plant collector struggling to turn Australian plants into money; see Endersby 2000 for more details.

68. Mangles had heard through his cousin Ellen (the wife of Western Australia's governor) about Molloy's interests in gardening and flowers (Lines 1996: 257–59). See also Stearn 1999: 48.

69. W. J. Hooker to Mueller, 9 April 1854, in Home et al. 1998: 177. For more on Mueller and his career, see Lucas 1995; Stevens 1997a; Moore 1997; Home et al. 1998; Home et al. 2002.

70. Oldfield to W. J. Hooker, 26 December 1855, KEW (KDC74).

71. Obituary of A. F. Oldfield, *Melbourne Argus*, 13 July 1889. Copy at Kew, KEW (KDC218): 239.

72. Oldfield to J. D. Hooker, 21 July 1861, KEW (KDC218). Oldfield seems to have

collected regularly for Mueller; see, e.g., Mueller to W. J. Hooker, 26 May 1861, KEW (KDC76); Mueller to Barlee, 28 February 1869, SROWA; Mueller to Bentham, 5 February 1866, KEW (KDC76).

73. Mueller told Bentham that Oldfield "had to give up his teachership in Tasmania on account of ill health"; Mueller to Bentham, 5 February 1866, KEW (KDC76).

74. Hooker's tribute to Oldfield, quoted in the obituary, had originally been made in the hope of securing him a pension; he asserted that "I can truly say that I know of no case of modest worth of heart, hand, and head more deserving of public recognition by the Governments of Australia than that of Mr. Augustus Oldfield." Obituary of A. F. Oldfield, *Melbourne Argus*, 13 July 1889. Copy at Kew, KEW (KDC218): 239.

75. Gunn to W. J. Hooker, 18 August 1832, in Burns and Skemp 1961: 21–23.

76. Gunn to W. J. Hooker, 15 November 1833, in Burns and Skemp 1961: 31–33.

77. Gaskell 1848 (1996): 113.

78. Cunningham to Colenso, 24 October 1838 and 4–11 December 1838, ATL (CP4).

79. Cunningham to Colenso, 9–17 January 1839, ATL (CP4).

80. Colenso to Cunningham, 1 March 1839, ATL (CP4).

81. Cunningham to Colenso, 11 April 1839, ATL (CP4). Paihia was the settlement where Colenso lived.

82. Colenso to Cunningham, 12 July 1839, ATL (CP4).

83. J. D. Hooker to W. J. Hooker, 7 September 1840, KEW (JDH/1/3).

84. Gunn to J. D. Hooker, 1840, KEW (JDH/2/10).

85. Gunn to J. D. Hooker, 4 July 1841, KEW (JDH/2/10).

86. Gunn to J. D. Hooker, 4 July 1841, KEW (JDH/2/10). A few months later, he regretted that he would be "omitting at present all reference to Bottled ale & other matters not strictly Botanical"; Gunn to J. D. Hooker, 8 September 1841, KEW (JDH/2/10). For Gunn and Hooker's collecting trips, see A. M. Buchanan 1990.

87. J. D. Hooker to Gunn, October 1844, ML (GC8).

88. Gunn to J. D. Hooker, 28 May 1845, KEW (KDC218). Colenso's letters expressed similar delight in hearing from Hooker, e.g., Colenso to J. D. Hooker, 17 May 1843, KEW (JDH/2/1/11); Colenso to J. D. Hooker, 6–10 August 1846, KEW (KDC73); Colenso to J. D. Hooker, 13 September 1890, KEW (JDH/2/1/11).

89. Tosh (1999: 109) argues that strong male-male friendships were so important partly because of the lack of opportunities for friendship with middle-class women (which was exacerbated by the fact that women were largely denied formal education, which often made them dull companions for men).

90. Kingsley 1848 (1994): 156. See also Michie 1999: 416–17; Adams 1999: 130–31.

91. Slinn 1999: 311–12. Tosh (1999: 110) believes that much homosexual feeling may have been sublimated in this ideal of male friendship, despite marriage remaining central to male status.

92. Thomas and Eves (1999: 3) stress the differences between letter writing and face-to-face contact as an essential component of colonial experience, noting that the distance of correspondence allowed relationships to be managed and renegotiated in different ways. See also Browne 1978.

93. Tosh (1999: 127) notes that such holidays gradually died out in the 1880s, when

family holidays became the norm; prior to that, such trips were considered a break from both work and home.

94. There are more than eighty surviving letters, spanning nearly sixty years of correspondence.

95. See n. 9 regarding letters of introduction.

96. Gunn to W. J. Hooker, 16 January 1836, in Burns and Skemp 1961: 51–52. Mrs. Gunn died on 25 June 1836 in Dublin, leaving Gunn with five children to raise; he remarried in 1841 (Burns and Skemp 1961: 63).

97. Gunn to W. J. Hooker, 2 September 1836, 10 November 1836, and 20 November 1836, in Burns and Skemp 1961: 55–56, 57–58, 61.

98. Gunn to W. J. Hooker, 10 November 1836, in Burns and Skemp 1961: 57–58.

99. Gunn to W. J. Hooker, 31 March 1837, in Burns and Skemp 1961: 62–66.

100. Gunn to J. D. Hooker, 29 April 1841, KEW (JDH/2/10).

101. J. D. Hooker to Gunn, 29 January 1854, ML (JDH).

102. Lines 1996: 64, 90.

103. Lines 1996: 243–44, 57.

104. Lines 1996: 280.

105. Gunn to W. J. Hooker, 31 March 1837, in Burns and Skemp 1961: 63; A. M. Buchanan 1990: 191.

106. Gunn to W. J. Hooker, 21 April 1838, in Burns and Skemp 1961: 75. Joseph Hooker briefly acknowledged her as being "among the many zealous collectors" of Tasmanian algae (1859: cxxvi).

107. By examining the labels on the surviving specimens in Gunn's herbarium (now part of the Tasmanian Herbarium in Hobart), Alex Buchanan (1990: 190–91) has identified about twenty people who supplied Gunn with specimens. Colenso, too, had contacts within New Zealand with whom he exchanged specimens and whom he mentioned in numerous letters; e.g., to W. J. Hooker, 14 February 1840, KEW (KDC73); to J. D. Hooker, 3 January 1865 and 30 June 1883, KEW (KDC174).

108. Archer to J. D. Hooker, 14 February 1860, KEW (KDC218). "Müller" was the common nineteenth-century spelling of Ferdinand von Mueller's name.

109. J. D. Hooker 1844–47: xii, 1851–53: iv.

110. Piper 1906: 17; Lyall to J. D. Hooker, 12 September 1858, KEW (KDC218).

111. Lyall to J. D. Hooker, 8 April 1859, KEW (KDC218).

112. W. J. Hooker to David Monro, 21 September 1854, KEW (KDC74).

113. R. Desmond 1994; Wright-St. Clair 2006; W. J. Hooker to D. Monro, 20 March 1853; D. Monro to W. J. Hooker, 5 May 1854; W. J. Hooker to D. Monro, 21 September 1854, KEW (KDC74); J. D. Hooker 1853: iv.

114. A. Secord 1994a.

115. Gunn to J. D. Hooker, 23 May 1841, KEW (JDH/2/10); Gunn to J. D. Hooker, 27 January 1845, KEW (KDC218). Georgiana Molloy was not part of Hooker's network; she corresponded directly only with Mangles.

116. The need for a good "botanical horse" is a reminder that a botanist's tools could be living ones. Gunn's horse was surely a technology, similar to a plant press or a vasculum, but perhaps Drummond's daughter Euphemia, Sapper Buttle, and "the shepherd" were of no greater significance than a horse to someone of Gunn's stature. The same

might easily be true of Gunn himself, when he was viewed from the distant metropolis. Latour's approach (1987) to studying science is a useful reminder that specimens, collecting tools, and horses are all essential characters in the colonial relationship; the collecting network as a whole—people, plants, bottles of beer, and horses—became a tool for Hooker.

117. Although Drummond collected in Western Australia, not Tasmania, Hooker's flora "brought together evidence of nearly 8000 flowering plants having been collected or observed in Australia" in order to assess "the affinities and distribution of all the Tasmanian species" (J. D. Hooker 1859: iii).

118. Thomas and Eves (1999: 1, 5–6) emphasize the importance of thinking about letters as artifacts, not merely as texts, and that absence was the most common occasion for letter writing.

119. Colenso to J. D. Hooker, 13 September 1890, KEW (JDH/2/1/4).

120. Harvey to J. D. Hooker, 2 June 1845, KEW (JDH/2/1/11).

121. For examples of such a view of colonial science, see essays in Reingold et al. 1987; MacLeod et al. 1988; MacLeod et al. 1994; Home et al. 1991; Jardine et al. 1996; Miller et al. 1996.

122. Zaheer Baber (1996) is one of a number of recent historians of colonial science who have also argued the need for a more dynamic model of the colonial/metropolitan relationship. See also Osborne 1994: 175.

CHAPTER FOUR

1. See Muir 1971; Wakeman 1973; Fox et al. 1973: 559–82; Anderson 1994: 1–18; J. A. Secord 2000a.

2. I am indebted to Anne Secord (2002) for encouraging me to consider illustration in this way.

3. Lewis 1992: 4–5.

4. J. D. Hooker, quoted in Anon. 1901: 553.

5. W. J. Hooker 1822.

6. W. J. Hooker to W. Hutton, 17 May 1837, APS (WH).

7. Anon. 1901: 553–54.

8. "Elephant folio" was a standard papermakers' term for a large sheet, measuring 28 × 23 inches (711 × 584 mm). By "Elephant folio," however, it is possible Greville may have been referring to a still larger size, known as "double Elephant," which was 40 × 26.5 inches (1016 × 673 mm).

9. Walters et al. 2001: 56; A. Secord 2002: 43–44. Henslow's surviving wall charts are in the Whipple Museum of the History of Science, Cambridge University (http://www.hps.cam.ac.uk/whipple/index.html).

10. Walters et al. 2001: 240–41.

11. J. D. Hooker, collecting instructions for Wilford, [1857], KEW (KCL/13/1); W. J. Hooker 1821.

12. Fitch to J. D. Hooker, 19 June 1845, KEW (JDH/2/1/8).

13. Bermingham 2000: x, 127.

14. Bermingham 2000: 215–17.

15. Ruskin 1857 (1971): 10.

16. Bermingham 2000: 106–9, 15–17.

17. L. Huxley 1918a: 8; Rajnai 1982; A. Secord 2007. A portrait of William Hooker by Cotman is now in the Victoria and Albert Museum, London.

18. L. Huxley 1918a: 4. In 1887, Hooker recorded that these pictures were "now in the Wallace Collection," but while they have several pictures from Turner's collection, the Cotmans are no longer among them.

19. Rajnai 1982: 10, 13; Bermingham 2000: 137–38.

20. Fitch 1869 (1995): 340.

21. Kemp 1996: 202–3.

22. Blunt et al. 1994: 244; Bermingham 2000: 202–3.

23. Barber 1980: 30–31.

24. [Brewster] 1833: 33.

25. The first photographic illustrations for a botanical book were done by a woman, Anna Atkins, who produced *Cyanotypes of British and Foreign Flowering Plants and Flowers* (1854) using a process invented by John Herschel. Anne Pratt published and illustrated numerous serious botanical works, but Jane Webb Loudon was perhaps the best-known female botanist, illustrator, and publisher of the nineteenth century, whose work was directly inspired by Lindley (Bermingham 2000: 212–15).

26. Stapf 1926: 39.

27. Gaskell 1864–66 (1987): 278–79.

28. Shteir 1997a: 237.

29. Anon. 1831: 317–18.

30. Morrell et al. 1981: 149, 50–55.

31. Shteir 1989, 1996, 1997a: 236.

32. Anne Secord (2002: 29) has made a similar point about the pleasure of looking at botanical illustrations, a focus which she argues escapes anachronistic historiographic models that create divisions between producers and consumers of botanical knowledge. She argues that models of exclusion are inadequate for investigating the complex relationships between different practitioners and their varied audiences.

33. J. D. Hooker to Gunn, 13 May 1844, ML (GC8). The "artist" was almost certainly the publisher, Lovell Reeve (who did his own lithography), and not Fitch, since it was only in the early 1850s that Hooker finally accepted that, given the subsidy, it was unreasonable to expect Fitch to work for nothing. For the *Flora Novae-Zelandiae* and *Flora Tasmaniae* Fitch was paid 30 shillings a plate: he had previously been expected to produce as many plates as the Hookers wanted for a fixed annual salary (which was still only £150 in the early 1850s); see Lewis 1992: 15.

34. J. D. Hooker to Gunn, October 1844, ML (GC8).

35. J. D. Hooker to W. Wilson, 28 August 1844, KEW (JDH/2/3/13). I am grateful to Anne Secord for bringing the Wilson letters that I cite in this chapter to my attention; see A. Secord 2002: 33–34.

36. Gunn to W. J. Hooker, 17 March 1849, in Burns and Skemp 1961: 119–21.

37. Gunn to W. J. Hooker, 15 November 1833, in Burns and Skemp 1961: 32.

38. Gunn to J. D. Hooker, 17 October 1845, KEW (KDC218).

39. W. Wilson to W. J. Hooker, 23 June 1831; quoted in A. Secord 2002: 35.

40. A. Secord 2002: 35.

41. Fitch 1869 (1995): 332.

42. J. D. Hooker 1870; Bentham 1865; Lindley 1862. The third rival flora that Hooker discussed, Charles Babington's *Manual of British Botany,* was not illustrated: its "unphilosophical" status was the product of Babington's classification, as I discuss in chapter 5.

43. J. D. Hooker to J. Hector, [after 23 December 1868], in Yaldwyn et al. 1998: 107.

44. Bentham 1865: v.

45. J. D. Hooker 1870: vi–vii.

46. Lewis 1992: 1.

47. Fitch 1869 (1995): 333.

48. Gunn to W. J. Hooker, 31 March 1837, in Burns and Skemp 1961: 62–66. The picture in question appeared as Tab. II in volume 1 of W. J. Hooker's periodical *Icones Plantarum* (1837).

49. Hill, quoted in Saunders 1995: 75–76.

50. Bentham 1865: xlvii-xlviii. The technique was still in use in the twentieth century; see Turrill 1963: 134.

51. Harvey to J. D. Hooker, 27 December 1844, KEW (JDH/2/1/11).

52. Hill, quoted in Saunders 1995: 75–76.

53. Fitch 1869 (1995): 339.

54. In *The Rhododendrons of Sikkim Himalaya,* 2d ed. (1849); quoted in Daston et al. 1992: 96–98.

55. H. C. Watson to W. J. Hooker, 26 April 1844; quoted in Egerton 2003: 98.

56. Lewis 1992: 1.

57. Fitch 1869 (1995): 339.

58. Fitch 1869 (1995): 339.

59. Bentham 1865: v.

60. Stapf 1926: 38.

61. Fitch 1869 (1995): 337.

62. Lewis 1992: 12–13.

63. J. D. Hooker 1855: i. On indigenous drawing traditions, see also Olarte 1993: 85–99.

64. J. D. Hooker 1855: iv.

65. J. D. Hooker, collecting instructions for Wilford, [1857], KEW (KCL/13/1).

66. J. D. Hooker to W. J. Hooker, 17 March 1840, in L. Huxley 1918a: 57. Joseph also mentioned his improved drawings in other letters, e.g., J. D. Hooker to W. J. Hooker, 6 June 1841, in L. Huxley 1918a: 61–62.

67. Lindley 1847: 133.

68. J. D. Hooker, collecting instructions for Wilford, [1857], KEW (KCL/13/1).

69. William Wilson to J. D. Hooker, 27 November 1844, KEW (KDC106).

70. Fitch 1869 (1995): 334.

71. William Wilson to J. D. Hooker, 13 September 1844, KEW (KDC106).

72. J. D. Hooker, collecting instructions for Wilford, [1857], KEW (KCL/13/1).

73. Stapf 1926: 38; Blunt et al. 1994: 265.

74. Daston 2001, 2004; Daston et al. 1992.

75. Robison 1841: 402–3.

76. William Wilson to J. D. Hooker, 13 September 1844, KEW (KDC106).

77. William Wilson to J. D. Hooker, 4 September 1844, KEW (KDC106).

78. Fitch 1869 (1995): 337.

79. Lorraine Daston and Peter Galison (1992: 96–98) argue that the demand for a single, typical form could not finally be reconciled with the demand for accurate depictions, which eventually led to a rejection of the kind of composite drawing Fitch produced.

80. My thinking on this point is indebted to the work of Jonathan Crary, who has noted that the word "observer" comes from the Latin *observare,* "to conform one's actions, to comply with" some set of rules. Crary defines an observer as "one who sees within a prescribed set of possibilities, one who is embedded in a system of conventions and limitations," and argues that these are far more than representational practices (1990: 5–6).

81. Quoted in Bermingham 2000: ix.

82. J. D. Hooker, collecting instructions for Wilford, [1857], KEW (KCL/13/1); J. D. Hooker to Gunn, October 1844, ML (GC8).

83. A. Secord 2002: 38–39.

84. Gill Saunders (1995: 97) notes that "typification" has long been an important function of botanical illustration. The *International Code of Botanical Nomenclature* even allows for an illustration to be designated as the type specimen; see *ICBN* (Vienna Code), chapter II, "Status, Typification, and Priority of Names Article," 9.1 (and art. 37.4) and 9.2; http://www.ibot.sav.sk/karolx/kod/0000Viennatitle.htm.

85. Collingwood 1858: 4–5.

86. Latour 1987: 223–24.

87. Daston et al. 1992: 87–89, 95. This idea of averaging across numerous samples parallels Crary's argument (which in turn draws on Foucault's analysis of the power inherent in statistics) that techniques of observation help to create and enforce a "normal" observing subject (1990: 16).

88. R. Wight, *Madras Journal of Science,* 15 October 1837; quoted in Saunders 1995: 7.

89. For a discussion of how scientific images shaped the understanding of a species, see J. Smith 2006.

90. Patricia Anderson (1994: 1–7) has argued that Victorian printing and pictures in particular were a new forum through which various middle-class groups attempted to exercise intellectual and moral leadership.

CHAPTER FIVE

1. I am indebted to J. B. S. Haldane for this observation, see *My Friend Mr. Leakey.*

2. Hooker's letter has not survived, but fortunately Colenso quoted Hooker's acerbic comment in his reply: Colenso to J. D. Hooker, 24 August 1854, KEW (KDC174).

3. See, e.g., Brockway 1979; Latour 1987; Miller 1996; Stevens 1997a; Drayton 2000; McOuat 2001.

4. Bentham 1865: xlvi.

5. W. J. Hooker to Monro, 4–18 October 1842, KEW (KDC74).

6. Agreement between W. J. Hooker and Purdie, 1843 April 14, KEW (KCL/11).

7. J. D. Hooker to Wilford, undated, KEW (KCL/13/1).

8. J. D. Hooker, collecting instructions for Wilford, [1857], KEW (KCL/13/1).

9. J. D. Hooker to W. H. Harvey, [before 21 June] 1845, KEW (JDH/2/3/5).

10. Lindley 1847: 134–35.

11. J. D. Hooker 1851–53: 5.

12. Allen 1986; Latour 1987: 225; Star et al. 1989: 410–11.

13. Larsen 1993: 331–32.

14. Stevens 1997a: 351.

15. Colenso to J. D. Hooker, 24 August 1854, KEW (KDC174).

16. J. D. Hooker 1870: vi-vii.

17. See J. D. Hooker to Mary Turner, 18 April 1843, in L. Huxley 1918a: 28–29.

18. J. D. Hooker to Bentham, 6 July 1844, KEW (JDH/2/3); J. D. Hooker et al. 1862–83.

19. J. D. Hooker 1851–53: 27–32.

20. J. D. Hooker 1851–53: 1.

21. J. D. Hooker 1851–53: 2.

22. Linnaeus 1751 (2005).

23. J. D. Hooker 1851–53: i, 2.

24. Bentham 1865: xlix.

25. J. D. Hooker to Hector, 4 September 1866, in Yaldwyn et al. 1998: 70. *Lecidae* (now spelled *Lecidia*) are a genus of lichens, while *Confervae* are a group of freshwater algae.

26. Gunn to W. J. Hooker, 15 November 1833, in Burns and Skemp 1961: 31–33. Hooker had sent Gunn Loudon's book at the Gunn's request: Gunn to W. J. Hooker, 1 July 1833, in Burns and Skemp 1961: 31.

27. Stevens 1994: 30–38, 40–41.

28. Stevens 1994: 207–8.

29. Mueller to W. J. Hooker, 5 April 1855, in Home et al. 1998: 215; Stevens 1997a: 350. Joachim Steetz (1804–62) was a lecturer at the Hamburg botanic garden; Conrad Gideon Scuchhardt (1829–92) was director of the botanic garden at Waldau Agricultural College (near Köningsberg). Neither visited Australia. See Home et al. 1998: 556, 59.

30. Gunn to J. D. Hooker, 20 August 1844, KEW (KDC218).

31. Ten years earlier he had noted the deficiencies in Loudon's *Encyclopaedia* and commented that "collecting plants for my Garden and arranging them . . . gives me a much more correct idea of natural affinities between the different genera than any book could give *me*"; Gunn to W. J. Hooker, 14 September 1834, in Burns and Skemp 1961: 39.

32. [Oliver] 1863b: 501. The review was ascribed to Oliver by Darwin in a letter to Hooker: C. Darwin to J. D. Hooker, 19 January 65, in Burkhardt et al. 2002: 29 and n. 7.

33. For more on Mueller and Bentham's relationship and the production of the *Flora*, see Moore 1997; Lucas 2003; and letters in Home et al. 1998, 2002.

34. Colenso to J. D. Hooker, 27 December 1884, KEW (KDC174); "sp. nov." are *species nova*, or new species.

35. Colenso to J. D. Hooker, 3 February 1852, KEW (BNZ).

36. J. D. Hooker 1853: vi–vii.

37. Colenso to J. D. Hooker, 3 February 1852, KEW (BNZ).

38. Colenso to J. D. Hooker, 3 February 1852, KEW (BNZ). The notion that there was a "proper magnitude" for the flora of an island of a particular size and latitude was part of prevailing theories of plant distribution that will be examined in chapter 6.

39. Gunn to J. D. Hooker, May 1844, ML (RCG). He later boasted: "You must enlarge your House to hold all my Collections, but I am most anxious that your Flora Tasmanica should beat all the rest"; Gunn to J. D. Hooker, 21 January 1847, KEW (KDC218).

40. Gunn complained that "half our plants are lost under the huge name 'Nov. Holl.'—I attach much value to our Tasmanian Flora as being that of the Southern extremity as it were of the great Australian Continent"; Gunn to J. D. Hooker, 20 August 1844, KEW (KDC218). "Nov. Holl." was of course "New Holland," the original Dutch name for Australia.

41. He also insisted that, although the ferns he named might "have been found in *other* countries, still they *were new* to this one"; Colenso to J. D. Hooker, 24 August 1854, KEW (KDC174).

42. Gunn to J. D. Hooker, 14 April 1845, in Burns and Skemp 1961: 111.

43. Gunn to J. D. Hooker, 28 May 1845, KEW (KDC218).

44. J. D. Hooker 1851–53: 1.

45. *Official Descriptive and Illustrated Catalogue of the Great Exhibition 1851*, 2:1000–1002.

46. R. Taylor, "(copy) a list of the articles which I sent to the *Royal Exhibition* 1851 May," ACCL (TMS).

47. Examples of economically valuable plant products include cotton, indigo, tea, and cinchona. For the economic value of botany at the time, see Lindley 1838; R. Desmond 1995: 191; Cain 1999; Drayton 2000.

48. Hooker received a copy in 1842, when the *Erebus* arrived in the Falkland Islands (L. Huxley 1918a: 132). Herbarium-making instructions are also to be found in Henfrey 1849: 234–35; and Bentham 1865: xlvii-xlviii.

49. Quoted from the fourth edition (1847), but the same instructions appeared in the first; Lindley 1847: 134.

50. Britten 1896: 133–34.

51. Britten 1896: 133.

52. Quoted from the fourth edition (1847), but the same instructions appeared in the first; Lindley 1847: 134.

53. Britten 1896: 132. Drayton (2000: 263–64) comments that that the size of William Hooker's herbarium sheets (16½ × 10½ inches) became the standard size throughout the Commonwealth.

54. Lindley 1847: 134–35; Bentham 1865: xlvii. By contrast, eighteenth-century herbarium sheets often had a range of related plants artistically laid out on a single sheet.

55. Quoted from the fourth edition (1847), but the same instructions appeared in the first; Lindley 1847: 134.

56. Britten 1896: 135–36.

57. Henfrey 1849: 234–35.

58. Lindley 1847: 134–35.

59. Britten 1896: 137.

60. The number of levels and the terms used to describe them varied from botanist to botanist throughout the period: some recognized Classes, Divisions, Tribes, and Subspecies in addition to the categories I list.

61. When Hooker declined the offer of the Glasgow chair of botany in 1845, his decision would almost certainly have been influenced by the fact that, unlike Edinburgh University, Glasgow had no herbarium; William Hooker had created his own and had taken it with him to Kew. The need to be able to use his father's herbarium was probably decisive in Joseph's decision to remain in London (Bellon 2000: 60).

62. J. D. Hooker to Bentham, 28 May 1852, KEW (JDH/2/3). For Edgeworth, see R. Desmond 1994.

63. J. D. Hooker to Bentham, 28 May 1852, KEW (JDH/2/3).

64. J. D. Hooker to Bentham, April 1853, KEW (JDH/2/3).

65. J. D. Hooker 1862: 278.

66. J. D. Hooker to Bentham, [1845], KEW (JDH/2/3).

67. J. D. Hooker 1851–53: xiii-xiv. *Lomaria procera* is now classified as part of the genus *Blechnum*.

68. J. D. Hooker to H. W. Bates, 26 March 1861, APS (JDH).

69. Colenso to J. D. Hooker, 24 August 1854, KEW (KDC174).

70. Colenso, ML (CGB).

71. J. D. Hooker 1851–53: 27–32.

72. J. D. Hooker to Bentham, [1846], (JDH/2/3).

73. Colenso quoted Hooker's words in his response: Colenso to J. D. Hooker, 24 August 1854, KEW (KDC174).

74. Colenso to J. D. Hooker, 3 February 1852, KEW (BNZ).

75. Hooker told Harvey that "*Erica Markaii* I never thought distinct from *tetralix*," because his herbarium contained "many dried intermediate states"; J. D. Hooker to Harvey, 7 August 1846, KEW (JDH/2/3/5).

76. J. D. Hooker to Darwin, [12 December 1843–11 January 1844], in Burkhardt et al. 1986. See J. D. Hooker 1844–47: 5–6, 232–33.

77. J. D. Hooker to Harvey, 19 February 1859, in L. Huxley 1918a: 470–71. See also Stevens 1997a: 347–48.

78. J. D. Hooker and Thomson 1855: 43 (emphasis added).

79. I am therefore led to dissent from Richard Bellon's (2000: 167–68) assertion that Hooker's "unshakable faith in the value of the general herbarium" was the product of the "two dominant intellectual fonts" in his life: his father's teaching and the ideas of Alexander von Humboldt. While these intellectual influences were undoubtedly important, I have tried to illustrate why I think the material practices of classification mattered even more.

80. Bellon also notes the central importance of the broad species concept for Hooker, but he and I have reached very different conclusions about its implications: Bellon regards Hooker as an advocate of specially created species prior to the mid-1850s, whereas I read him as agnostic on the matter of how species were created while remaining preoccupied with the question of stable species. See Bellon 2006: 1–7.

81. Asa Gray to Darwin, 30 June 1855, in Burkhardt et al. 1989: 362.

82. [Henfrey] 1857: 60, 72–73.

83. [Forbes] 1852: 388–89.

84. Forbes had hoped to become a gentlemanly naturalist, but his income fell short of his ambitions and he was forced to take a wide variety of jobs to earn enough to live on. He supplemented his income from the Geological Survey and as professor of botany at University College London with a great deal of paid writing, such as this piece in the *Westminster* (Mills 1984: 373). He also reviewed extensively for the *Literary Gazette* (Browne 1981: 212–13).

85. Drayton 2000: 148.

86. [Forbes] 1852: 389–90.

87. I would, nevertheless, resist calling this a professionalizing campaign for the reasons outlined in chapter 1: despite the common desire for better funding, there were still important differences within this group.

88. J. D. Hooker 1853: viii. His decision to include the phrase "however they originated or were created" is significant, as I shall argue in a later chapter.

89. J. D. Hooker 1853: xvii.

90. J. D. Hooker 1853: xvi–xvii.

91. Colenso to J. D. Hooker, 24 August 1854, KEW (KDC174). Colenso's quotation comes from "Epistle I: Of the Nature and State of Man, with Respect to the Universe," section VII.

92. J. D. Hooker 1853: xiv.

93. J. D. Hooker and Thomson 1855: 11.

94. J. D. Hooker to Bentham, [?1852], KEW (JDH/2/3).

95. J. D. Hooker to Bentham, 20 September 1854, KEW (JDH/2/3).

96. J. D. Hooker to Bentham, [?1852], KEW (JDH/2/3).

97. J. D. Hooker to Harvey, 19 February 1859, in L. Huxley 1918a: 470–71.

98. J. D. Hooker 1851–53: xiv.

99. For the colonial floras project, see Galloway 1998: 55. See also the anonymous review "New Colonial Floras," probably by Daniel Oliver ([Oliver] 1863b).

100. Latour 1987: 249–52; Schaffer 1997, 1995.

101. However, this subsidy is also a reminder of my earlier point about the unease such colonial patronage would have caused Hooker.

102. J. D. Hooker to Gunn, October 1844, ML (GC8); Galloway 1998: 50.

103. [Oliver] 1863b: 497.

104. Knight to J. D. Hooker, 7 November 1866, KEW (KDC175).

105. Colenso 1883: 32–33.

106. Colenso sent Hooker a copy of "On Nomenclature" the day it appeared; Colenso to J. D. Hooker, 29 October 1883, KEW (KDC174).

107. One weakness of Latour's theory is his tendency to focus too closely on those at the center of his cycles. For example, his description of the astronomer Tycho Brahe as sitting at "*the end* of a long network" is a curious one; a network is precisely a structure that has no end (1987: 227; emphasis added). Latour's choice of words implies that the activities at the center provide the cycle with its real goals, and despite his rhetoric about paying equal attention to all the actants within a network, there is sometimes a degree of "center-centrism" in Latour's work.

108. Michel Callon (1985) argues that some elements within a network are stable

enough to serve as a basis for negotiation, whereas others are labile, allowing certain actors to redefine both the meaning of the elements and thus the actors' roles. In this case, the living plants form a relatively stable resource that can be negotiated over, but the classifying practices are fluid, allowing Hooker to redefine himself and the Kew herbarium as an "obligatory point of passage," thus ensuring that the colonial collectors are enrolled in his philosophical program.

109. Blainey 1966; Chambers 1991; Endersby 2000.

110. C. Darwin to J. D. Hooker, 22 August [1857], in Burkhardt et al. 1990: 443.

111. C. Darwin to J. D. Hooker, 1 August [1857], in Burkhardt et al. 1990: 438. The *Oxford English Dictionary* gives this as the first recorded use of the terms, though clearly they were already familiar enough to be used without explanation.

112. C. Darwin to J. D. Hooker, 23 February [1858], in Burkhardt et al. 1991: 31.

113. J. D. Hooker to C. Darwin, [25] February 1858, in Burkhardt et al. 1991: 35.

114. J. D. Hooker to Bentham, [?1854], KEW (JDH/2/3). Part of this letter is also quoted by L. Huxley (1918a: 473), who gives its date as 1853. However, the letter itself is undated, and as Hooker explicitly says, he is at work on the introductory essay of the *Flora Indica* (published in 1855), and so the letter is more likely to have been written in 1854 or later.

115. As Richard Bellon notes, Hooker's Indian trip was necessitated by his failure to win the Edinburgh chair of botany. Hooker told a friend: "I had a resource in the determination I formed then & there of recruiting my crippled credit by again serving abroad"; J. D. Hooker to Elizabeth [Rigby] Eastlake, 29 July 1848, KEW (IL); also quoted in Bellon 2001: 55.

116. J. D. Hooker to Bentham, 14 September 1851, KEW (JDH/2/3). Johann Friedrich Klotzsch (1805–60) was a mycologist and keeper of the Royal Herbarium in Berlin (L. Huxley 1918a: 25; Brummitt et al. 1992).

117. J. D. Hooker to Asa Gray, 21 June 1854; quoted in Stevens 1997b: 353. William Dunlop Brackenridge (1810–93) was the superintendent of the US National Botanic Garden in Washington.

118. The last recorded collection of specimens was sent in 1898: Colenso to J. D. Hooker, 1 January 1898, KEW (JDH/2/1/11). In 1898, Colenso sent a contribution of £2.2.0 toward the medal the Linnean Society was planning in honor of the man he still referred to as "my very dear friend Sir Joseph D. Hooker"; Colenso to [Secretary, Linn. Soc.], 15 February 1898, LSL (BC).

CHAPTER SIX

1. Hooker's Indian travels are not discussed in detail in this book; an account of them has been given in Bellon 2000. See also Arnold 2005.

2. For the impact of taxonomic instability on the status of botany, see Stevens 1994: 212; Ritvo 1997b: 336–39.

3. Gledhill 2002: 16–17.

4. The generic name *Cardamine* is derived from the Greek word κρδαμον (*kardamon*), which referred to an Indian spice (however, the cardamom pods which we use in cooking are derived from various plants, none of which are *Cardamine* or closely related to it).

5. Linnaeus 1751 (2005): 217.

6. Linnaeus 1751 (2005).

7. Stearn 1971; Schiebinger 1996; Koerner 1999; Drayton 2000.

8. Stevens 1994; Drayton 2000; Gledhill 2002.

9. Stevens 1994; Drayton 2000: 19; Spary 2000. For Adanson, see Zirkle 1946: 107; Farley 1982: 12; Browne 1989: 597; Allen 1994: 35; Stevens 1994. For Antoine-Laurent and the rest of the de Jussieu family, see Stevens 1994. For the importance of the Jardin des plantes as a model for Kew, see Drayton 1993, 2000. For the de Candolles, see Green 1909 (1967): 16–17; G. Nelson 1978; Browne 1983; Rehbock 1983; Stevens 1984, 1994; D. H. Nicolson 1991. For Brown, see Mabberley 1985.

10. Gledhill 2002: 19.

11. Bowler 1992: 164.

12. White 1999; Allen 1994: 40; Drayton 2000: 141–42.

13. [Brewster] 1833: 69.

14. Lindley 1835: vi; see also Stevens 1994: 210.

15. [Brewster] 1833: 68–69.

16. Brown 1830 (1960); Mabberley 1985; Stevens 1994: 100.

17. The claim that the Linnaean system was disused by the 1830s or 1840s has been made by many historians, including Morton 1981: 364–75; Allen 1986: 4–5; Foucault 1989: 125–65; Bowler 1992: 258–60; Larsen 1993: 167; Saunders 1995: 89; Shteir 1997a: 31; Stearn 1999: 34.

18. [Lindley] 1833b.

19. Stapf 1926; Blunt et al. 1994: 213.

20. Anon. 1834: 342.

21. Lindley 1834a: iii-iv.

22. Anon. 1834: 342.

23. Lindley 1830; quoted in [Dixon] 1851.

24. [Dixon] 1851.

25. [Dixon] 1851. The reviewer's contrast between the catalog and the herbarium reinforces the point made in the previous chapter about the importance of the latter as a mark of the professed botanist's status.

26. J. D. Hooker 1851–53: 312.

27. J. D. Hooker 1851–53: 312.

28. Lindley 1847: 124.

29. [Lankester] 1846a: 573.

30. [Luxford] 1850.

31. [Luxford] 1850: 38; R. Desmond 1994: 218. This James Drummond appears not to have been related to the James Drummond who collected plants in Western Australia.

32. Quoted in [Luxford] 1850: 41.

33. [Luxford] 1850: 41, 58.

34. Stevens 1994: 212.

35. [Luxford] 1850: 46.

36. [Luxford] 1850: 46.

37. At various times Newman published and/or edited *The Field*, *The Phytologist*,

The Entomological Magazine, and *The Zoologist*, and he was a founder of the Entomological Society. In 1840 he was expelled from the Religious Society of Friends (Quakers) for marrying "a person who is not of our Society" but was reinstated in 1865 after informing Southwark Monthly Meeting that "he had ever remained through these years of separation a 'Friend' in principle" (*Dictionary of Quaker Biography*). See also R. Desmond 1994: 515. The septenary system was first outlined in Newman's *Sphinx Vespiformis* (1832) and further developed in his *System of Nature* (1843).

38. [Luxford] 1850: 61.

39. [Luxford] 1850: 62.

40. There has, as yet, been no detailed study of quinarianism and its impacts; however, discussions of the system can be found in Winsor 1976; Browne 1979, 1983; di Gregorio 1982; Rehbock 1983; A. Desmond 1985; O'Hara 1991; Ospovat 1995: 101; McOuat 1996, 2001; Ritvo 1997a; Coggon 2002.

41. MacLeay 1825: 63; Ospovat 1995: 107.

42. J. D. Hooker to W. J. Hooker, 7 March 1843, in L. Huxley 1918a: 84. The family's name is now spelled MacLeay, but the alternative spelling McLeay was still common during the early nineteenth century.

43. J. D. Hooker to W. J. Hooker, 7 March 1843, in L. Huxley 1918a: 83–84.

44. J. D. Hooker to W. J. Hooker, 25 November 1842, in L. Huxley 1918a: 132.

45. MacLeay 1825: 48–49. My thanks to Dr. Sachiko Kusukawa for the translation.

46. Ospovat 1995: 107–8.

47. McOuat 1996: 482–85.

48. A. Desmond 1985: 158–62; McOuat 1996: 482, 2001: 16–17.

49. See [Haworth] 1825a: 200–202, 1825b: 105–6, 183–84; MacLeay 1830. See also McOuat 1996: 482–85.

50. Baskerville 1839: v. Thomas Baskerville was a Canterbury doctor (R. Desmond 1994).

51. Baskerville 1839: v–vii, 1.

52. Knight to Grey, 28 September 1852, ACCL (GLNZ21); R. Desmond 1994: 444. "Sowerby" refers to James de Carle Sowerby (1787–1871), eldest son and collaborator of James Sowerby (1757–1822), the artist of the original plates for J. E. Smith's *English Botany* (1790–1814). The book's reputation largely rested on its illustrations and was often known (to Smith's extreme irritation) as "Sowerby's English Botany"). The younger Sowerby retained the original Linnaean classification in the cheap 1847 edition he produced, which was probably the one Lynd used.

53. McMillan 1980. See Swainson 1834: 196 for an example of his promotion of quinarianism.

54. The manuscript was given to J. H. Maiden, then director of the Sydney botanic gardens, by Oldfield's brother sometime after Augustus Oldfield's death on 22 May 1889. A note on the manuscript by Maiden says it was "probably intended for publication" since it appears to be a fair copy, but there is no evidence as to the date it was written beyond the fact that many of its pages bear the paper manufacturer's watermark: "E Towgood 1865."

55. A. F. Oldfield, RBGS (OSM): lvii–lxi.

56. These echoes of Hooker's views are not surprising; Oldfield was familiar with

Hooker's writings and quoted from them in the manuscript; A. F. Oldfield, RBGS (OSM): 23–28, 34–37.

57. For *Naturphilosophie*, see Jardine 1996.

58. A. F. Oldfield, RBGS (OSM): xiii, xxvii–xxviii, lvi–lvii, 87–92, 95–100. Oken's and Hooker's attitudes to *Naturphilosophie* are discussed in chapter 7.

59. A. F. Oldfield, RBGS (OSM): 684–761.

60. [?Oliver] 1863a: 37–38. The review was probably by Daniel Oliver, who was responsible for the journal's coverage of flowering plants.

61. Stevens 1997a: 348.

62. Colenso to J. D. Hooker, 24 August 1854, KEW (KDC174).

63. Colenso to J. D. Hooker, 6–10 August 1846, KEW (KDC73). A fern's caudex is the swollen stem base from which next year's growth begins; "coriaceous" means "leathery"; "sori" is the plural of "sorus," the organs on the underside of fern fronds that produce the spores; an epiphyte grows on another plant but is not parasitic on it.

64. Colenso to J. D. Hooker, 24 August 1854, KEW (KDC174).

65. Bowler 1992: 260.

66. [Bentham] 1856: 518.

67. [Bentham] 1856: 517.

68. Watson 1860: 24. As far as I have been able to discover, "Fanny Ficklemind" was invented by Watson, presumably in a reference to the midcentury fashion for "crazy patchwork." I am grateful to Linda Cluckie for this suggestion.

69. Watson 1860: 24.

70. J. D. Hooker 1851–53: i.

71. [Brewster] 1833: 68–69. For Alexander von Humboldt, see chapter 1.

72. [Brewster] 1833: 68–69.

73. Colenso listed the contents of his botanical library in a letter to William Hooker in 1842. It included several books that used the Linnaean system, e.g., James Rattray's *Botanical Chart; or, Concise Introduction to the Linnaean System of Botany* (Glasgow, c. 1830). See Colenso to W. J. Hooker, 1 December 1842, HBM (CL).

74. L. Huxley 1918a: 22.

75. J. D. Hooker to William Munro, September 1858, in L. Huxley 1918a: 469. For more on Munro, see Chichester 2004.

76. J. D. Hooker 1853: xxvi.

77. J. D. Hooker to A. Gray, 21 September 1853; quoted in D. Porter 1993: 13.

78. J. D. Hooker and Thomson 1855: 8–9.

79. J. D. Hooker and Thomson 1855: 9.

80. J. D. Hooker and Thomson 1855: 35–36.

81. Colenso to J. D. Hooker, 22 January 1883, KEW (JDH/2/1/4).

82. J. D. Hooker to James Hector, 27 January 50, in Yaldwyn et al. 1998.

83. Colenso to J. D. Hooker, 11 September 1865, 22 January 1883, and 15 June 1883, KEW (KDC174).

84. Colenso to J. D. Hooker, 9 April 1864, KEW (KDC174).

85. J. D. Hooker 1853: viii.

86. Strickland 1841: 185–86; McOuat 1996: 494–95.

87. McOuat 1996: 476. Strickland's rules are also discussed in Ritvo 1997b: 340–43.

88. The zoological committee was granted £10 for its work while, in yet another example of the priorities of the BAAS, a rival committee on the nomenclature of stars, led by Herschel and Whewell, received a grant of £32 (McOuat 1996: 506).

89. McOuat 1996: 511. In this, they built on the work of Linnaeus, who argued that "only genuine botanists have the ability to apply names to plants" (1751 [2005]: 169).

90. Strickland, quoted in McOuat 1996: 512. For Strickland, the local naturalists were mainly those in Britain's provinces, who had many features in common with the colonial naturalists.

91. [Seemann] 1863: 327 (emphasis added).

92. Hooker regarded Strickland's rules as useful but limited since they were devised for zoological, rather than botanical, purposes. As he would later write: "With every wish to bind ourselves by the canons (most of which are excellent) laid down by the British Association for nomenclature in Natural History, we have, in common with every botanist who has tried to do so, been obliged to set them aside in many instances" (J. D. Hooker and Thomson 1855: 43).

93. As McOuat (2001) has noted, the British Museum used its power and influence in just this way, to settle destabilizing controversies over classification.

CHAPTER SEVEN

1. Heyck 1982: 27–29, 30–31; Cantor et al. 2004: 20–29.

2. The literature on book history and its relation to the sciences in the Victorian period is large and growing rapidly; useful introductions and overviews include Yeo 1991; Cooter et al. 1994; Anderson 1994; Lightman 1997; Johns 2002; J. A. Secord 2000a; Fyfe 2004; Cantor et al. 2004.

3. Jardine 2000b: 396–98.

4. Johns 2002: 66.

5. [Dixon] 1851: 6, 10, 14–16.

6. Brock et al. 1984: 107.

7. Anon. 1850a.

8. Mabberley 2004.

9. Anon. 1831: 317.

10. Stapf 1926; Brock et al. 1984; Blunt et al. 1994: 211–13.

11. J. D. Hooker to Gunn, 13 May 1844, ML (GC8).

12. W. J. Hooker to Dawson Turner, 1 April 1844, in L. Huxley 1918a: 171.

13. Dance 2004.

14. Lewis 1992: 19.

15. Baulch 1961: xviii.

16. Gunn to J. D. Hooker, 26 September 1844, KEW (KDC218).

17. Colenso to J. D. Hooker, 24 August 1854, KEW (KDC174). The "little V.D.L. publication" was Colenso 1845.

18. Petersen 1966.

19. Hamlin 1971: 696.

20. By the time he died in 1896, Mueller had published 1,400 items, many of them taxonomic. Forty year earlier, Hooker had already been complaining that Mueller's work

had appeared in 30 or 40 different British, German, and Australian periodicals (Stevens 1997a: 354–55). As far as Hooker was concerned, this astonishing output was largely the product of egotism; as he wrote to James Hector, Mueller "is so devoured of vanity and jealousy of Colonial notoriety" that he had "blasted his reputation by insatiable greed of glory & hasty descriptions (& some bad) of endless plants all to catch the priority of nomenclature"; J. D. Hooker to Hector, 28 May 1867, in Yaldwyn et al. 1998: 80.

21. J. D. Hooker to Bentham, 8–9 August 1859, KEW (JDH/2/3).

22. Stevens 1997a: 358–59.

23. J. D. Hooker and Thomson 1855: vi-vii.

24. They first began to devise the rule in the 1860s (Stevens 1991).

25. Bentham 1878; quoted in Stevens 1991: 162. Bentham's original training as a lawyer is apparent from his language, and his attitude to nomenclature may well have been influenced by the prominent role accorded to custom and tradition in British law.

26. Hooker's and Bentham's fellow lumper Asa Gray also followed the Kew Rule in the Harvard Herbarium (Stevens 1991: 157).

27. Stevens 1997a: 355.

28. Colenso to B. D. Jackson, 29 December 1884, LSL (BC).

29. Colenso to B. D. Jackson, 15 June 1885, LSL (BC).

30. Colenso to B. D. Jackson, 15 June 1885, LSL (BC).

31. Lindley 1858: 3, 5.

32. Lindley 1858: 5.

33. Lindley 1858: 8.

34. Colenso 1883.

35. Colenso 1883: 9. He was quoting his own "Essay on the Maori Races" (1865). Colenso may have been the first white New Zealander to use the term *pakeha* to describe himself and his fellow colonists.

36. Colenso 1883: 9.

37. Colenso 1883: 16–21.

38. Colenso 1883: 21.

39. Colenso 1883: 22–23.

40. McKenzie 2002: 202–3.

41. Colenso 1883: 29.

42. Colenso 1883: 30–31.

43. Linnaeus 1751 (2005): 172.

44. J. Lindley (no source given); quoted in Colenso 1883: 30.

45. J. D. Hooker to C. Darwin, [11 April 1857], in Burkhardt et al. 1990: 369.

46. C. J. F. Bunbury to Mrs. C. Lyell, 29 April 1846, in Bunbury 1906a: 234.

47. J. D. Hooker 1851–53: xx-xxi.

48. J. D. Hooker 1853: xx.

49. J. D. Hooker to Bentham, July 1855, KEW (JDH/2/3). Part of this letter is quoted in L. Huxley 1918a: 374.

50. Another reason Hooker's work became more speculative during the 1850s was that the rigid Baconianism of the early decades of the nineteenth century, which eschewed hypotheses, was giving way to views that allowed more scope for speculation and the role of the theorist's imagination (Yeo 1985 [2001]: 267).

51. C. Darwin to J. D. Hooker, 13 June [1850], in Burkhardt et al. 1988: 344.

52. Whewell's copy is in the Trinity College Library, Cambridge, while Darwin's annotated copy is in the Cambridge University Library. In a letter to the botanist William Munro, an expert on Indian grasses, Hooker mentioned that he and Thomson had given away 120 copies of the *Flora Indica* and had another 130 to sell; J. D. Hooker to Munro, 21 December 1856, in L. Huxley 1918a: 358.

53. J. D. Hooker and Thomson 1855: 19.

54. A similar point could be made about the comparisons between the Indian and New Zealand floras that Hooker and Thomson made: the breadth of their experience was a reminder both to the colonial botanists of how inadequate their parochial local knowledge was and to the gentlemen of the BAAS of the need to pay naturalists on voyages (J. D. Hooker and Thomson 1855: 26). Hooker also compared the Australian and Indian floras in the *Flora Tasmaniae* (J. D. Hooker 1860: i, xxviii–xxix, l, liii–liv, ciii).

55. For natural history's rival traditions, see Sloan 1996: 297–301.

56. Rehbock 1983: 10; Yeo 1985 (2001); Smith 1994.

57. J. A. Secord 1997: ix-xvi, 2000b.

58. J. A. Secord 1997: xxi-xxiii. Whewell's review of the *Principles* expressed some reservations, but nevertheless he saw Lyell as a philosopher able to generalize about the vast quantities of facts that the geologists had assembled and thus render them intelligible to the general reader (Yeo 1993: 99).

59. [Seemann] 1867.

60. J. D. Hooker 1859: xxv.

61. J. D. Hooker and Thomson 1855: 12–13.

62. J. D. Hooker and Thomson 1855: 14. Letters between Hooker and Whewell discussing the revisions (written in April and May 1857) are in the archives at Kew and Trinity College, Cambridge. Whewell acknowledged Hooker in print: Whewell 1857. The following year, Hooker told Darwin: "We went to the Whewells at Trinity Lodge for 4 days & my wife enjoyed it much" (J. D. Hooker to Darwin, 22 December 1858, in Burkhardt et al. 1991: 219). Hooker's respect for Whewell is also evident in a letter to Asa Gray in which Hooker commented on how much he liked the philosopher's "self-confident style" of writing (J. D. Hooker to Asa Gray, 29 March 1857, in L. Huxley 1918a: 478).

63. L. Huxley 1918a: 207–8. Hooker's friendship with Forbes would also have strengthened his interest in Whewell's work, as Forbes was strongly influenced by the latter (Rehbock 1983: 69–70). Browne argues persuasively that Darwin, Forbes, Wallace, and Hooker were all directly influenced by Lyell's thinking on these topics, and that his sense of geological history as process inspired them to think of distribution in the same terms rather than as a static pattern (Browne 1983: 104–5).

64. J. D. Hooker 1856b: 248–49. The emphasis on "bold original ideas" as the key to developing laws suggests Whewell's influence.

65. J. D. Hooker 1856b: 249.

66. Whewell 1857: vol. 3, 451. Also quoted in J. A. Secord 1997: xxiii.

67. These groupings are necessarily fluid, not least because key naturalists like Forbes and Darwin worked on both plants and animals; nevertheless, they prove to be

useful analytic tools. While the linguistic fault lines in zoology were rather different, it is interesting to note that Rupke (1994: 161) argues that Richard Owen's work and audiences were effectively divided between the French example of Cuvier's comparative anatomy and the German alternative of Lorenz Oken's *Naturphilosophie*.

68. For Schleiden, see Farley 1982: 48; Lenoir 1989: 114, 95, 242; Chadarevian 1993.

69. M. Nicolson 1987: 169; Kinch 1980: 97; Egerton 2003: 24–26.

70. Egerton 2003: 27–28.

71. [Lindley] 1831: 262. The question mark in parentheses is Lindley's.

72. [Lindley] 1831: 262. The philosophical professors in question were John Stevens Henslow (Cambridge), Robert Graham (Edinburgh), William Hooker (Glasgow), and, of course, Lindley himself (University College London).

73. [Lindley] 1831: 262.

74. [Lankester] 1846a: 573.

75. McOuat 1996: 492; Morrell et al. 1981: 138; English 2004.

76. See A. Desmond 1989 for more on the relationship between French scientific and political radicalism.

77. [Lankester] 1846a: 573.

78. J. D. Hooker and Thomson 1855: 36–37.

79. Lenoir 1989. These divisions have been challenged by Robert J. Richards (1992: 48, 59–60, 124), but I will not discuss his critique since it does not bear directly on my argument here.

80. However, Owen was also indebted to Cuvier's functionalist anatomy, and the relative influence of *Naturphilosophie* and functionalism on his thought waxed and waned throughout his career. See Rupke 1994 for a fuller discussion.

81. Lindley 1834b: 27.

82. [Luxford] 1850: 58. For more on Carl Gustav Carus, Goethe, and Owen's attitudes to their ideas, see Rupke 1994: 161–219.

83. In 1852, the *Westminster*'s review of his collected works acknowledged that "although Germany and France have applied Goethe's morphological ideas with great success, yet England—true to her anti-metaphysical instinct . . . has been very chary of giving them admission, because the real philosophic method which underlies them is not appreciated" ([Lewes] 1852: 490).

84. Anon. 1847a: 783. This controversial translation of Oken's book was produced by the Ray Society at Richard Owen's instigation (E. Richards 1987: 163–65). For the controversy surrounding the *Vestiges*, see J. A. Secord 2000b. Although contempt for *Naturphilosophie* was a recurring theme of Lindley's *Gardeners' Chronicle*, the journal was not invariably anti-German, referring reverentially to Goethe whenever issues of "morphology" were discussed.

85. Anon. 1850b. The book under review was Carl Wilhelm von Nägeli's *Gattungen einzelliger Algen* (Kinds of Single-Cell Algae, 1849).

86. L. Huxley 1918a: 425–26.

87. L. Huxley 1918a: 402.

88. Lenoir 1989.

89. E. Newman 1848: 156.

90. [Forbes] 1851: 5. "Vestigianism" was, of course, the philosophy of the anonymous *Vestiges*, itself considered synonymous with unphilosophical speculation by many reputable men of science. For example, Darwin described it as "that strange unphilosophical, but capitally-written book, the Vestiges" (Darwin to William Fox, [24 April 1845], in Burkhardt et al. 1987: 181). And he later admitted that "I am almost as unorthodox about species as the Vestiges itself, though I hope not *quite* so unphilosophical" (Darwin to Huxley, 2 September [1854], in Burkhardt et al. 1989: 213).

91. Quoted in E. Richards 1987: 162–63.

92. [Lankester] 1848: 973–974.

93. J. D. Hooker to Darwin, 3 February 1849, in Burkhardt et al. 1988: 204. I am indebted to Rebecca Stott for drawing this letter to my attention. For a detailed discussion of Strickland's campaign, see McOuat 1996.

94. Yeo 1993: 127–28. Morrell and Thackray (1981: 268–70) also note this division of labor.

95. Forbes 1843: 18. See also Alborn 1996.

96. See J. Smith 1994 for a fuller account of the broad cultural impact of Baconian ideas and language.

97. Forbes 1843: 22, 20.

98. For the pedigree of the idea of ascending via induction, see Oldroyd 1986: 142–67.

99. Alborn 1996: 94–97, 103, 10.

100. J. D. Hooker 1859: vii.

101. J. D. Hooker 1853: xii.

102. J. D. Hooker to Gray, 28 July 1855, GHA.

103. See C. Darwin to J. D. Hooker, 6 November [1855], in Burkhardt et al. 1989: 493–94n4.

104. [?Lindley] 1855.

105. J. D. Hooker 1859: ii.

106. C. Darwin to J. D. Hooker, 3 January [1860], in Burkhardt et al. 1993: 6–7.

107. C. J. F. Bunbury to Leonora Pertz, 9 March 1860, and C. J. F. Bunbury to Katherine Lyell, 17 March 1860, in Bunbury 1906b: 154–155.

108. [Carpenter] 1860: 385; Watson 1860: 33.

109. J. D. Hooker and Thomson 1855: 42–43.

CHAPTER EIGHT

1. Browne 1979, 1980, 1983, 1996; Rehbock 1983; Drayton 1993, 2000.

2. J. D. Hooker to Gunn, 12 August 1846, ML (GC8).

3. Hooker told Gunn that "there are still things in Brown's herb. from Mt W^{n}. which have been gathered by no one else"; J. D. Hooker to Gunn, 12 August 1846, ML (GC8). "Mt W^{n}" was Mount Wellington, which overlooks Hobart. Hooker commented on Brown's meanness with specimens in several letters: e.g., "we *never* saw Brown give away a plant though he daily receives them, it is quite a disease with him"; J. D. Hooker to Gunn, 13 May 1844, ML (GC8).

4. Forbes was particularly interested in the notion of representative species and emphasized their importance (Rehbock 1983: 156–57).

5. Browne 1983: 138.

6. J. D. Hooker to Gunn, 17 April 1851, ML (GC8).

7. J. D. Hooker 1853: 4.

8. Gunn subscribed to the *Gardeners' Chronicle.* Gunn to J. D. Hooker, 4 July 1841, KEW (JDH/2/10). Colenso does not mention the journal, but because he corresponded with Lindley (who edited it), he would almost certainly have also received copies.

9. J. D. Hooker 1856a: 192.

10. J. D. Hooker 1851–53: 5. Larson (1986: 455–56) argues that the study of plant distribution had its origins in the increasingly common local floras of the late eighteenth century, which normally began with a detailed physical description of the country covered, together with a general description of the character of its flora, before listing all the species found in the region.

11. Colenso to J. D. Hooker, 24 August 1854, KEW (KDC174).

12. J. D. Hooker to C. Darwin, [2–6 April 1845], in Burkhardt et al. 1987: 167–68.

13. Hooker and Thomson "deplored the defective geographical nomenclature adopted in almost every work treating of the Natural History of India, and the fact that 'E. Ind.' or 'Ind. Or.' is considered in most cases sufficiently definite information as to the native place of any production found between Ceylon and Tibet, or Cabul and Singapur" (J. D. Hooker and Thomson 1855: 2).

14. J. D. Hooker 1853: xix.

15. Oldfield to W. J. Hooker, 20 August 55, KEW (KDC74).

16. A. F. Oldfield, RBGS (OSM): 792.

17. Gunn to J. D. Hooker, 22 December 1845, KEW (KDC218).

18. Colenso to J. D. Hooker, 3 February 1852, KEW (BNZ).

19. Cunningham to Colenso, 11 April 1839, ATL (CP4).

20. A. F. Oldfield, RBGS (OSM): 800–801, 809.

21. J. D. Hooker 1853: i.

22. J. D. Hooker 1853: vii.

23. [Brewster] 1833: 41.

24. [Brewster] 1833: 41.

25. L. Huxley 1918a: 79–81; R. Desmond 1999: 70–75. J. D. Hooker continued to take an interest in the tussock grass for many years. He sent Sir George Grey, governor of New Zealand, a packet of its seeds in 1865; J. D. Hooker to Grey, 17 May 1865, ACCL (GL21).

26. As Drayton (2000: 107–8, 19–20) has shown, the empire's botanic gardens were intended to boost Britain's economy and to demonstrate the efficient, benevolent nature of British rule; and, of more direct military relevance, Kew's networks could provide practical information on geography and population. See also Brockway 1979; McCracken 1997.

27. Browne 1992: 460–61; Drayton 2000: 126–27.

28. Drayton 2000: 192–93; R. Desmond 1995: 192–93.

29. The *Manual* also included a letter that discussed the need for naval officers in particular to make notes on useful timbers as they traveled, since "timber for masting and ship-building purposes is annually becoming more scarce"; W. J. Hooker 1851: 425, 36.

30. W. J. Hooker 1851: 416.

31. L. Huxley 1918a: 65.

32. R. Desmond 1999: 75.

33. [J. Bell] 1863: 507–8.

34. Drayton 2000: 207–11.

35. J. D. Hooker to Gunn, [August 1840], ML (GC8). The letter is undated other than "Saturday" but can be dated from another letter in the same archive that is dated "19 August" and apologizes for not "delivering the accompanying letter"—which appears to be the undated one quoted. This probability is strengthened because, although the *Erebus* was in Tasmania twice (from 16 August to 12 November 1840 and then from 6 April to early July 1841), Hooker and Gunn only saw each other regularly on the first visit in 1840 (Gunn was in Launceston during the second visit and they never met again) (L. Huxley 1918a: 105–9, 112–20).

36. E.g., J. D. Hooker and Thomson 1855: 36; J. D. Hooker 1856b: 215, 1867 (1984): 60.

37. For Sabine, see Good 2004.

38. Cawood (1979: 514) suggests that, in the context of 1840s politics, Herschel's term was almost certainly a deliberate political pun. On the importance of mapping within Humboldt's work, see M. Nicolson 1987: 180–81.

39. Herschel 1845: xxxiii-xxxiv.

40. Brown's *Botany of Terra Australis* appeared several years before Humboldt announced his own ideas on botanical arithmetic; see Brown 1814: 5–6; Mabberley 1985: 191, 356–58.

41. My thanks to Dr. Sachiko Kusukawa for the translation. In 1833 the *Edinburgh Review* quoted Humboldt's description ([Brewster] 1833: 68); ten years later, Edward Forbes (1843: 21) quoted it in his inaugural lecture as professor of botany at King's College, London; and another decade on, the *Athenaeum* mourned: "It was for a long time the reproach of this country that the *facile princeps botanicorum*—Robert Brown—was amongst us and we knew him not" ([Lankester] 1854).

42. [Brewster] 1835: 380.

43. J. D. Hooker to W. J. Hooker, 3 February 1840, in L. Huxley 1918a: 162. Brown had tried to calculate the areas of the Southern Hemisphere floras that were richest in genera, assuming these must be the original centers from which southern families—especially the Proteaceae—originated. His efforts were the starting point for Hooker's own attempts to understand the relationships between the southern floras (Browne 1983: 62).

44. J. D. Hooker to W. J. Hooker, 25 November 1842, in L. Huxley 1918a: 79–80.

45. [Holland] 1858. Holland was a noted physician and a distant relative of Darwin's; he frequently reviewed scientific works for the *Quarterly Review*. For the importance of numerical values and mathematics, see also Schaffer 1997.

46. Herschel 1830 (1966): 175–76.

47. Darwin to Asa Gray, 25 August [1856], in Burkhardt et al. 1990: 208.

48. Darwin to J. D. Hooker, 31 March [1844], in Burkhardt et al. 1987: 24.

49. Browne 1983: 68–74.

50. Egerton 2003: 28.

51. J. D. Hooker 1856b: 248–49.

52. Murchison 1846: xxxv.

53. Forbes 1846.

54. Browne 1983: 115–16.

55. Here, as elsewhere, my historiographic approach is influenced by Jardine 2000a.

56. For more on Buffon and Willdenow, see Kinch 1980: 97; Egerton 2003: 26–27.

57. G. Nelson 1978: 269–72. Gareth Nelson credits Buffon with founding modern biogeography, a claim that James Larson (1986: 448) thinks is overstated. However, the details of Larson's argument are not relevant to my discussion here.

58. Such species are now referred to as "disjunct," but the term was not used in the period. I have referred to them as sporadic because Hooker used the term in *Flora Novae-Zelandiae:* "while the instances are rare of sporadic species, as such are called which are found in small numbers in widely sundered localities" (1853: x). He was drawing on Augustin-Pyramus de Candolle's use of *"sporadique"* (Browne 1983: 111–13). However, "disconnected" would also be a legitimate shorthand term, since Hooker translated Alphonse de Candolle's phrase *"Espèces disjointes"* as "disconnected (*disjointes*) species" (1856b: 116–17).

59. Browne 1983: 112.

60. Proponents of multiple centers included Alphonse de Candolle (son of Augustin-Pyramus), who suggested that a single focus was impossible but was unsure whether the various original members of a species appeared simultaneously within their region or gradually over time. In Hooker's review of de Candolle's work he noted that the latter listed several disconnected species and argued that there was "no plausible explanation" for them, a view Hooker disputed (J. D. Hooker 1856b: 116–17). Philip Lutely Sclater was another advocate of multiple centers who based his view on the assumption that creation was fixed and static, with no changes having ever occurred (Kinch 1980: 110–11). Louis Agassiz also supported multiple centers of creation and coined a theory of the "autochthonal" origin of all the members of a species, simultaneously, across its entire modern range. The entomologist Lacordaire assumed that multiple creations were the only solution to disconnected species (Browne 1983: 111–13, 39).

61. A. F. Oldfield, RBGS (OSM): 849–51. Spontaneous generation was not an especially outlandish theory at the time; see Strick 2000.

62. A. F. Oldfield, RBGS (OSM): 849–50.

63. Taylor 1866. The printed book bears no date; however, 1866 is given in Bagnall 1970: 1011.

64. Oldfield cited Sclater's views on multiple creations (n. 62), claiming that they "so fully agree with these [i.e., his own] views that I here insert them," and quoted a lengthy section from Sclater's work on Madagascar (Taylor 1866: 19–20, 25–26).

65. Taylor 1866.

66. J. D. Hooker to A. Gray, 26 January 1854, in L. Huxley 1918a: 473–74.

67. Karl Sprengel was one naturalist who took the latter view. See K. Sprengel and A.-P. de Candolle, *Elements of the Philosophy of Plants* (Edinburgh, 1821), 283; quoted in Browne 1983: 113.

68. A.-P. de Candolle, *Géographie,* 415–16; quoted in G. Nelson 1978: 284.

69. Browne 1983: 116.

70. Forbes 1846: 336.

71. As Michael Kinch (1980: 113) argues, Lyell became the founder of a historical geological explanation of distribution and the *Principles* were crucial to disseminating this view. Browne (1983: 102–4) also argues for the centrality of Lyell's work.

72. Forbes 1846: 349.

73. Browne 1983: 117.

74. Hooker acknowledged Forbes's "Connexion" essay, as he had earlier done in the *Flora Antarctica* (J. D. Hooker 1851–53: xxii, 1844–47: 210).

75. Henfrey 1852: 378, 82–83.

76. J. D. Hooker 1856b: 118. In 1865, ten years after Forbes's death, the *Popular Science Review* praised his work (Coultas 1865: 37).

77. [Bentham] 1856: 492. Forbes had died two years earlier, in 1854.

78. [Bentham] 1856: 490–91.

79. [Bentham] 1856: 492–93.

80. Bentham noted that comparisons of different floras were often impossible, "owing to the great diversity of views entertained as to what constitutes a species" ([Bentham] 1856: 493).

81. Darwin to J. D. Hooker, [10 September 1845], in Burkhardt et al. 1987: 253.

82. Browne 1979, 1983.

83. Browne 1983: 198–99.

84. J. D. Hooker 1867 (1984): 60; Browne 1983: 133; Turrill 1953: 121.

85. J. D. Hooker to Darwin, [26 June or 3 July 1856], in Burkhardt et al. 1990: 156.

86. J. D. Hooker 1844–47: xi-xii.

87. J. D. Hooker 1851–53: xix.

88. J. D. Hooker 1859: l, 1844–47: 211–12.

89. J. D. Hooker 1859: lxxxix.

90. Malcolm Nicolson (1987: 167) has drawn on Erik Nordenskiöld's work to argue that the nineteenth century had two distinct distributional traditions: a floristic one (i.e., studies of species and their distribution) and a "vegetational geography" one (studies of plant communities).

91. J. A. Secord 1989: 76–79.

92. J. A. Secord 1989: 64–65.

93. As Latour (1999: 54–55) has observed, a great deal has to happen to nature before mathematics can describe it.

94. Browne 1983: 113.

95. J. D. Hooker 1859: xciv.

96. See L. Huxley 1918a: 506.

97. As Latour notes, "in losing the forest, we win knowledge over it. In a beautiful contradiction, the English word 'oversight' exactly captures the two meanings of this domination by sight, since it means at once looking at something from above and ignoring it" (1999: 38).

98. Drayton 2000: 204.

99. Kinch (1980: 91) also argues that the origin of species was central to nineteenth-century interest in biogeography. Larson (1986: 447n1) criticizes Janet Browne for anachronistically imposing nineteenth-century concerns onto the eighteenth century.

100. Browne 1983: 159.

101. J. D. Hooker 1887: 5. Also quoted in L. Huxley 1918a: 5.

CHAPTER NINE

1. Morgan 1994: 21.

2. Dickens 1838: 8.

3. A. R. Smith 1847: 3, 1–2. Pippins are a kind of apple.

4. Brander 1975: 17; Best 1979: 268–78.

5. Galton 1874.

6. Morgan 1994: 26–29.

7. Daunton 1989: 132; Corfield 2000: 174.

8. Craik 1856 (2005): 9.

9. Mitchell 1983: 40–41.

10. Craik 1856 (2005): 36.

11. Craik 1856 (2005): 13.

12. Anon. 1866: 42–43; Mitchell 1983: 44.

13. Mitchell 1983: 50–51.

14. Craik 1856 (2005): 320.

15. Morgan 1994: 133–37.

16. Morrell et al. 1981.

17. J. D. Hooker to Darwin, 28 September 1846, in Burkhardt et al. 1987: 342.

18. For fuller discussions of the Forbes/Watson dispute, see Browne 1983: 74; Rehbock 1983: 176.

19. J. D. Hooker to C. Darwin, [7 May 1863], in Burkhardt et al. 1999: 387.

20. J. D. Hooker to Darwin, [before 3 September 1846], in Burkhardt et al. 1987: 336–37.

21. In a letter to C. C. Babington (29 October 1841); quoted in Egerton 2003: 131–32.

22. H. C. Watson, "Remarks on the Distinction of Species in Nature, and in Books," *London Journal of Botany* (1843): 613–22; quoted in Egerton 2003: 140–41.

23. Watson 1859: 12–14.

24. J. D. Hooker 1853: xxvi (emphasis added).

25. Watson 1859: 13–14.

26. Anon. 1847b: 491. The review was anonymous but its style and sentiments suggest it was written by Lindley.

27. Watson 1847: 465.

28. J. D. Hooker to Darwin, 28 September 1846, in Burkhardt et al. 1987: 342.

29. Forbes to J. D. Hooker, 31 October 1846, KEW (JDH/2/1/8). The words in parentheses are a marginal addition by Forbes. Part of his letter is also quoted in Rehbock 1983. Watson's work was acknowledged in Forbes 1846: 342.

30. Murchison 1846: xxxiii.

31. Browne (1983: 107–8) notes that, in contrast to geologically minded naturalists (who saw the patterns of distribution in terms of arrangements in time), Watson approached his work in topographic terms, simply looking for observable, measurable ratios of species to genera, etc., in space.

32. Forbes had not published any noticeably philosophical, botanical works since becoming professor of botany at King's College, London, and Rehbock (1983: 175–77) argues

persuasively that Forbes's "Connexion" essay was in part intended to justify his occupying a botanical chair.

33. Watson 1847: 465.

34. Darwin to J. D. Hooker, 3 September 1846, in Burkhardt et al. 1987: 339.

35. J. D. Hooker to Darwin, 28 September 1846, in Burkhardt et al. 1987: 342.

36. J. D. Hooker to Darwin, [before 3 September 1846], in Burkhardt et al. 1987: 337.

37. Watson 1859: 8.

38. [Lindley] 1859: 911. Originally anonymous, Hooker attributed it to Lindley in a letter to Darwin, [21 November 1859], in Burkhardt et al. 1991: 383.

39. [Lindley] 1859: 911.

40. Watson 1845: 108.

41. Egerton 2003: 3–4.

42. Prideaux, *Strictures on the Conduct of Hewett Watson, F.L.S., in His Capacity of Editor of the Phrenological Journal* (1840); quoted in Egerton 2003: 63.

43. Watson, "Review of Thurmann's *Essai de phytostatique*," *Hooker's Journal of Botany and Kew Garden's Miscellany* 2 (1850): 187–92, 3 (1850): 918–21; quoted in Egerton 2003: 109.

44. C. C. Babington to J. H. Balfour, 23 January 1846; quoted in Egerton 2003: 130.

45. Quoted in Egerton 2003: 222.

46. H. C. Watson to A. de Candolle, 4 September 1859; quoted in Egerton 2003: 223.

47. Egerton 2003: 223–24.

48. Anon. 1847b: 491. It was in direct response to the *Gardeners' Chronicle*'s "vituperative notice" that Watson launched his attack on Lindley in which the latter was described as kin to "Mrs Fanny Ficklemind" (Watson 1860: 6–7).

49. Darwin to J. D. Hooker, [20 November 1859], in Burkhardt et al. 1991: 382.

50. J. D. Hooker to Darwin, [21 November 1859], in Burkhardt et al. 1991: 383.

51. Browne 1979: 64. Outram (1984: 135–36) notes that one of *Naturphilosophie*'s basic principles was the idea of endless polarities as the primary explanatory mechanism for the ways in which the few basic underlying forces of nature played themselves out in living objects. This association of Forbes's idea with *Naturphilosophie* and Oken would have made it doubly unpalatable to Hooker and Darwin.

52. For more details, see Browne 1983; Rehbock 1983.

53. Quoted in Rehbock 1983: 103–4.

54. Darwin to J. D. Hooker, 7 July [1854], in Burkhardt et al. 1989: 201. His reference was to "animal magnetism," an occult force capable of moving tables at events such as séances. Spiritualism, and associated phenomena such as mesmerism, were regarded as distinctly unpalatable by most philosophical naturalists; Huxley also compared the *Vestiges*' theory to the "table-turning folly" (T. H. Huxley 1854: 425). For mesmerism, see Winter 1998.

55. Darwin to J. D. Hooker, [3 September 1845], in Burkhardt et al. 1987: 250.

56. J. D. Hooker to Darwin, 2 February 1846, in Burkhardt et al. 1987: 296.

57. Darwin to J. D. Hooker, [13 March 1846], in Burkhardt et al. 1987: 300.

58. Rudwick (1985: 113–14) mentions the importance of courtesy and friendship in maintaining or restoring civility during controversies.

59. Anon. 1833: 442. The *Westminster* refers simply to "Murray's Physiology of

Plants," and its author simply as "Mr Murray," but the volume in question was probably John Murray's anonymously published *Syllabus of Six Lectures on Physiology of Plants* (1833). His later *Strictures on Morphology: Its Unwarrantable Assumptions, and Atheistical Tendency* (London, 1845) was owned by Hooker, who lent his copy to Darwin; see J. D. Hooker to Darwin, [23] March 1845, in Burkhardt et al. 1986: 162–65. Murray was a popular scientific writer; see R. Desmond 1994.

60. C. Darwin to J. D. Hooker, 30 May [1860], in Burkhardt et al. 1993: 232.

61. [Day] 1839: 81.

62. Anon. 1849: 48–49.

63. Anon. 1839: vi-vii.

64. J. D. Hooker to W. J. Hooker, 3 February 1840, in L. Huxley 1918a: 162.

65. W. J. Hooker to J. D. Hooker, [?1840], in L. Huxley 1918a: 162.

66. C. J. F. Bunbury to Mrs. C. Lyell, 29 April 1846, in Bunbury 1906a: 233–34.

67. Anon. 1839: vi-vii.

68. Anon. 1839: 109.

69. A. R. Smith 1847: 51–52.

70. Anon. 1849: 91.

71. Robinson 1850: xxxvi–xxxvii. Whewell's 1837 BAAS address also noted that the "manly vigour of discussion" at the Geological Society should be "tempered always by mutual respect and by good manners"; quoted in Rudwick 1985: 234.

72. [Forbes] 1852: 395.

73. Browne 1981: 206; Gay et al. 1997: 428–31.

74. Allen 1994: 96. Forbes's mocking "Sonnet to the British Public" (sung by the Red Lions) is quoted in Allen 1994: 76.

75. The manliness of fieldwork was explicitly seen within the Geological Survey as character building and morally improving as well as instilling "a hearty love of song and good cheer" (R. Porter 1978: 827). Rudwick (1985: 40–41) also notes that geological fieldwork was respectable for gentlemen because it encompassed aspects of the manly exercises (such as riding and hunting) that were a traditional component of gentlemanliness. The Red Lions also organized expeditions together, though these generally bypassed healthy exercise in favor of heading straight to a tavern (Gay et al. 1997: 432–33).

76. J. A. Secord 1986b: 240–41.

77. Wilson et al. 1861: 247–50; Mills 1984: 375. Andrew Ramsay's diary (Imperial College Archives, London, cited in Gay et al. 1997: 431–35) mentions picking up prostitutes. Unmarried young men like the Red Lions were prostitutes' largest group of clients, and their behavior was sufficiently common to attract relatively little overt censure, even though it was not approved of, and there was often strong peer-group pressure to join in (Tosh 1999: 108). See also Mason 1995.

78. Wilson et al. 1861: 380. See also Gay et al. 1997.

79. H. C. Watson to Sir Walter Calverley Trevelyan, 8 October 1838; quoted in Egerton 2003: 49.

80. H. C. Watson to W. J. Hooker, 27 August 1834; quoted in Egerton 2003: 100.

81. Egerton 1980: 189.

82. For Watson's views, see Egerton 2003: 111–13, 47–48. For the political implications of Lamarckianism, see A. Desmond 1989.

83. For example, Forbes opposed both the 1832 Reform Bill and the Chartist movement (Mills 1984: 375). Forbes had also satirized phrenology while a student in 1834 (Rehbock 1983: 176–77). See also Gay et al. 1997: 435–36. For the significance of the Philosophical Club, see MacLeod 1983: 74–75; Waller 2001.

84. Edward Forbes to J. D. Hooker, 11 May [1847/48?], KEW (JDH/2/1/8).

85. MacLeod 1983: 74–75.

86. C. Darwin to J. D. Hooker, 15 [May 1855], in Burkhardt et al. 1989: 330–31.

87. Hartog 2004.

88. Rupke has noted the importance of London's numerous clubs in Owen's career. The most important and prestigious of these was the Literary Club, also known as Dr. Johnson's Club or more often simply as "The Club." It had a fixed number of only forty members who dined together fortnightly at the Thatched House in St. James's Street. Owen was elected at the unusually early age of forty in 1845; Hooker was not elected until 1878 (after being president of the Royal Society). See Rupke 1994: 55–59.

89. Lytton 1836: 1.152.

90. Yeo 1993: 135–37. Whewell, for example, stressed the importance of scientific societies, where a "collision with other minds" ensured that speculations were tested (Yeo 2002: 58).

91. Tosh 1999: 127–34.

92. R. A. Buchanan 1983: 417–20.

93. J. A. Secord 1997: xi.

94. J. D. Hooker to Asa Gray, [n.d.], in L. Huxley 1918b: 43.

95. Hooker quoted in Turrill 1963: 197. C. Darwin to J. D. Hooker, 25 and 26 January 1862, in Burkhardt et al. 1997: 48; A. Desmond 2001: 7.

96. Diary entry, 27 January 1868, in Bunbury 1906b: 154–55.

97. J. D. Hooker to T. H. Huxley, 4 January 1861, in L. Huxley 1918b: 59. See also Turrill 1963: 197–99.

98. Darwin to J. D. Hooker, 27 [June 1854], in Burkhardt et al. 1989: 197. Both Darwin and Hooker were present at the births of their children, a practice that had become common among middle-class men. Prince Albert was with Queen Victoria when she first gave birth in 1841, and the tenor of the press commentary makes it clear that his action was widely perceived as commonplace and laudable (Tosh 1999: 82).

99. Darwin to J. D. Hooker, 27 May [1855], in Burkhardt et al. 1989: 338. For an account of the seed experiments, see Darwin to *Gardeners' Chronicle*, 21 May [1855], in Burkhardt et al. 1989: 331–34; Browne 1995: 517–21, 27–29.

100. Darwin to J. D. Hooker, 5 June [1855], in Burkhardt et al. 1989: 344.

101. J. D. Hooker to Darwin, [6–9 June 1855], in Burkhardt et al. 1989: 345.

102. The same ideals of scientific courtesy may also help explain Colenso's tolerance of Hooker's criticisms, assuming that he had absorbed such ideals from Cunningham and/or Hooker himself.

103. Bellon 2001: 51–56.

104. Whewell 1845.

105. Gaskell 1864–66 (1987): 117–24. I am indebted to Clare Pettitt for drawing these passages to my attention.

106. [Forbes] 1852: 397.

107. When Marner first arrives in the village of Raveloe, Eliot says: "He had inherited from his mother some acquaintance with medicinal herbs and their preparation" and so he had an "inherited delight to wander through the fields in search of foxglove and dandelion and coltsfoot." However, concerned that his misfortunes are a punishment for his sins, he worries whether his healing knowledge is godly or not, and neglects his herbs until the child Eppie comes into his life. Her company and pleasure in nature lead him to feel forgiven for his past sins, and so "Silas began to look for the once familiar herbs again" (Eliot 1861 [1985]: 57, 185). For various views of Marner's femininity, see Rignall 2001: 400–404. On sensibility as a feminine attribute, see Tosh 1999: 44. It is also no coincidence that Marner is a weaver, as they were among the most prominent of the early-nineteenth-century artisan botanists; see A. Secord 1994a.

108. Quoted in Morgan 1994: 122.

109. Morgan 1994: 130–31.

110. Anon. 1849: 97–98.

111. [Day] 1836: 72.

112. [T. Bell] 1857.

113. Yeo 1993: 30–31; White 2003.

114. C. Darwin to J. D. Hooker, 5 July [1856], in Burkhardt et al. 1990: 171.

115. Forbes 1843: 23. Morrell and Thackray (1981) have noted that the BAAS avoided discussion of the practical or commercial applications of science. Whewell was particularly opposed to any notion of promoting science on the basis of its utility, and Yeo emphasizes that he was far from unusual in his views; Lyell, Lyon Playfair, and others all agreed that even in cases where the usefulness of particular kinds of knowledge was unavoidable, abstract scientific principles should be regarded as preeminent (Yeo 1993: 226–27).

116. J. D. Hooker to Darwin, 28 September 1846, in Burkhardt et al. 1987: 342.

117. Browne 1979: 62.

118. Camerini 1996: 47–48.

119. Colenso 1883: 32–33.

120. Allen 1994: 169–71. Ritvo (1997b: 344–46) notes the connection between proliferating names and commercial considerations implied by the term "species-monger," arguing that it implied both "social condescension as well as scientific disapprobation." See also Ritvo 1997a. McOuat (2001: 18) has noted that Robert Grant made the same connection when he noted that "objects of zoology are scarcely of any appreciable value until they are identified and names are assigned to them by competent authorities" but that "being properly named, [a specimen's] value may be raised to 30, 40, 50 guineas."

121. A. F. Oldfield, RBGS (OSM): 27.

122. Colenso 1883: 32.

123. J. A. Secord 1997: xii.

124. Morgan 1994: 136–37.

125. I. Ashe, *Medical Education and Medical Interests* (London: Longman, 1868); quoted in Morgan 1994.

126. Anon. 1849: 35.

127. For gentlemen of science, see Morrell et al. 1981; Rudwick 1985; Shapin 1994.

128. Anon. 1849: 1.

129. Diary entry, 27 January 1868, in Bunbury 1906b: 154–55.

130. [Bentham] 1856: 492.

131. Anon. 1849: 34.

132. Anon. 1849: 36–37.

133. J. D. Hooker 1853: xix.

134. J. D. Hooker and Thomson 1855: 20–21.

135. J. D. Hooker to Harvey, [13] October 1856, KEW (JDH/2/3/5).

136. J. D. Hooker to Harvey, December 1856, KEW (JDH/2/3/5). Part of this letter is also quoted in L. Huxley 1918a: 371.

137. Best 1979: 268–70.

CHAPTER TEN

1. [?Masters] 1876: 1.

2. J. D. Hooker 1872b: 634.

3. [?Masters] 1876: 1.

4. Hall et al. 2000.

5. Hoyles 1991: 169–86.

6. Anon. 1874: 333.

7. R. Desmond 1995: 233; Anon. 1878.

8. Mr. T. Cave, MP; quoted in Anon. 1879.

9. *The Garden* (21 September 1878); quoted in Blunt 1978: 167.

10. Quoted in [?Masters] 1872c: 939.

11. J. D. Hooker to H. W. Bates, [?10 May 1863], APS (JDH).

12. I am thus unable to agree with Richard Bellon's assessment that Hooker completed "the transition from rootless neophyte to stable professional in the spring of 1855" (2001: 57).

13. J. D. Hooker to H. W. Bates, [?10 May 1863], APS (JDH).

14. Mr. W. E. Forster, MP; quoted in Anon. 1879.

15. Parsons 1854.

16. J. D. Hooker to C. Darwin, 15 September 1863, in Burkhardt et al. 1999: 629.

17. J. D. Hooker to W. J. Hooker, 18 May 1843, in L. Huxley 1918a: 165 (emphasis added).

18. L. Huxley 1918b: 507–17.

19. An honor that was followed by Knight Commander of the Star of India (1879), Grand Commander of the Star of India (1897), and the Order of Merit (1907); Endersby 2004.

20. Bellon 2001.

21. J. D. Hooker 1868: 4.

22. J. D. Hooker to R. Lingen, 19 February 1872, KEW (KG5).

23. Hooker 1868; quoted in R. Desmond 1995: 226.

24. J. D. Hooker to C. Darwin, [3 November 1865], in Burkhardt et al. 2002: 292.

25. MacLeod 1974: 45–47.

26. J. D. Hooker 1868: 22, 28–29.

27. Rupke 1994: 29–43.

28. MacLeod 1974: 46.

29. MacLeod 1974: 48–49.

30. For the St. Petersburg congress, see Rendle 1935: 36.

31. J. D. Hooker to Ayrton, 27 March 1869, in Lubbock 1872: 73.

32. There is extensive correspondence about this clerk and his qualifications in the government's Blue Book on the subject; Lubbock 1872: 1–34.

33. MacLeod 1974: 53. Macleod gives Smith's first name as Robert, but Robert Smith was the unsatisfactory clerk whom Ayrton foisted on Hooker.

34. Sir Douglas Strutt Galton (1822–99) was a sanitary engineer who made numerous improvements to barracks and hospitals (Vetch 2004).

35. MacLeod 1974: 54.

36. J. D. H. to Treasury, 1 May 1872, in Lubbock 1872: 23–24. The warrant (dated 29 October 1867 and signed by George Russell) is included in the Blue Book: Lubbock 1872: 27.

37. R. Callendar to J. D. Hooker, [n.d.], in Lubbock 1872: 28.

38. J. D. Hooker to R. Callendar, [n.d.], in Lubbock 1872: 28.

39. J. D. Hooker to H. Barkly, 8 April 1872, CUL (HFC).

40. J. D. Hooker to C. Darwin, 20 October 1871, CUL (DAR) 103: 85–86, 87–92.

41. C. Darwin to J. D. Hooker, 20 October [1871], CUL (DAR) 94: 209–10.

42. J. D. Hooker to C. Darwin, 31 October 1871, CUL (DAR) 103: 93–95. Hooker used almost the same words in a letter to his friend James Hector in New Zealand, describing Ayrton's response "a document which I officially declare to be a tissue of misrepresentation & misstatements & evasions *intended* to mislead the P.M.: & he further in a private letter to the P.M. made a statement which I officially pronounce to be a *deliberate falsehood*"; J. D. Hooker to Hector, 14 December 1871, in Yaldwyn et al. 1998: 144.

43. J. D. Hooker to A. West, 30 October 1871, KEW (KG5).

44. The signatories included Lyell, Darwin, George Bentham, Henry Holland, Spottiswoode, Huxley, and Tyndall (Lyell et al. 1872).

45. Lyell et al. 1872. The letter is also quoted in Lubbock 1872: 28.

46. *Nature* listed some of the bodies that supported Hooker, including a meeting of "leading botanists and horticulturalists" at the *Gardeners' Chronicle* offices; the Council of the Royal Horticultural Society; the Council of the Royal Botanical Society; and the Council of the Meteorological Society (*Nature* 143 [25 July 1872]: 249). Another list of botanical and similar societies, including foreign ones, appeared in *Gardeners' Chronicle* 30 (27 July 1872). The Royal Society of Arts and Sciences of Mauritius also expressed support (*Nature* 161 [28 November 1872]: 71).

47. MacLeod 1974: 58–59, 62–63.

48. [?Lockyer] 1872b: 209.

49. Anon. 1872h.

50. Anon. 1872d: 169.

51. Quoted in [?Masters] 1872b.

52. Quoted in [?Masters] 1872b.

53. Anon. 1871: 112.

54. Anon. 1872e: 239.

55. Anon. 1872i: 31.

56. Anon. 1872f: 56.

57. Boulger et al. 2004.

58. Anon. 1872a: 939.

59. [?Masters] 1872c: 967.

60. [?Masters] 1872a: 933.

61. [?Masters] 1872d: 1001.

62. A. S. Ayrton, 8 August 1872; quoted in Anon. 1872b: 1100.

63. A. S. Ayrton, Memorandum, in Lubbock 1872: 51–64.

64. *Hansard*, 8 August 1872; quoted in MacLeod 1974: 67.

65. Lyell et al. 1872.

66. W. E. Gladstone, 8 August 1872; quoted in Anon. 1872b: 1100–1101.

67. Anon. 1872d: 169–70.

68. A. S. Ayrton, 8 August 1872; quoted in Anon. 1872b: 1100.

69. Anon. 1872h.

70. Barlow 1958: 105.

71. Anon. 1872h.

72. Lyell et al. 1872.

73. J. D. Hooker to Hector, 14 December 1871, in Yaldwyn et al. 1998: 144–45.

74. [?Lockyer] 1872c: 401.

75. J. D. Hooker to H. Barkly, 9 September 1872, CUL (HFC).

76. MacLeod 1974: 68–69.

77. J. D. Hooker to C. Darwin, 14 August 1873, CUL (DAR) 103: 167–68.

78. MacLeod 1974: 71.

79. [?Lockyer] 1872c: 401.

80. Anon. 1872d: 170.

81. Anon. 1872g: 66.

82. J. D. Hooker to Hector, 19 September 1872, in Yaldwyn et al. 1998: 148.

83. Lubbock 1872: 63–64.

84. Owen 1872a: 169–70; Drayton 2000: 217.

85. Owen 1872a: 172.

86. Owen 1872a: 173.

87. Owen 1872a: 173–74; Drayton 2000: 219.

88. [?Lockyer] 1872a: 280.

89. [?Masters] 1872e: 1066.

90. Anonymous editorial comment preceding Harvey et al. 1858 (1870): 97. See Drayton 2000: 199–201. For the background to the original memorial, see Rupke 1994: 97–102.

91. MacLeod 1974: 52–53.

92. Harvey et al. 1858 (1870): 98.

93. R. Owen, "Papers Relating to the Enlargement of the British Museum," *Parliamentary Papers*, 1859; quoted in Rupke 1994: 103–4.

94. A. Desmond 1982: 17; Rupke 1994: 104.

95. Owen (1872a: 172–73) refers to Kew throughout as "the Botanical Department under the Commissioners of Works," presumably to flatter Ayrton and irritate Hooker.

96. J. D. Hooker 1872a: 516–17.

97. Devonshire Commission 1872: Q. 6747, 438.

98. Devonshire Commission 1872: Q. 6745.

99. Devonshire Commission 1872: Q. 6732, Q. 6733, 437.

100. C. Darwin to J. D. Hooker, 27 October [1872], CUL (DAR) 94: 235–36.

101. J. D. Hooker to C. Darwin, 29 October 1872, CUL (DAR) 103: 128–29.

102. Owen 1872b: 6.

103. King 1872: 84.

104. Carruthers 1872b: 26.

105. They included Charles Cardale Babington, John Hutton Balfour, Thomas Thomson, Charles Darwin, and George Bentham,

106. Berkeley et al. 1873: 212.

107. Law 1873: 243.

108. Thiselton-Dyer 1873: 244.

109. Thomason 2004.

110. Thiselton-Dyer 1873: 244–45.

111. See [Thomson] 1865 for a review of Jordan's *Diagnoses d'espèces nouvelles ou méconnues* (1864).

112. Carruthers 1872a: 449–50.

113. J. D. Hooker to W. J. Hooker, 11 April 1849, Kew, India Letters: 158. Quoted in Bellon 2001: 69–70.

114. Drayton 2000: 197–99.

115. W. J. Hooker; quoted in Allan 1967: 217.

116. Hooker's wife, Francis, told Darwin that "Joseph has this morning received the offer of the Directorship"; F. H. Hooker to C. Darwin, 13 September [1865], in Burkhardt et al. 2002: 232.

117. Richard Drayton (2000: 197) makes the point that the arrangement Banks made was already unacceptable in the 1840s, when William Hooker first offered the government his herbarium.

118. [?Masters] 1872a: 933.

119. Lyell et al. 1872.

120. *Pall Mall Gazette*, reprinted in Anon. 1872c: 1068.

121. Lyell et al. 1872.

122. Anon. 1872h.

123. MacLeod 1974: 71.

124. Alborn 2004.

125. Tyndall to Derby, 2 April 1872; quoted in MacLeod 1974: 58–59.

126. J. D. Hooker 1872b: 634–35.

127. J. D. Hooker 1872b: 634–35.

128. J. D. Hooker 1872b: 634–35.

129. Richard Drayton (1999: 148–53) has christened this theory, which both Hookers shared, "the Whig Theory of Landscape."

130. [?Masters] 1875: 332.

131. Quoted in Blunt 1978: 165.

132. R. Desmond 1995: 228.

133. [?Masters] 1876: 2.

134. R. Desmond 1995: 228–30.

135. J. D. Hooker to C. Darwin, 29 October 1872, CUL (DAR) 103: 128–29.

136. Anon. 1878.

137. *Richmond and Twickenham Times*, 15 September 1877; quoted in R. Desmond 1995: 236.

138. Memo, 29 October 1877, "Admission of the Public, 1853–1925"; quoted in R. Desmond 1995: 236.

139. Anon. 1878.

140. "Kew Gardens. Forenoon Opening, 1866–1889," Kew Archives (1875), fol. 144.

141. [?Robinson], *The Garden*, 13 April 1878; quoted in Blunt 1978: 167.

142. Sir Trevor Lawrence, MP; quoted in Anon. 1879.

143. Gerard Noel; quoted in Anon. 1879.

144. [?Masters] 1873: 1631–32.

145. Drayton 2000: 207–11.

146. Devonshire Commission 1872: Q. 6670, 435.

147. L. Huxley 1918b: 140.

148. Drayton 2000: 220.

149. As Paul White (2003: 33–34) has argued in the case of Thomas Huxley, good manners played a vital part in the informal world of the clubs that dominated scientific London; codes of etiquette needed to be observed if science was to be successfully used as a route to gentlemanliness.

150. MacLeod 1974: 70–71.

151. J. D. Hooker, annual report 1877; quoted in Anon. 1878.

152. Blunt 1978: 170.

CONCLUSION

1. J. D. Hooker; quoted in Anon. 1901: 552–55.

2. See A. Desmond 1994, 1997; White 2003; A. Desmond et al. 1991; Browne 1995, 2002; Camerini 1996; Raby 2001; Slotten 2004.

3. For an overview of these approaches, see Golinski 1998 (2005).

4. J. D. Hooker et al. 1862–83; Gledhill 2002.

5. C. Darwin to executors, 20 December 1881, DAR (CD library—\Index Kewensis\ tom. 1).

6. Basalla 1967. An overview of the debates is provided in Drayton 1999. Some critics of these approaches have used Bruno Latour's ideas to provide an alternative model which is more dynamic and less Eurocentric and emphasizes active processes rather than static inequalities. See Latour 1988; Stemerding 1991; Miller 1996; Spary 2000.

7. J. D. Hooker 1859: ii.

8. J. D. Hooker 1859: iv.

9. Allan 1967: 194–95, 200; Hull 1973: 86; Bowler 1989: 193; Stevens 1997a: 346.

10. L. Huxley 1918a: 481; Turrill 1963: 76; A. Desmond and Moore 1991: 373, 415; Ruse 1999: 138–41; A. Desmond 1999: 208; Bellon 2000.

11. Among his more significant shorter essays are J. D. Hooker 1851, 1856b, 1862, 1867 (1984).

12. See, e.g., Browne 2002: 22–37.

13. Browne 1978: 361–62.

14. J. D. Hooker 1853: viii.

15. J. D. Hooker 1853: xii.

16. [T. Bell] 1857.

17. C. Lyell to C. J. F. Bunbury, 30 April 1856, in Bunbury 1906b: 212.

18. C. J. F. Bunbury to C. Lyell, 12 May 1856, in Bunbury 1906b: 92.

19. This is a reference to Fontanelle's theory of the plurality of worlds, as elaborated by Christiaan Huygens: since the sun was like other stars, they probably had planets that were likely to be inhabited, and those inhabitants were likely to be rational inhabitants like us. Whewell's *Dialogue on the Plurality of Worlds* (1854) argued against this on the basis that the sun appeared to be *un*like other stars, so the earth was probably unique (Knight 1981: 42–43).

20. J. D. Hooker 1853: xxvi.

21. Darwin to J. D. Hooker, 20 October 1858, in Burkhardt and Smith 1991: 174.

22. J. D. Hooker to Asa Gray, 21 October 1858; quoted in D. Porter 1993: 32–33.

23. Schaffer 1984.

24. As McOuat (1996) has shown, Darwin was an enthusiastic supporter of Hugh Strickland's attempts to impose a stable classificatory system.

25. J. D. Hooker 1853: i–ii.

26. J. D. Hooker and Thomson 1855: 19.

27. J. D. Hooker 1859: iv (emphasis added).

28. J. D. Hooker 1856b: 255.

29. J. D. Hooker 1859: iii–iv. While he was writing the *Flora Tasmaniae* essay, Hooker described the Indian botanist George Thwaites as having been "a devoted variationist"; J. D. Hooker to Bentham, 17 July [1859], KEW (JDH/2/3). Parts of this letter are quoted in L. Huxley 1918a: 484. George Henry Kendrick Thwaites worked at the botanic gardens, Peradeniya, Ceylon, from 1849 to 1880 (R. Desmond 1994).

30. C. J. F. Bunbury to Leonora Pertz, 9 March 1860, in Bunbury 1906b: 154.

31. Bellon 2006: 4.

32. J. D. Hooker to Darwin, 14 September 1845, in Burkhardt et al. 1987: 250.

33. J. D. Hooker to Asa Gray, 21 October 1858; quoted in D. Porter 1993: 32–33.

34. J. D. Hooker to Asa Gray, 21 October 1858; quoted in D. Porter 1993: 32–33.

35. J. D. Hooker 1862: 278–79.

36. C. Darwin 1859 (1985): 399–404.

37. Bellon (2006: 3) has also reached this conclusion, noting that Darwin regularly modified his ideas in response to Hooker's comments and criticisms.

38. Browne 1983: 135. See also Darwin to J. D. Hooker, 15 November [1856], in Burkhardt et al. 1990: 271–72; and J. D. Hooker to Darwin, [16 November 1856], in Burkhardt et al. 1990: 272–73.

39. I am indebted to members of Leeds University's Division of History and Philosophy of Science for their comments on an earlier version of this argument and in particular to J. R. R. Christie and M. J. S. Hodge, who helped me to clarify the relationship between ambiguity and equivocation.

40. J. D. Hooker to Darwin, 12 December 1843–11 January 1844, in Burkhardt et al. 1987.

41. Darwin to J. D. Hooker, [11 January 1844], in Burkhardt et al. 1987.

42. J. D. Hooker to J. S. Henslow, [March 1860], in L. Huxley 1918a: 535.

BIBLIOGRAPHY

PRIMARY SOURCES: MANUSCRIPTS AND ARCHIVES

ACCL: Auckland Central City Library, Manuscripts and Archives.

- ACCL (GL21): Grey, Sir George. Grey Letters.
- ACCL (GLNZ21): Grey, Sir George. Grey New Zealand Letters.
- ACCL (TL): Taylor, Richard. Taylor Letters (Grey Collection, GNZ MSS 297/4).
- ACCL (TMS): Taylor, Richard. MS notes on New Zealand and its Native inhabitants (GNZ MSS 297/25).

APS: American Philosophical Society, Philadelphia.

- APS (JDH): Hooker, Joseph Dalton. J. D. Hooker Papers (B/H76).
- APS (WH): Hutton, William. William Hutton Papers (B/H978).

ATL: Alexander Turnbull Library, Manuscripts and Archives. National Library of New Zealand, Wellington.

- ATL (CP1): Colenso, William. Colenso Papers, vol. 1 (MS-0585).
- ATL (CP4): Colenso, William. Colenso Papers, vol. 4 (MS-0585).

AUR: Anderson University Records, Glasgow.

- AUR (SR): Student Records: Popular Evening Classes Attendance Lists.

CUL: Cambridge University Library Archives.

- CUL (DAR): Darwin, Charles Robert. Correspondence and other papers, 1821–82 (Ms DAR).
- CUL (HFC): Hooker, Joseph Dalton. Sir Joseph Hooker, family correspondence and papers (Ms Add. 9537).

GHA: Gray Herbarium Archives. Harvard University, Cambridge, MA.

HBM: Hawke's Bay Museum, Library and Archives. Napier, New Zealand.

- HBM (CL): Colenso, William. Colenso Letters.

KEW: Archives, Royal Botanic Gardens, Kew.

- KEW (BNZ): Colenso, William. Botany of New Zealand manuscripts.
- KEW (IL): Various authors. India Letters.
- KEW (JDH/1/1): Hooker, Joseph Dalton. *Antarctic Journal.*
- KEW (JDH/1/3): Hooker, Joseph Dalton. Letters and Journal.
- KEW (JDH/2/1/4): Various authors. Letters to J. D. Hooker.

KEW (JDH/2/1/8): Various authors. Letters to J. D. Hooker (ENG-GAG).
KEW (JDH/2/1/11): Various authors. Letters to J. D. Hooker (HAR).
KEW (JDH/2/3): Hooker, Joseph Dalton. Letters from J. D. Hooker (BEN-BUR).
KEW (JDH/2/3/5): Hooker, Joseph Dalton. Letters from J. D. Hooker (GRA-HAR).
KEW (JDH/2/3/13): Hooker, Joseph Dalton. Letters from J. D. Hooker (PAL-WRI).
KEW (JDH/2/10): Various authors. J. D. Hooker Correspondence Received.
KEW (KCL/1/1): Various authors. Kew Collector's List, vol. 1.
KEW (KCL/9/1): Various authors. Kew Collector's List, vol. 3.
KEW (KCL/11): Various authors. Kew Collector's List, vol. 5.
KEW (KCL/13/1): Various authors. Kew Collector's List, vol. 8.
KEW (KDC73): Various authors. Kew Director's Correspondence, vol. 73.
KEW (KDC74): Various authors. Kew Director's Correspondence, vol. 74.
KEW (KDC76): Various authors. Kew Director's Correspondence, vol. 76.
KEW (KDC106): Various authors. Kew Director's Correspondence, vol. 106.
KEW (KDC174): Various authors. Kew Director's Correspondence, vol. 174.
KEW (KDC175): Various authors. Kew Director's Correspondence, vol. 175.
KEW (KDC218): Various authors. Kew Director's Correspondence, vol. 218.
KEW (KG5): Various authors. Hooker/Ayrton Controversy.

LSL: Linnean Society of London, Library.
LSL (Orchids): Archer, William. Original Drawings of Tasmanian Orchideae.
LSL (BC): Various authors. Bound Correspondence.

ML: Mitchell and Dixson Libraries, State Library of New South Wales, Sydney.
ML (CGB): Colenso, William. *Glossarium Botanicum: Novae Zelandiae* (ML MSS 3109).
ML (GC8): Various authors. Gunn Correspondence.
ML (JDH): Hooker, Joseph Dalton. Letters to R. C. Gunn.
ML (RCG): Gunn, Ronald Campbell. Gunn Papers.

RBGS: Archives, Royal Botanic Gardens, Sydney.
RBGS (OSM): Oldfield, Augustus Frederick. Species Manuscripts.
RBGS (MRC): Manuscript Records of Royal Botanic Gardens Sydney Correspondence.

SLV: Manuscripts Collection, State Library of Victoria, Melbourne.
SLV (RCG): Gunn, Ronald Campbell. Letters from R. C. Gunn.

SROWA: State Records Office of Western Australia, Perth.

UGR: University of Glasgow Records.
UGR (CCI): Class Catalogues and Indexes.
UGR (JAS): Dr. Jeffray's Anatomy Students Register.

UTA: University of Tasmania Archives.
UTA (Journal): Archer, William. *Journal.*

PRIMARY SOURCES: PUBLISHED

For reviews and articles that were originally published anonymously, authors' names are enclosed in square brackets; uncertain or speculative attributions are indicated with a question mark (e.g., [?Lindley, John]). Attributions for articles in the *Athenaeum* are taken from *The Athenaeum Index of Reviews and Reviewers* (http://web.soi.city.

ac.uk/~asp/v2/home.html); all others are from *The Wellesley Index to Victorian Periodicals, 1824–1900*, unless stated otherwise in a note.

Anon. 1831. "Review of W. J. Hooker *Botanical Miscellany*." *Monthly Review or Literary Journal* 2, no. 3: 317–26.

Anon. 1833. "John Murray's *The Physiology of Plants* and James Main's *Illustrations of Vegetable Physiology*." *Westminster Review* 19, no. 38: 431–42.

Anon. 1834. "Review of Lindley *Ladies' Botany*." *Monthly Review or Literary Journal* 2, no. 3: 342–53.

Anon. 1839. *Advice to a Young Gentleman, on Entering Society, by the Author of "The Laws of Etiquette."* London.

Anon. 1847a. "Review of Oken's 'Elements of Philosophy.'" *Gardeners' Chronicle*, p. 783.

Anon. 1847b. "Review of Watson's *Cybele Britannica*." *Gardeners' Chronicle*, p. 491.

Anon. 1849. *The English Gentleman: His Principles, His Feelings, His Manners, His Pursuits*. London: G. Bell.

Anon. 1850a. Editorial. *Gardeners' Magazine of Botany, Horticulture, Floriculture, and Natural Science* 1, no. 1: 1–2.

Anon. 1850b. "Review of Nägeli's *Gattungen einzelliger Algen*." *Gardeners' Chronicle*, p. 103.

Anon. 1866. "The Author of *John Halifax*." *British Quarterly Review* 44: 32–58.

Anon. 1871. "Nice Little Holiday Tasks (for Little Ministers)." *Punch: or the London Charivari* 61: 112.

Anon. 1872a. "An Allegory." *Gardeners' Chronicle*, no. 28: 938–39.

Anon. 1872b. "Ayrton versus Hooker (reprinted from *The Times*)." *Gardeners' Chronicle*, no. 33: 1100–1101.

Anon. 1872c. "The Kew Gardens Case." *Gardeners' Chronicle*, no. 32: 1068.

Anon. 1872d. "Mr. Ayrton Again." *Saturday Review* 34, no. 876: 169–70.

Anon. 1872e. "The Noble Savage among the Antiquaries." *Punch: or the London Charivari* 62: 239.

Anon. 1872f. "Punch's Essence of Parliament (August 10)." *Punch: or the London Charivari* 62: 56.

Anon. 1872g. "Punch's Essence of Parliament (August 17)." *Punch: or the London Charivari* 62: 66.

Anon. 1872h. "Report of Parliamentary Debate on the Kew Gardens Question." *Spectator* 45, no. 2302: 997.

Anon. 1872i. "Song by a 'Noble Savage.'" *Punch: or the London Charivari* 62: 31.

Anon. 1874. "Report of Richmond (Surrey) Select Vestry Meeting." *Gardeners' Chronicle*, n.s., 2, no. 37: 333–34.

Anon. 1878a. "Kew Gardens." *Times*, no. 29351: 9.

Anon. 1878b. "Kew Gardens (Hooker's Annual Report for 1877)." *Times*, p. 9.

Anon. 1879. "Parliamentary Intelligence, Petitions: Kew Gardens." *Times*, p. 6.

Anon. 1901. "The Opening of the New Botanical Department at the Glasgow University." *Annals of Botany* 15, no. 59: 551–58.

Babington, Charles Cardale. 1843. *Manual of British Botany, Containing the Flowering Plants and Ferns, Arranged according to the Natural Orders*. London: John Van Voorst.

Balfour, John H. 1849. *A Manual of Botany: Being an Introduction to the Study of the Structure, Physiology and Classification of Plants*. London: John Joseph Griffin.

Baskerville, Thomas. 1839. *Affinities of Plants: With Some Observations upon Progressive Development*. London: Taylor and Walton.

[Bell, James]. 1863. "Chinchona Cultivation in India." *Edinburgh Review or Critical Journal* 118, no. 242: 507–22.

[Bell, Thomas]. 1857. "Review of JD Hooker's *Botany of the Antarctic Voyage*, Part II: *Flora Novae-Zelandiae*." *Athenaeum*, no. 1530: 242–43.

[Bentham, George]. 1856. "Review of De Candolle's *Geographical Botany* and Other Works." *Edinburgh Review or Critical Journal* 104, no. 212: 490–518.

Bentham, George. 1865. *Handbook of the British Flora: A Description of the Flowering Plants and Ferns Indigenous to, or Naturalized in, the British Isles*. London: Lovell Reeve.

Berkeley, Miles, Charles Cardale Babington, et al. 1873. "The National Herbaria." *Nature* 7, no. 168: 212–13.

[Brewster, David]. 1833. "Memoir and Correspondence of the Late Sir James Edward Smith." *Edinburgh Review or Critical Journal* 57, no. 115: 39–69.

———. 1835. "Reports of the British Association for the Advancement of Science." *Edinburgh Review or Critical Journal* 60, no. 122: 363–94.

Britten, James. 1896. "Flowering Plants and Ferns." In *Notes on Collecting and Preserving Natural-History Objects*, edited by John Ellor Taylor, 117–37. London: Gibbings.

Brown, Robert. 1814. *General Remarks, Geographical and Systematical, on the Botany of Terra Australis*. London: William Bulmer's.

———. 1830 (1960). *Prodromus Florae Novae Hollandiae et Insulae Van Diemen*. New York: H. R. Engelmann (J. Cramer) and Wheldon and Wesley. Originally published 1810.

Browning, John. 1881. *Priced List of Microscopes*. London: John Browning.

Bunbury, Charles James Fox. 1906a. *The Life of Sir Charles J. F. Bunbury*. Vol. 1. London: John Murray.

———. 1906b. *The Life of Sir Charles J. F. Bunbury*. Vol. 2. London: John Murray.

Carpenter, W. B. 1860. "The Theory of Development in Nature." *British and Foreign Medico-Chirurgical Review* 25 (April): 367–404.

Carruthers, William. 1872a. "Botanical Museums." *Nature* 7, no. 153: 449–52.

———. 1872b. "Letter to the Editor (Reply to Hooker)." *Nature* 7, no. 159: 26.

Colenso, William. 1845. *Classification and Description of Some Newly-Discovered Ferns Collected in the Northern Island of New Zealand, in . . . 1841–42*. Launceston, Van Diemen's Land: Office of the *Launceston Examiner*.

———. 1865. "Essay on the Maori Races." *Transactions of the New Zealand Institute* 1: 5–75.

———. 1883. "On Nomenclature." In *Three Literary Papers*, 1–35. Wellington: Privately printed.

Collingwood, Cuthbert. 1858. *On the Scope and Tendency of Botanical Study: An Inaugural Address Delivered before the Liverpool Royal Infirmary School of Medicine, May 3, 1858*. London: Longman.

Coultas, Harland. 1865. "On the Origin of the Local Floras of Great Britain and Ireland." *Popular Science Review* 4, no. 13: 28–39.

Craik, Dinah Maria Mulock. 1856 (2005). *John Halifax, Gentleman*. Stroud: Nonsuch Publishing.

Darwin, Charles. 1859 (1985). *On the Origin of Species by Means of Natural Selection: Or the Preservation of Favoured Races in the Struggle for Life*. Harmondsworth, UK: Penguin Books.

[Day, Charles William]. 1836. *Hints on Etiquette and the Usages of Society*. London: Longman, Rees, Orme, Brown, Green, and Longman.

———. 1839. *Hints on Etiquette and the Usages of society*. London: Longman, Rees, Orme, Brown, Green, and Longman.

Devonshire Commission. Royal Commission on Scientific Instruction and the Advancement of Science. 1872. *First, Supplementary, and Second Reports, with Minutes of Evidence and Appendices*. Vol. 1. London: House of Commons.

Dickens, Charles. 1838. *Sketches of Young Gentlemen: Dedicated to the Young Ladies*. London: Chapman and Hall.

Dieffenbach, Ernst. 1843. *Travels in New Zealand*. London: John Murray.

[Dixon, Edmund Saul]. 1851. "Gardening: Reviews of *Transactions of the Horticultural Society of London*, and Other Titles." *Quarterly Review* 89, no. 177: 1–32.

Eliot, George. 1861 (1985). *Silas Marner: The Weaver of Raveloe*. Harmondsworth, UK: Penguin Books.

———. 1866 (1995). *Felix Holt: The Radical*. Harmondsworth, UK: Penguin Books.

———. 1872 (1994). *Middlemarch*. Harmondsworth, UK: Penguin Books.

Fielding, Henry. 1749. *History of Tom Jones, a Foundling*. London: A. Millar.

Fitch, Walter Hood. 1869 (1995). "Botanical Drawing." In *The Art of Botanical Illustration*, edited by Wilfrid Blunt and William T. Stearn, 332–40. London: Antique Collector's Club.

Forbes, Edward. 1843. *An Inaugural Lecture on Botany: Considered as a Science and as a Branch of Medical Education*. London: John Van Voorst.

———. 1845. "On the Distribution of Endemic Plants, More Especially Those of the British Islands, Considered with Regard to Geological Changes." *Report of the British Association for the Advancement of Science*, 15th Meeting, Cambridge, June 1845: 67–68.

———. 1846. "On the Connexion between the Distribution of the Existing Fauna and Flora of the British Isles, and the Geological Changes Which Have Affected Their Area, Especially during the Epoch of the Northern Drift." *Memoirs of the Geological Survey* 1: 336–432.

[Forbes, Edward]. 1851. "Review of Sedgwick's *Discourse on the Studies of the University of Cambridge*." *Literary Gazette, and Journal of Belles Lettres*, no. 1772 (4 January): 5–7.

———. 1852. "Plants and Botanists (Reviews of Lindley's *Vegetable Kingdom* and Other Books)." *Westminster Review* 58, no. 2: 385–98.

Galton, Francis. 1874. *English Men of Science: Their Nature and Nurture*. London: Macmillan.

Gaskell, Elizabeth Cleghorn. 1848 (1996). *Mary Barton: A Tale of Manchester Life*. Harmondsworth, UK: Penguin Books.

———. 1864–66 (1987). *Wives and Daughters: An Every-Day Story*. Oxford: Oxford University Press.

Harvey, William Henry, Arthur Henfrey, et al. 1858 (1870). "Natural History Museum." *Nature* 2, no. 31: 97–98.

[Haworth, Adrian]. 1825a. "A Few Observations on the Natural Distribution of Animated Nature. By a Fellow of the Linnean Society." *Philosophical Magazine* 62: 200–202.

———. 1825b. "A New Binary Arrangement of the Brachyurous Crustacea." *Philosophical Magazine* 65: 105–6, 183–84.

Henfrey, Arthur. 1849. *The Rudiments of Botany: A Familiar Introduction to the Study of Plants.* London: John Van Voorst.

———. 1852. *The Vegetation of Europe, Its Conditions and Causes.* London: John Van Voorst.

———. 1857. *An Elementary Course of Botany: Structural, Physiological and Systematic.* London: John Van Voorst.

[Henfrey, Arthur]. 1857. "Review of Bradbury's *Ferns of Great Britain &c. Nature-Printed* and Other Books." *Quarterly Review* 101, no. 201: 57–79.

Herschel, John Frederick William. 1830 (1966). *A Preliminary Discourse on the Study of Natural Philosophy.* New York: Johnson Reprint Corp.

———. 1845. "Presidential Address to the British Association for the Advancement of Science." In *Report of the British Association for the Advancement of Science,* xxvii–xliv. London: John Murray.

[Holland, Henry]. 1858. "The Progress and Spirit of Physical Science." *Edinburgh Review or Critical Journal* 108, no. 219: 71–104.

Hooker, Joseph Dalton. 1844–47. *Flora Antarctica.* Vol. 1 of *The Botany of the Antarctic Voyage.* London: Reeve Brothers.

———. 1851. "On the Vegetation of the Galapagos Archipelago, as Compared with That of Some Other Tropical Islands and of the Continent of America." *Transactions of the Linnean Society of London* 20: 235–62.

———. 1851–53. *Flora Novae-Zelandiae.* Vol. 2 of *The Botany of the Antarctic Voyage.* London: Lovell Reeve.

———. 1853. *Introductory Essay to the "Flora Novae-Zelandiae."* London: Lovell Reeve.

———. 1855. *Illustrations of Himalayan Plants: Chiefly Selected from Drawings Made for the Late J. F. Cathcart, Esq.re of the Bengal Civil Service. The Descriptions and Analyses by J. D. Hooker; the Plates Executed by W. H. Fitch.* London: Lovell Reeve.

———. 1856a. "Botanical Geography." *Gardeners' Chronicle,* no. 12 (22 March 1856): 192–93.

———. 1856b. "Notice of Alphonse de Candolle's *Géographie botanique raisonée.*" *Hooker's Kew Journal of Botany* 8: 54–64, 82–88, 112–21, 151–57, 181–91, 214–19, 248–56.

———. 1859. "On the Flora of Australia" (introductory essay to *Flora Tasmaniae*). London: Lovell Reeve.

———. 1860. *Flora Tasmaniae.* Vol. 3 of *The Botany of the Antarctic Voyage.* London: Lovell Reeve.

———. 1862. "Outlines of the Distribution of Arctic Plants." *Transactions of the Linnean Society of London* 23: 251–348.

———. 1864–67. *Handbook of the New Zealand Flora: A Systematic Description of the*

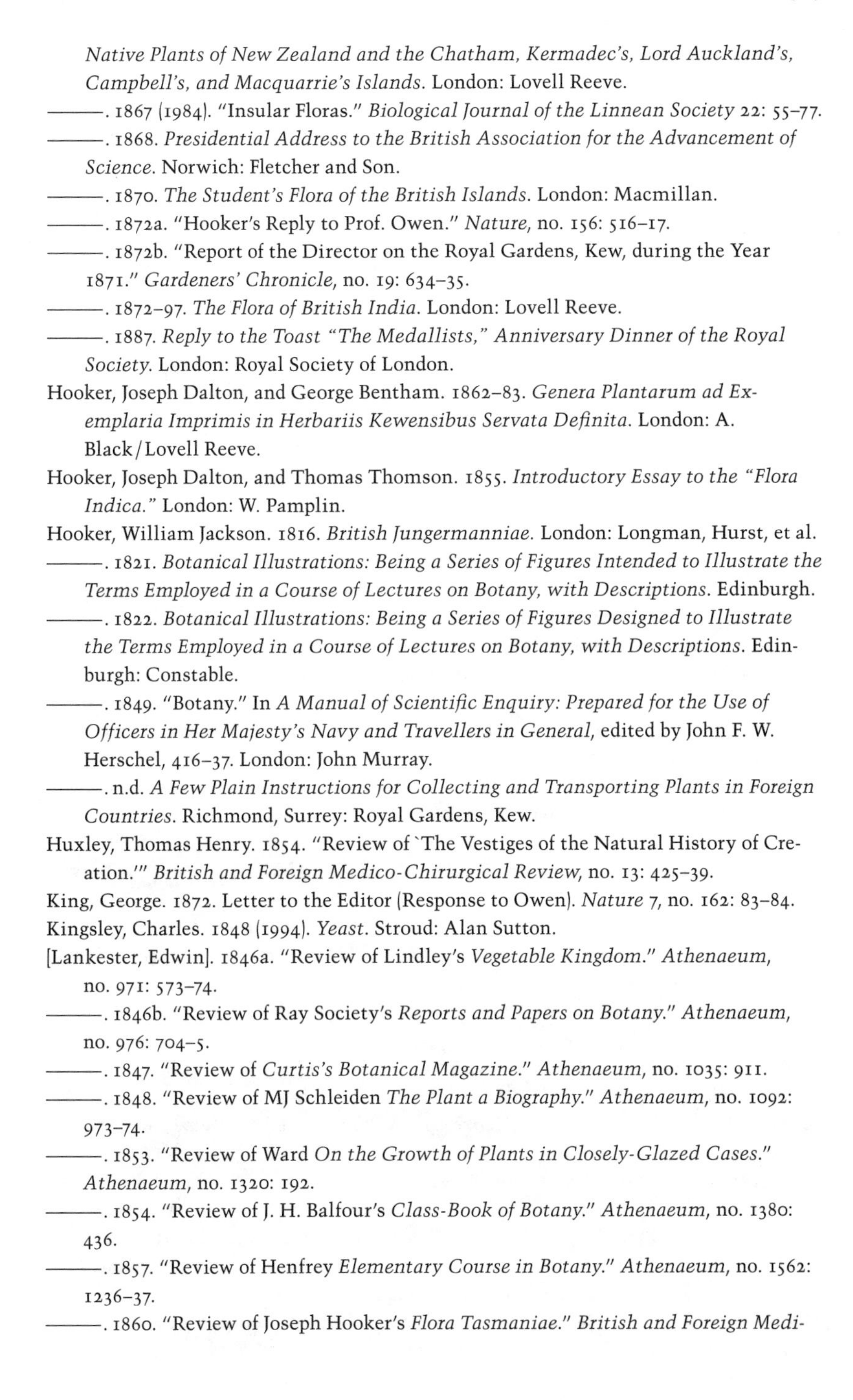

Native Plants of New Zealand and the Chatham, Kermadec's, Lord Auckland's, Campbell's, and Macquarrie's Islands. London: Lovell Reeve.

———. 1867 (1984). "Insular Floras." *Biological Journal of the Linnean Society* 22: 55–77.

———. 1868. *Presidential Address to the British Association for the Advancement of Science*. Norwich: Fletcher and Son.

———. 1870. *The Student's Flora of the British Islands*. London: Macmillan.

———. 1872a. "Hooker's Reply to Prof. Owen." *Nature*, no. 156: 516–17.

———. 1872b. "Report of the Director on the Royal Gardens, Kew, during the Year 1871." *Gardeners' Chronicle*, no. 19: 634–35.

———. 1872–97. *The Flora of British India*. London: Lovell Reeve.

———. 1887. *Reply to the Toast "The Medallists," Anniversary Dinner of the Royal Society*. London: Royal Society of London.

Hooker, Joseph Dalton, and George Bentham. 1862–83. *Genera Plantarum ad Exemplaria Imprimis in Herbariis Kewensibus Servata Definita*. London: A. Black / Lovell Reeve.

Hooker, Joseph Dalton, and Thomas Thomson. 1855. *Introductory Essay to the "Flora Indica."* London: W. Pamplin.

Hooker, William Jackson. 1816. *British Jungermanniae*. London: Longman, Hurst, et al.

———. 1821. *Botanical Illustrations: Being a Series of Figures Intended to Illustrate the Terms Employed in a Course of Lectures on Botany, with Descriptions*. Edinburgh.

———. 1822. *Botanical Illustrations: Being a Series of Figures Designed to Illustrate the Terms Employed in a Course of Lectures on Botany, with Descriptions*. Edinburgh: Constable.

———. 1849. "Botany." In *A Manual of Scientific Enquiry: Prepared for the Use of Officers in Her Majesty's Navy and Travellers in General*, edited by John F. W. Herschel, 416–37. London: John Murray.

———. n.d. *A Few Plain Instructions for Collecting and Transporting Plants in Foreign Countries*. Richmond, Surrey: Royal Gardens, Kew.

Huxley, Thomas Henry. 1854. "Review of 'The Vestiges of the Natural History of Creation.'" *British and Foreign Medico-Chirurgical Review*, no. 13: 425–39.

King, George. 1872. Letter to the Editor (Response to Owen). *Nature* 7, no. 162: 83–84.

Kingsley, Charles. 1848 (1994). *Yeast*. Stroud: Alan Sutton.

[Lankester, Edwin]. 1846a. "Review of Lindley's *Vegetable Kingdom*." *Athenaeum*, no. 971: 573–74.

———. 1846b. "Review of Ray Society's *Reports and Papers on Botany*." *Athenaeum*, no. 976: 704–5.

———. 1847. "Review of *Curtis's Botanical Magazine*." *Athenaeum*, no. 1035: 911.

———. 1848. "Review of MJ Schleiden *The Plant a Biography*." *Athenaeum*, no. 1092: 973–74.

———. 1853. "Review of Ward *On the Growth of Plants in Closely-Glazed Cases*." *Athenaeum*, no. 1320: 192.

———. 1854. "Review of J. H. Balfour's *Class-Book of Botany*." *Athenaeum*, no. 1380: 436.

———. 1857. "Review of Henfrey *Elementary Course in Botany*." *Athenaeum*, no. 1562: 1236–37.

———. 1860. "Review of Joseph Hooker's *Flora Tasmaniae*." *British and Foreign Medi-

co-Chirurgical Review, or Quarterly Journal of Practical Medicine and Surgery 13 (April): 385.

Law, William. 1873. Letter to the Editor (National Herbaria). *Nature* 7, no. 170: 243.

[Leifchild, John R.]. 1859. "Review of Forbes *Outlines of the Natural History of Europe.*" *Athenaeum*, no. 1669: 535–37.

[Lewes, George Henry]. 1852. "Goethe as a Man of Science." *Westminster Review* 58, no. 2: 479–506.

Lindley, John. 1830. *An Introduction to the Natural System of Botany: or, a Systematic View of the Organisation, Natural Affinities, and Geographical Distribution of the Whole Vegetable Kingdom.* London: Longman.

[Lindley, John]. 1831. "Review of *New Elements of Botany* by A. Richard." *Athenaeum*, no. 182: 262.

———. 1833a. "Review of J. Forbes Royle *Illustrations of the Botany and Other Branches of the Natural History of the Himalayan Mountains.*" *Athenaeum*, no. 319: 829.

———. 1833b. "Review of Thomas Castle *A Synopsis of Systematic Botany.*" *Athenaeum*, no. 321: 870.

Lindley, John. 1834a. *Ladies' Botany: or a Familiar Introduction to the Study of the Natural System of Botany.* Vol. 1. London: James Ridgway and Sons.

———. 1834b. "On the Principal Questions at Present Debated in the Philosophy of Botany." In *Report of the British Association for the Advancement of Science*, 27–57. London: John Murray.

[Lindley, John]. 1834c. "Review of Daubeny *Inaugural Lecture on the Study of Botany.*" *Athenaeum*, no. 344: 401–2.

Lindley, John. 1835. *A Synopsis of the British Flora.* London: Longman, Rees, Orme, Brown, Green, and Longman.

———. 1838. "Report to the Committee Appointed by the Lords of the Treasury in January 1838 to Inquire into the Management, &c of the Royal Gardens at Kew." London: House of Commons.

———. 1847. *The Elements of Botany: Structural and Physiological.* London: Bradbury and Evans.

[?Lindley, John]. 1855. "Hooker's *Flora Indica.*" *Gardeners' Chronicle*, no. 44: 723.

Lindley, John. 1858. *Descriptive Botany: or, The Art of Describing Plants Correctly in Scientific Language.* London.

[Lindley, John]. 1859. "Review of Watson's `Cybele Britannica.'" *Gardeners' Chronicle*, no. 46 (12 November): 911–12.

Lindley, John. 1862. *School Botany, Descriptive Botany, and Vegetable Physiology; or, the Rudiments of Botanical Science.* 12th ed. London: Bradbury, Evans.

[?Lockyer, Norman]. 1872a. "Notes (Hooker and Ayrton, July 11)." *Nature* 6, no. 141: 209.

———. 1872b. "Notes (Hooker and Ayrton, August 8)." *Nature* 6, no. 145: 280–81.

———. 1872c. "Notes (Hooker and Ayrton, September 12)." *Nature* 6, no. 150: 401.

Lubbock, John, ed. 1872. *Copies "Of Papers Relating to Changes Introduced into the Administration of the Office of Works Affecting the Direction and Management of the Gardens at Kew."* London: House of Commons.

[Luxford, George]. 1850. "Natural Systems of Botany: Review of Lindley's *Vegetable Kingdom* and Other Works." *Westminster Review* 54, no. 1: 38–65.

Lyell, Charles, Charles Darwin, et al. 1872. "Letter to W. E. Gladstone." *Nature* 6, no. 141: 211–16.

Lytton, Edward Bulwer. 1836. *England and the English*. London: Richard Bentley.

MacLeay, William Sharp. 1825. "Remarks on the Identity of Certain General Laws Which Have Been Lately Observed to Regulate the Natural Distribution of Insects and Fungi." *Transactions of the Linnean Society of London* 14, pt. 1: 46–68.

———. 1830. *A Letter on the Dying Struggle of the Dichotomous System*. London: Richard Taylor.

[?Masters, Maxwell Tylden]. 1872a. Editorial (July 13). *Gardeners' Chronicle*, no. 28: 933–34.

———. 1872b. "Mr. Ayrton and Dr. Hooker." *Gardeners' Chronicle*, no. 28: 939.

———. 1872c. Editorial (July 20). *Gardeners' Chronicle*, no. 29: 967–68.

———. 1872d. Editorial (July 27). *Gardeners' Chronicle*, no. 30: 1001–2.

———. 1872e. Editorial (August 10). *Gardeners' Chronicle*, no. 32: 1065–66.

———. 1873. Editorial (December 6). *Gardeners' Chronicle*, no. 49: 1631.

———. 1875. Editorial (Kew Director's Report). *Gardeners' Chronicle*, n.s., 4, no. 89: 332.

———. 1876. "The Royal Botanic Gardens, Kew." *Gardeners' Chronicle*, n.s., 6, no. 136: 1–12.

Murchison, Roderick Impey. 1846. "Presidential Address to the British Association for the Advancement of Science." In *Report of the British Association for the Advancement of Science*, xxvii–xliii. London: John Murray.

Newman, Edward. 1848. "Notice of 'The Principles of Nature, Her Divine Revelations, and a Voice to Mankind.' By and through Andrew Jackson Davis, the Poughkeepsie Seer and Clairvoyant." *Phytologist: A Popular Botanical Miscellany* 3: 149–57.

Newman, J. 1845. *Catalogue of Philosophical Instruments: Manufactured and Sold by J. Newman*. London: J. Newman.

[?Oliver, Daniel]. 1863a. "British Floras." *Natural History Review: A Quarterly Journal of Biological Science*, no. 9: 34–39.

[Oliver, Daniel]. 1863b. "New Colonial Floras." *Natural History Review: A Quarterly Journal of Biological Science* 3, no. 12: 497–507.

[Owen, Richard]. 1860. "Review: 'On the Origin of Species by Means of Natural Selection' and Other Works." *Edinburgh Review* 111, no. 226 (April): 487–532.

Owen, Richard. 1872a. Letter to the Editor (Response to Hooker). *Nature* 7, no. 158: 5–6.

———. 1872b. "Statement Relative to the Botanical Departments Respectively under the Trustees of the British Museum and the Commissioners of Works." In *Copies "Of Papers Relating to Changes Introduced into the Administration of the Office of Works Affecting the Direction and Management of the Gardens at Kew,"* edited by John Lubbock, 169–75. London: House of Commons.

Parsons, William, Earl of Rosse. 1854. "Address Delivered before the Royal Society." *Proceedings of the Royal Society of London* 7: 247–63.

Rees, Abraham, ed. 1819. *The Cyclopaeida: or Universal Dictionary of Arts, Sciences, and Literature*. London: Longman.

Robinson, Rev. Thomas Romney. 1850. "Presidential Address to the British Association for the Advancement of Science, 1849." In *Annual Reports of the British Association for the Advancement of Science*, xxix–xlvi. London: John Murray.

Robison, John. 1841. "Camera Lucida for Drawing Flowers." *Edinburgh New Philosophical Journal* 30, no. 60: 402–4.

Ruskin, John. 1857 (1971). *The Elements of Drawing*. New York: Dover.

[Seemann, Berthold Carl]. 1863. "Review of John T. Boswell-Syme (ed.), *English Botany*." *Athenaeum*, no. 1872: 327–28.

———. 1867. "Review of Alexander Silver, *Outlines of Elementary Botany*." *Athenaeum*, no. 1994: 56.

Smiles, Samuel. 1859 (2002). *Self-Help: With Illustrations of Character, Conduct and Perseverance*. Oxford: Oxford University Press.

Smith, Albert Richard. 1847. *The Natural History of the Gent*. London: D. Bogue.

Steward, J. H. 1867. *Catalogue of Optical Instruments, Clocks, Watches and Jewellery*. London: J. H. Steward.

Strickland, Hugh E. 1841. "On the True Method of Discovering the Natural System in Zoology and Botany." *Annals and Magazine of Natural History* 6, no. 36: 184–94.

Swainson, William. 1834. *A Preliminary Discourse on the Study of Natural History*. London: Longman.

Taylor, Rev. Richard. 1848. *A Leaf from the Natural History of New Zealand*. Wellington: Robert Stokes, at the office of the "New Zealand Spectator and Cook's Strait Guardian," Manners-Street; Te Aro & J. Williamson, "New Zealander Office," Auckland.

———. 1866. *The Age of New Zealand*. Auckland: Geo. T. Chapman, Bookseller and Stationer.

Thiselton-Dyer, William. 1873. Letter to the Editor (National Herbaria). *Nature* 7, no. 170: 243–44.

[Thomson, Thomas]. 1865. "Species and Subspecies." *Natural History Review: A Quarterly Journal of Biological Science* 5: 226–42.

Ward, Nathaniel Bagshaw. 1852. *On the Growth of Plants in Closely Glazed Cases*. London: John Van Voorst.

Watson, Hewett C. 1845. "On the Theory of 'Progressive Development,' Applied in Explanation of the Origin and Transmutation of Species." *Phytologist: A Popular Botanical Miscellany* 2: 40–47, 61–68, 108–13, 225–28.

———. 1847. *Cybele Britannica; or British Plants and Their Geographical Relations*. Vol. 1. London: Longmans.

———. 1859. *Cybele Britannica; or British Plants and Their Geographical Relations*. Vol. 4. London: Longmans.

———. 1860. *Part First of a Supplement to the "Cybele Britannica": To Be Continued Occasionally as a Record of Progressive Knowledge concerning the Distribution of Plants in Britain*. London: Printed for Private Distribution.

Whewell, William. 1845. *Of a Liberal Education in General; and with Particular Reference to the Leading Studies of the University of Cambridge*. London: John W. Parker.

———. 1857. *History of the Inductive Sciences from the Earliest to the Present Time*. London: J. W. Parker.

Wilson, George, and Archibald Geikie. 1861. *Memoir of Edward Forbes*. London: Macmillan and Edmonston.

SECONDARY SOURCES

Adams, James Eli. 1999. "Victorian Sexualities." In *A Companion to Victorian Literature and Culture*, edited by Herbert F. Tucker, 125–38. Oxford: Blackwell.

Alborn, Timothy L. 1996. "The Business of Induction: Industry and Genius in the Language of British Scientific Reform, 1820–1840." *History of Science* 34: 91–115.

———. 2004. "Lubbock, John, First Baron Avebury (1834–1913)." In *Oxford Dictionary of National Biography*, Oxford University Press, http://www.oxforddnb.com/view/article/34618 (accessed 30 August 2006).

Allan, Mea. 1967. *The Hookers of Kew, 1785–1911*. London: Michael Joseph.

Allen, David Elliston. 1959. "The History of the Vasculum." *Proceedings of the Botanical Society of the British Isles* 3: 135–50.

———. 1965. "Some Further Light on the History of the Vasculum." *Proceedings of the Botanical Society of the British Isles* 6: 105–12.

———. 1969. *The Victorian Fern Craze*. London: Hutchinson.

———. 1986. *The Botanists: A History of the Botanical Society of the British Isles through a Hundred and Fifty Years*. Winchester: St. Paul's Bibliographies.

———. 1994. *The Naturalist in Britain: A Social History*. Princeton, NJ: Princeton University Press.

———. 1996. "Tastes and Crazes." In *Cultures of Natural History*, edited by Nicholas Jardine, James A. Secord, and Emma Spary, 394–407. Cambridge: Cambridge University Press.

———. 2001a. "Early Professionals in British Natural History." In *Naturalists and Society: The Culture of Natural History in Britain, 1700–1900*, 1–12. Aldershot: Ashgate / Variorum.

———. 2001b. *Walking the Swards: Medical Education and the Rise and Spread of the Botanical Field Class*. Aldershot: Ashgate<ts>/</ts>Variorum.

———. 2003. "Bentham's *Handbook of the British Flora*: From Controversy to Cult." *Archives of Natural History* 30, no. 2: 224–36.

Anderson, Patricia J. 1994. *The Printed Image and the Transformation of Popular Culture, 1790–1860*. Oxford: Clarendon Press.

Arnold, David. 2005. "Envisioning the Tropics: Joseph Hooker in India and the Himalayas, 1848–1850." In *Tropical Visions in an Age of Empire*, edited by Felix Driver and Luciana Martins, 137–55. Chicago: University of Chicago Press.

Baber, Zaheer. 1996. *The Science of Empire: Scientific Knowledge, Civilization, and Colonial Rule in India*. Albany: State University of New York Press.

Bagnall, Austin Graham. 1970. *New Zealand National Bibliography to the Year 1960*. Vol. 1, *To 1889*. Wellington, NZ: A. R. Shearer, Government Printer.

Bagnall, Austin Graham, and G. C. Peterson. 1948. *William Colenso, Printer, Missionary, Botanist, Explorer, Politician: His Life and Journeys*. Wellington, NZ: A. H. and A. W. Reed.

Barber, Lynn. 1980. *The Heyday of Natural History, 1820–1870*. London: Jonathan Cape.

Barker, R. M., and W. R. Barker. 1990. "Botanical Contributions Overlooked: The Role and Recognition of Collectors, Horticulturalists, Explorers and Others in the Early Documentation of the Australian Flora." In *A History of Systematic Botany in Australia*, edited by P. S. Short, 37–85. South Yarra, Victoria: Australian Systematic Botany Society.

Barlow, Nora, ed. 1958. *The Autobiography of Charles Darwin*. London: Collins.

Barton, Ruth. 1998. "Huxley, Lubbock, and Half a Dozen Others: Professionals and Gentlemen in the Formation of the X Club, 1851–1864." *Isis* 89, no. 3: 410–44.

———. 2000. "Haast and the Moa: Reversing the Tyranny of Distance." *Pacific Science* 54, no. 3: 251–63.

———. 2003. "'Men of Science': Language, Identity and Professionalization in the Mid-Victorian Scientific Community." *History of Science* 41: 73–119.

Basalla, George. 1967. "The Spread of Western Science." *Science* 156: 611–22.

Baulch, W. 1961. "Ronald Campbell Gunn." In *Van Diemen's Land Correspondents: Letters from R. C. Gunn, R. W. Lawrence, Jorgen Jorgenson, Sir John Franklin and Others to Sir William J. Hooker, 1827–1849*, edited by T. E. Burns and J. R. Skemp, xiii–xix. Launceston, UK: Queen Victoria Museum.

Belich, James. 1986. *The Victorian Interpretation of Racial Conflict: The Maori, the British, and the New Zealand Wars*. Montreal: McGill-Queen's University Press.

Bellon, Richard. 2000. "Joseph Hooker and the Progress of Botany, 1845–65." PhD diss., University of Washington.

———. 2001. "Joseph Hooker's Ideals for a Professional Man of Science." *Journal of the History of Biology* 34: 51–82.

———. 2005. "A Question of Merit: John Hutton Balfour, Joseph Hooker and the 'Concussion' over the Edinburgh Chair of Botany." *Studies in the History and Philosophy of the Biological and Biomedical Sciences* 36: 25–54.

———. 2006. "Joseph Hooker Takes a 'Fixed Post': Transmutation and the 'Present Unsatisfactory State of Systematic Botany,' 1844–1860." *Journal of the History of Biology* 39: 1–39.

Bermingham, Ann. 2000. *Learning to Draw: Studies in the Cultural History of a Polite and Useful Art*. New Haven, CT: Yale University Press.

Best, Geoffrey. 1979. *Mid-Victorian Britain, 1851–75*. London: Fontana Press.

Blainey, Geoffrey. 1966. *The Tyranny of Distance: How Distance Shaped Australia's History*. Melbourne: Sun Books.

Blunt, Wilfrid. 1978. *In for a Penny: A Prospect of Kew Gardens; Their Flora, Fauna and Falballas*. London: Hamish Hamilton.

Blunt, Wilfrid, and William T. Stearn. 1994. *The Art of Botanical Illustration*. London: Antique Collector's Club.

Bonneuil, Christophe. 2002. "The Manufacture of Species: Kew Gardens, the Empire, and the Standardisation of Taxonomic Practice in Late 19th Century Botany." In *Instruments, Travel, and Science: Itineraries of Precision from the 17th to the 20th Century*, edited by Marie-Noelle Bourguet, Christian Licoppe, and H. Otto Sibum, 189–215. London: Routledge.

Boulger, G. S., and William T. Stearn. 2004. "Masters, Maxwell Tylden (1833–1907)." In *Oxford Dictionary of National Biography*, Oxford University Press, http://www.oxforddnb.com/view/article/34928 (accessed 30 August 2006).

Bowler, Peter J. 1989. *Evolution: The History of an Idea*. Berkeley and Los Angeles: University of California Press.

———. 1992. *The Fontana History of the Environmental Sciences*. London: Fontana Press.

Brander, Michael. 1975. *The Victorian Gentleman*. London: Gordon Cremonesi.

Briggs, Asa. 1990. *Victorian Cities*. Harmondsworth, UK: Penguin Books.

Brock, W. H. 1984. "Brewster as a Scientific Journalist." In *"Martyr of Science": Sir David Brewster, 1781–1863*, edited by A. D Morrison-Low and John R. R. Christie, 37–44. Edinburgh: Royal Scottish Museum.

———. 2004. "Taylor, Richard (1781–1858)." In *Oxford Dictionary of National Biography*, edited by Lawrence Goldman. Oxford: Oxford University Press.

Brock, William Hodson, and Arthur Jack Meadows. 1984. *The Lamp of Learning: Taylor and Francis and the Development of Science Publishing*. London: Taylor and Francis.

Brockway, Lucille. 1979. *Science and Colonial Expansion: The Role of the British Royal Botanic Gardens*. New York: Academic Press.

Browne, Janet. 1979. "C. R. Darwin and J. D. Hooker: Episodes in the History of Plant Geography, 1840–1860." PhD diss., Imperial College, University of London.

———. 1980. "Darwin's Botanical Arithmetic and the 'Principle of Divergence,' 1854–1858." *Journal of the History of Biology* 13, no. 1 (Spring): 53–89.

———. 1981. "The Making of the *Memoir* of Edward Forbes, F.R.S." *Archives of Natural History* 10, no. 2: 205–19.

———. 1983. *The Secular Ark: Studies in the History of Biogeography*. New Haven, CT: Yale University Press.

———. 1989. "Botany for Gentlemen: Erasmus Darwin and *The Loves of the Plants*." *Isis* 80, no. 304: 593–621.

———. 1992. "A Science of Empire: British Biogeography before Darwin." *Revue d'histoire des sciences* 45, no. 4: 453–75.

———. 1995. *Charles Darwin: Voyaging*. London: Jonathan Cape.

———. 1996. "Biogeography and Empire." In *Cultures of Natural History*, edited by Nicholas Jardine, James A. Secord, and Emma Spary, 305–21. Cambridge: Cambridge University Press.

———. 2002. *Charles Darwin: The Power of Place*. London: Jonathan Cape.

Brummitt, R. K., and C. E. Powell. 1992. *Authors of Plant Names: A List of Authors of Scientific Names of Plants, with Recommended Standard Forms of Their Names, Including Abbreviations*. London: Royal Botanic Gardens, Kew.

Brunner, Bernd. 2003. *The Ocean at Home: An Illustrated History of the Aquarium*. New York: Princeton Architectural Press.

Buchanan, A. M. 1990. "Ronald Campbell Gunn (1808–1881)." In *A History of Systematic Botany in Australia*, edited by P. S. Short, 179–92. South Yarra, Victoria: Australian Systematic Botany Society.

Buchanan, R. A. 1983. "Gentlemen Engineers: The Making of a Profession." *Victorian Studies* 26, no. 4: 407–29.

Burkhardt, Frederick, and Sydney Smith, eds. 1986. *The Correspondence of Charles Darwin*. Vol. 2, *1837–1843*. Cambridge: Cambridge University Press.

———. 1987. *The Correspondence of Charles Darwin*. Vol. 3, *1844–1846*. Cambridge: Cambridge University Press.

———. 1988. *The Correspondence of Charles Darwin*. Vol. 4, *1847–1850*. Cambridge: Cambridge University Press.

———. 1989. *The Correspondence of Charles Darwin*. Vol. 5, *1851–1855*. Cambridge: Cambridge University Press.

———. 1990. *The Correspondence of Charles Darwin*. Vol. 6, *1856–1857*. Cambridge: Cambridge University Press.

———. 1991. *The Correspondence of Charles Darwin*. Vol. 7, *1858–1859*. Cambridge: Cambridge University Press.

Burkhardt, Frederick, Janet Browne, Duncan M. Porter, et al., eds. 1993. *The Correspondence of Charles Darwin*. Vol. 8, *1860*. Cambridge: Cambridge University Press.

Burkhardt, Frederick, Joy Harvey, Duncan M. Porter, et al., eds. 1997. *The Correspondence of Charles Darwin*. Vol. 10, *1862*. Cambridge: Cambridge University Press.

Burkhardt, Frederick, Duncan M. Porter, Sheila Ann Dean, et al., eds. 1999. *The Correspondence of Charles Darwin*. Vol. 11, *1863*. Cambridge: Cambridge University Press.

———. 2002. *The Correspondence of Charles Darwin*. Vol. 13, *1865*. Cambridge: Cambridge University Press.

Burns, T. E., and J. R. Skemp, eds. 1961. *Van Diemen's Land Correspondents: Letters from R. C. Gunn, R. W. Lawrence, Jorgen Jorgenson, Sir John Franklin and Others to Sir William J. Hooker, 1827–1849*. Launceston, UK: Queen Victoria Museum.

———. 2006. "Gunn, Ronald Campbell (1808–1881)." In *Australian Dictionary of Biography*, online ed., Australian National University, http://www.adb.online.anu.edu.au/biogs/A010448b.htm.

Cain, P. J. 1999. "Economics and Empire: The Metropolitan Context." In *The Oxford History of the British Empire*, vol. 3, *The Nineteenth Century*, edited by Andrew Porter, 31–52. Oxford: Oxford University Press.

Callon, Michel. 1985. "Some Elements of a Sociology of Translation: Domestication of the Scallops and the Fishermen of St. Brieuc Bay." In *Power, Action and Belief*, edited by John Law, 196–233. London: Routledge and Kegan Paul.

Camerini, Jane. 1996. "Wallace in the Field." *Osiris* 11: 44–65.

———. 1997. "Remains of the Day: Early Victorians in the Field." In *Victorian Science in Context*, edited by Bernard Lightman, 354–77. Chicago: University of Chicago Press.

Cantor, Geoffrey, Gowan Dawson, et al., eds. 2004. *Science in the Nineteenth Century Periodical*. Cambridge: Cambridge University Press.

Carter, Tom. 1984. *The Victorian Garden*. London: Bell and Hyman.

Cawood, J. 1979. "The Magnetic Crusade: Science and Politics in Early Victorian Britain." *Isis* 70: 492–518.

Chadarevian, Soraya de. 1993. "Instruments, Illustrations, Skills, and Laboratories in Nineteenth Century German Botany." In *Non-verbal Communication in Science prior to 1900*, edited by R. G. Mazzolini, 529–50. Florence, Italy: Olschki.

Chambers, David Wade. 1991. "Does Distance Tyrannize Science?" In *International Science and National Scientific Identity: Australia between Britain and America*, edited by R. W. Home and S. G. Kohlstedt, 19–38. Dordrecht: Kluwer Academic.

Chichester, H. M. 2004. "Munro, William (1818–80)." Revised by Andrew Grout. In *Oxford Dictionary of National Biography*, Oxford University Press, http://www.oxforddnb.com/view/article/19550 (accessed 1 August 2006).

Clark, Manning, ed. 1957. *Sources of Australian History*. Melbourne: Oxford University Press.

Coggon, Jennifer. 2002. "Quinarianism after Darwin's *Origin:* The Circular System of William Hincks." *Journal of the History of Biology* 35: 5–42.

Coleman, D. C. 1958. *The British Paper Industry, 1495–1860: A Study in Industrial Growth*. Oxford: Clarendon Press.

Colquhoun, Kate. 2003. *A Thing in Disguise: The Visionary Life of Joseph Paxton*. London: Fourth Estate.

Comrie, J. D. 1932. *History of Scottish Medicine*. London: Wellcome History of Medicine Museum.

Cooter, Roger, and Stephen Pumfrey. 1994. "Separate Spheres and Public Places: Reflections on the History of Science Popularization and Science in Popular Culture." *History of Science* 32: 237–67.

Corfield, Penelope F. 2000. *Power and the Professions in Britain, 1700–1850*. London: Routledge.

Crary, Jonathan. 1990. *Techniques of the Observer: On Vision and Modernity in the Nineteenth Century*. Cambridge, MA: MIT Press.

Crosby, Alfred W. 1986. *Ecological Imperialism: The Biological Expansion of Europe, 900–1900*. Cambridge: Cambridge University Press.

Crowther, M. Anne, and Brenda White. 1988. *On Soul and Conscience: The Medical Expert and Crime*. Aberdeen: Aberdeen University Press.

Dance, S. Peter. 2004. "Reeve, Lovell Augustus (1814–1865)." In *Oxford Dictionary of National Biography*, edited by Lawrence Goldman. Oxford: Oxford University Press.

Darwin, Francis. 1888. *The Life and Letters of Charles Darwin*. Vol. 2. London: John Murray.

Daston, Lorraine. 2001. "Scientific Objectivity with and without Words." In *Little Tools of Knowledge: Historical Essays on Academic and Bureaucratic Practices*, edited by Peter Becker and William Clark, 259–84. Ann Arbor: University of Michigan Press.

———. 2004. "Attention and the Values of Nature in the Enlightenment." In *The Moral Authority of Nature*, edited by Lorraine Daston and Fernando Vidal, 100–126. Chicago: University of Chicago Press.

Daston, Lorraine, and Peter Galison. 1992. "The Image of Objectivity." *Representations* 40: 81–128.

Daunton, Martin. 1989. "'Gentlemanly Capitalism' and British Industry, 1820–1914." *Past and Present*, no. 122: 119–58.

Desmond, Adrian. 1982. *Archetypes and Ancestors: Palaeontology in Victorian London, 1850–1875*. London: Blond and Briggs.

———. 1985. "The Making of Institutional Zoology in London, 1822–1836." *History of Science* 23: 153–85, 224–50.

———. 1989. *The Politics of Evolution: Morphology, Medicine, and Reform in Radical London*. Chicago: University of Chicago Press.

———. 1994. *Huxley: The Devil's Disciple*. London: Michael Joseph.

———. 1997. *Huxley: Evolution's High Priest*. London: Michael Joseph.

———. 2001. "Redefining the X Axis: 'Professionals,' 'Amateurs' and the Making of Mid-Victorian Biology—a Progress Report." *Journal of the History of Biology* 34: 3–50.

Desmond, Adrian, and James Moore. 1991. *Darwin*. London: Michael Joseph.

Desmond, Ray. 1994. *Dictionary of British and Irish Botanists and Horticulturalists, Including Plant Collectors, Flower Painters and Garden Designers*. London: Taylor and Francis and Natural History Museum.

———. 1995. *Kew: A History of the Royal Botanic Gardens*. London: Harvill Press.

———. 1999. *Sir Joseph Dalton Hooker: Traveller and Plant Collector.* Woodbridge, Suffolk: Antique Collector's Club.

Dettelbach, Michael. 1996. "Humboldtian Science." In *Cultures of Natural History,* edited by Nicholas Jardine, James A. Secord, and Emma Spary, 287–304. Cambridge: Cambridge University Press.

———. 2005. "The Stimulations of Travel: Humboldt's Physiological Construction of the Tropics." In *Tropical Visions in an Age of Empire,* edited by F. Driver and L. Martins, 43–58. Chicago: University of Chicago Press.

"Dictionary of Quaker Biography." n.d. Unpublished, London. ©Library of the Religious Society of Friends.

di Gregorio, Mario A. 1982. "In Search of the Natural System: Problems of Zoological Classification in Victorian Britain." *History and Philosophy of the Life Sciences* 4, no. 2: 225–54.

DiNito, Andrea, and David Winter. 1999. *The Pressed Plant: The Art of Botanical Specimens, Nature Prints, and Sun Pictures.* New York: Stewart, Tabori, and Chang.

Dow, Derek, and Michael Moss. 1988. "The Medical Curriculum at Glasgow in the Early Nineteenth Century." *History of Universities* 7: 227–57.

Drayton, Richard H. 1993. "Imperial Science and a Scientific Empire: Kew Gardens and the Uses of Nature, 1772–1903." PhD diss., Yale University.

———. 1999. "Science, Medicine, and the British Empire." In *Historiography,* edited by Robin W. Winks, 264–76. Oxford: Oxford University Press.

———. 2000. *Nature's Government: Science, Imperial Britain, and the "Improvement" of the World.* New Haven, CT: Yale University Press.

Driver, Felix, and Martins Luciana, eds. 2005. *Tropical Visions in an Age of Empire.* Chicago: University of Chicago Press.

Egerton, Frank N. 1980. "Hewett Cottrell Watson." In *Dictionary of Scientific Biography,* edited by Charles Coulston Gillispie, 189–91. New York: Charles Scribner and Sons.

———. 2003. *Hewett Cottrell Watson: Victorian Plant Ecologist and Evolutionist.* Aldershot, UK: Ashgate.

Elliott, Brent. 1986. *Victorian Gardens.* London: B. T. Batsford.

Endersby, Jim. 2000. "A Garden Enclosed: Botanical Barter in Sydney, 1818–1839." *British Journal of the History of Science* 33, no. 118: 313–34.

———. 2001. "'From Having No Herbarium': Local Knowledge vs. Metropolitan Expertise; Joseph Hooker's Australasian Correspondence with William Colenso and Ronald Gunn." *Pacific Science* 55, no. 4: 343–58.

———. 2002. "Putting Plants in Their Place: Joseph Hooker's Philosophical Botany, 1838–65." PhD diss., University of Cambridge.

———. 2004. "Hooker, Sir Joseph Dalton (1817–1911)." In *Oxford Dictionary of National Biography,* Oxford University Press, http://www.oxforddnb.com/view/article/33970 (accessed 21 June 2007).

———. 2005. "Classifying Sciences: Systematics and Status in Mid-Victorian Natural History." In *The Organisation of Knowledge in Victorian Britain,* edited by M. Daunton, 61–85. Oxford: British Academy / Oxford University Press. 2005.

English, Mary P. 2004. "Lankester, Edwin (1814–1874)." In *Oxford Dictionary of National Biography,* Oxford University Press, http://www.oxforddnb.com/view/article/16054 (accessed 22 February 2007).

Farley, John. 1982. *Gametes and Spores: Ideas about Sexual Reproduction, 1750–1914.* Baltimore, MD: Johns Hopkins University Press.

Fichman, Martin. 1977. "Wallace: Zoogeography and the Problem of Land Bridges." *Journal of the History of Biology* 10, no. 1: 45–63.

FitzGerald, Sylvia. 2004. "Hooker, Sir William Jackson (1785–1865)." In *Oxford Dictionary of National Biography,* edited by Lawrence Goldman. Oxford: Oxford University Press.

Ford, Brian J. 1985. *Single Lens: The Story of the Simple Microscope.* London: Heinemann.

Foucault, Michel. 1989. *The Order of Things: An Archaeology of the Human Sciences.* London: Tavistock/Routledge.

Fox, Celina, and Michael Wolff. 1973. "Pictures from the Magazines." In *The Victorian City: Images and Realities,* edited by Harold James Dyos and Michael Wolff, 559–82. London: Routledge and Kegan Paul.

Fraser, Angus 2004. "Turner, Dawson (1775–1858)." In *Oxford Dictionary of National Biography,* edited by Lawrence Goldman. Oxford: Oxford University Press.

Fyfe, Aileen. 2004. *Science and Salvation: Evangelical Popular Science Publishing in Victorian Britain.* Chicago: University of Chicago Press.

Galloway, D. J. 1990. "Charles Knight." In *The Dictionary of New Zealand Biography,* 228–29. Wellington: Allen and Unwin.

———. 1998. "Joseph Hooker, Charles Knight, and the Commissioning of New Zealand's First Popular Flora: Hooker's *Handbook of the New Zealand Flora* (1864–1867)." *Tuhinga,* no. 10: 31–62.

Gates, Barbara T. 1998. *Kindred Nature: Victorian and Edwardian Women Embrace the Living World.* Chicago: University of Chicago Press.

Gay, Hannah, and John W. Gay. 1997. "Brothers in Science: Science and Fraternal Culture in Nineteenth-Century Britain." *History of Science* 35: 425–53.

Geyer-Kordesch, Johanna, and Fiona Macdonald. 1999. *Physicians and Surgeons in Glasgow: The History of the Royal College of Physicians and Surgeons in Glasgow, 1599–1858.* London: Hambledon Press.

Gilbert, Lionel. 1986. *The Royal Botanic Gardens, Sydney: A History, 1816–1985.* Melbourne: Oxford University Press.

Gillham, Nicholas Wright. 2001. *A Life of Sir Francis Galton: From African Exploration to the Birth of Eugenics.* Oxford: Oxford University Press.

Gilmour, Robin. 1993. *The Victorian Period: The Intellectual and Cultural Context of English Literature, 1830–1890.* London: Longman.

Gledhill, David. 2002. *The Names of Plants.* Cambridge: Cambridge University Press.

Glenn, Rewa. 1950. *The Botanical Explorers of New Zealand.* Wellington: A. H. and A. W. Reed.

Golinski, Jan. 1998 (2005). *Making Natural Knowledge: Constructivism and the History of Science.* Chicago: University of Chicago Press.

Good, Gregory A. 2004. "Sabine, Sir Edward (1788–1883)." In *Oxford Dictionary of National Biography,* Oxford University Press, http://www.oxforddnb.com/view/article/24436 (accessed 22 February 2007).

Gooday, Graeme N. J. 1997. "Instrumentation and Interpretation: Managing and Representing the Working Environments of Victorian Experimental Science." In *Victorian Science in Context,* edited by Bernard Lightman, 409–37. Chicago: University of Chicago Press.

Gooding, David, Trevor Pinch, et al., eds. 1989. *The Uses of Experiment: Studies in the Natural Sciences.* Cambridge: Cambridge University Press.

Green, J. Reynolds. 1909 (1967). *A History of Botany, 1860–1900: Being a Continuation of Sachs "History of Botany, 1530–1860."* New York: Russell and Russell.

Grove, Richard H. 1995. *Green Imperialism: Colonial Expansion, Tropical Island Edens, and the Origins of Environmentalism, 1600–1860.* Cambridge: Cambridge University Press.

Hall, Catherine. 2002. *Civilising Subjects: Metropole and Colony in the English Imagination, 1830–1867.* Chicago: University of Chicago Press.

Hall, Catherine, Keith McClelland, et al., eds. 2000. *Defining the Victorian Nation: Class, Race, Gender and the Reform Act of 1867.* Cambridge: Cambridge University Press.

Hamilton, David. 1981. *The Healers: A History of Medicine in Scotland.* Edinburgh: Canongate.

Hamlin, B. G. 1971. "The Bryophyte Collections of William Colenso in the Dominion Museum, Wellington." *New Zealand Journal of Botany* 9: 695–98.

Hartog, P. J. 2004. "Rue, Warren de la (1815–1889)." Revised by A. J. Meadows. In *Oxford Dictionary of National Biography,* Oxford University Press, http://www.oxforddnb.com/view/article/7447 (accessed 16 August 2006).

Heyck, T. W. 1982. *The Transformation of Intellectual Life in Victorian England.* London: Croom Helm.

Home, R. W., and S. G. Kohlstedt, eds. 1991. *International Science and National Scientific Identity: Australia between Britain and America.* Dordrecht: Kluwer Academic.

Home, R. W., A. M. Lucas, et al., eds. 1998. *Regardfully Yours: Selected Correspondence of Ferdinand von Mueller.* Vol. 1, *1840–1859.* Bern: Peter Lang.

———. 2002. *Regardfully Yours: Selected Correspondence of Ferdinand von Mueller.* Vol. 2, *1860–75.* Bern: Peter Lang.

Hoppen, K. Theodore. 1998. *The Mid-Victorian Generation, 1846–1886.* Oxford: Oxford University Press.

Hoyles, Martin. 1991. *The Story of Gardening.* London: Journeyman.

Hull, David L., ed. 1973. *Darwin and His Critics: The Reception of Darwin's Theory of Evolution by the Scientific Community.* Cambridge, MA: Harvard University Press.

Huxley, Leonard. 1918a. *Life and Letters of Joseph Dalton Hooker.* Vol. 1. London: John Murray.

———. 1918b. *Life and Letters of Joseph Dalton Hooker.* Vol. 2. London: John Murray.

Jardine, Nicholas. 1996. "*Naturphilosophie* and the Kingdoms of Nature." In *Cultures of Natural History,* edited by Nicholas Jardine, James A. Secord, and Emma Spary, 230–45. Cambridge: Cambridge University Press.

———. 2000a. "Books, Texts and the Making of Knowledge." In *Books and the Sciences in History,* edited by Nicholas Jardine and Marina Frasca-Spada, 393–407. Cambridge: Cambridge University Press.

———. 2000b. *The Scenes of Inquiry: On the Reality of Questions in the Sciences.* Oxford: Oxford University Press.

Jardine, Nicholas, James A. Secord, and Emma Spary, eds. 1996. *Cultures of Natural History.* Cambridge: Cambridge University Press.

Johns, Adrian. 2002. "The Book of Nature and the Nature of the Book." In *The Book History Reader*, edited by David Finkelstein and Alistair McCleery, 59–76. London: Routledge.

Keeney, Elizabeth B. 1992. *The Botanizers: Amateur Scientists in Nineteenth-Century America.*Chapel Hill: University of North Carolina Press.

Kemp, Martin. 1996. "'Implanted in Our Natures': Humans, Plants, and the Stories of Art." In *Visions of Empire: Voyages, Botany, and Representations of Nature*, edited by David Philip Miller and Peter Hans Reill, 197–229. Cambridge: Cambridge University Press.

Kinch, Michael Paul. 1980. "Geographical Distribution and the Origin of Life: The Development of Early Nineteenth-Century Explanations." *Journal of the History of Biology* 13, no. 1: 91–119.

Knight, David. 1981. *Ordering the World: A History of Classifying Man*. London: Burnett Books / André Deutsch.

Koerner, Lisbet. 1999. *Linnaeus: Nature and Nation*. Cambridge, MA: Harvard University Press.

Kuhn, Thomas S. 1970 (1975). *The Structure of Scientific Revolutions*. Chicago: University of Chicago Press.

Larsen, Anne. 1993. "Not Since Noah: The English Scientific Zoologists and the Craft of Collecting, 1800–1840." PhD diss., Princeton University.

———. 1996. "Equipment for the Field." In *Cultures of Natural History*, edited by Nicholas Jardine, James A. Secord, and Emma Spary, 358–77. Cambridge: Cambridge University Press.

Larson, James. 1986. "Not without a Plan: Geography and Natural History in the Late Eighteenth Century." *Journal of the History of Biology* 19, no. 3: 447–88.

Latour, Bruno. 1987. *Science in Action: How to Follow Scientists and Engineers through Society*. Cambridge, MA: Harvard University Press.

———. 1988. *The Pasteurization of France*. Cambridge, MA: Harvard University Press.

———. 1999. "Circulating Reference: Sampling the Soil in the Amazon Forest." In *Pandora's Hope: Essays on the Reality of Science Studies*, edited by Bruno Latour, 24–79. Cambridge, MA: Harvard University Press.

Latour, Bruno, and Steve Woolgar. 1979. *Laboratory Life: The Social Construction of Scientific Facts*. Beverly Hills, CA: Sage Publications.

Lenoir, Timothy. 1989. *The Strategy of Life: Teleology and Mechanics in Nineteenth-Century German Biology*. Chicago: University of Chicago Press.

Lewis, Jan. 1992. *Walter Hood Fitch: A Celebration*. London: HMSO / Royal Botanic Gardens, Kew.

Lightman, Bernard. 1997. "'The Voices of Nature': Popularizing Victorian Science." In *Victorian Science in Context*, edited by Bernard Lightman, 187–211. Chicago: University of Chicago Press.

Limoges, Camille. 1980. "The Development of the Muséum d'Histoire Naturelle of Paris, c. 1800–1914." In *The Organization of Science and Technology in France, 1808–1914*, edited by Robert Fox and George Weisz, 211–40. Cambridge: Cambridge University Press.

Lines, William J. 1996. *An All Consuming Passion: Origins, Modernity and the Australian Life of Georgiana Molloy*. Berkeley and Los Angeles: University of California Press.

Linnaeus, Carolus. 1751 (2005). *Philosophia Botanica*. Edited by Stephen Freer. Oxford: Oxford University Press.

Loudon, Irvine. 1992. "Medical Practitioners, 1750–1850, and the Period of Medical Reform in Britain." In *Medicine in Society: Historical Essays*, edited by Andrew Wear, 219–47. Cambridge: Cambridge University Press.

Lucas, A. M. 1995. "Letters, Shipwrecks and Taxonomic Confusion: Establishing a Reputation from Australia." *Historical Records of Australian Science* 10, no. 3: 207–21.

———. 2003. "Assistance at a Distance: George Bentham, Ferdinand von Mueller and the Production of *Flora australiensis*." *Archives of Natural History* 30, no. 2: 255–81.

Mabberley, D. J. 1985. *Jupiter Botanicus: Robert Brown of the British Museum*. Braunschweig: J. Cramer, in association with the British Museum (Natural History).

———. 2004. "Henfrey, Arthur (bap. 1820, d. 1859)." In *Oxford Dictionary of National Biography*, edited by Lawrence Goldman. Oxford: Oxford University Press.

Macintyre, Stuart. 1999. *A Concise History of Australia*. Cambridge: Cambridge University Press.

Mackay, David. 1990. "William Colenso." In *The Dictionary of New Zealand Biography*, 87–89. Wellington: Allen and Unwin.

MacLeod, Roy. 1971. "The Royal Society and the Government Grant: Notes on the Administration of Scientific Research, 1849–1914." *Historical Journal* 14, no. 2: 323–58.

———. 1974. "The Ayrton Incident: A Commentary on the Relations between Science and Government in England, 1870–1873." In *Science and Values: Patterns of Tradition and Change*, edited by Arnold Thackray and Everett Mendelsohn, 45–78. New York: Humanities Press.

———. 1983. "Whigs and Savants: Reflections on the Reform Movement in the Royal Society, 1830–48." In *Metropolis and Province: Science in British Culture, 1780–1850*, edited by Ian Inkster and Jack Morrell, 55–90. London: Hutchinson.

MacLeod, Roy, and Philip F. Rehbock, eds. 1988. *Nature in Its Greatest Extent: Western Science in the Pacific*. Honolulu: University of Hawai'i Press.

———. 1994. *Darwin's Laboratory: Evolutionary Theory and Natural History in the Pacific*. Honolulu: University of Hawai'i Press.

Magee, Gary Bryan. 1997. *Productivity and Performance in the Paper Industry: Labour, Capital, and Technology in Britain and America, 1860–1914*. Cambridge: Cambridge University Press.

Maiden, J. H. 1909. "Records of Tasmanian Botanists." *Proceedings of the Royal Society of Tasmania*, pp. 9–29.

Mandl, George T. 1985. *Three Hundred Years in Paper*. [London: 22 Hollen St., Wardour St. W1V 4BX, Clifton Hill].

Marchand, Leslie Alexis. 1971. *The Athenaeum: A Mirror of Victorian Culture*. New York: Octagon Books.

Martin, Stephen. 1993. *A New Land: European Perceptions of Australia, 1788–1850*. Sydney: Allen and Unwin.

Mason, Michael. 1995. *The Making of Victorian Sexuality*. Oxford: Oxford University Press.

McCord, Norman. 1991. *British History, 1815–1906*. Oxford: Oxford University Press.

McCracken, Donal. 1997. *Gardens of Empire: Botanical Institutions of the Victorian British Empire.* London: Leicester University Press.

McKenzie, Donald Francis. 2002. "The Sociology of a Text: Orality, Literacy and Print in Early New Zealand." In *The Book History Reader*, edited by David Finkelstein and Alistair McCleery, 189–215. London: Routledge.

McLean, Denis. 2006. "Dieffenbach, Johann Karl Ernst, 1811–1855." In *Dictionary of New Zealand Biography*, http://www.dnzb.govt.nz/ (updated 7 April 2006).

McMillan, Nora F. 1980. "William Swainson." In *Dictionary of Scientific Biography*, edited by Charles Coulston Gillispie, 167–68. New York: Charles Scribner and Sons.

McMinn, W. G. 1970. *Allan Cunningham: Botanist and Explorer.* Melbourne: Melbourne University Press.

McOuat, Gordon R. 1996. "Species, Rules and Meaning: The Politics of Language and the Ends of Definitions in 19th Century Natural History." *Studies in the History and Philosophy of Science* 27, no. 4: 473–519.

———. 2001. "Cataloguing Power: Delineating 'Competent Naturalists' and the Meaning of Species in the British Museum." *British Journal of the History of Science* 34, no. 120: 1–28.

Michie, Helena. 1999. "Under Victorian Skins: The Bodies Beneath." In *A Companion to Victorian Literature and Culture*, edited by Herbert F. Tucker, 407–24. Oxford: Blackwell.

Miller, David Philip. 1996. "Joseph Banks, Empire and "Centres of Calculation" in Late Hanoverian London." In *Visions of Empire: Voyages, Botany, and Representations of Nature*, edited by David Philip Miller and Peter Hans Reill, 21–37. Cambridge: Cambridge University Press.

Miller, David Philip, and Peter Hans Reill, eds. 1996. *Visions of Empire: Voyages, Botany, and Representations of Nature.* Cambridge: Cambridge University Press.

Mills, Eric L. 1984. "A View of Edward Forbes, Naturalist." *Archives of Natural History* 11, no. 3: 365–93.

Mitchell, Sally. 1983. *Dinah Mulock Craik.* Boston: Twayne.

Moore, James. 1997. "Green Gold: The Riches of Baron Ferdinand von Mueller." *Historical Records of Australian Science* 11, no. 3: 371–88.

Morgan, Marjorie. 1994. *Manners, Morals and Class in England, 1774–1858.* Basingstoke, UK: Macmillan.

Morrell, Jack. 1996. "Professionalisation." In *Companion to the History of Modern Science*, edited by R. C. Olby, G. N. Cantor, J. R. R. Christie, and M. J. S. Hodge, 980–89. London: Routledge.

———. 1997. "Individualism and the Structure of British Science in 1830." In *Science, Culture and Politics in Britain, 1750–1870*, 183–204. Aldershot, UK: Ashgate/Variorum.

Morrell, Jack, and Arnold Thackray. 1981. *Gentlemen of Science: Early Years of the British Association for the Advancement of Science.* Oxford: Oxford University Press.

Morrison-Low, A. D. 2004. "Brewster, Sir David (1781–1868)." In *Oxford Dictionary of National Biography*, edited by Lawrence Goldman. Oxford: Oxford University Press.

Morton, A. G. 1981. *History of Botanical Science: An Account of the Development of Botany from Ancient Times to the Present Day.* London: Academic Press.

Morus, Iwan Rhys. 2005. *When Physics Became King.* Chicago: University of Chicago Press.

Moyal, Ann Mozley. 1976. *Scientists in Nineteenth Century Australia: A Documentary History.* Stanmore, NSW: Cassell Australia.

Muir, Percy Horace. 1971. *Victorian Illustrated Books.* New York: Praeger.

Nelson, E. Charles. 1990. "James and Thomas Drummond: Their Scottish Origins and Curatorships in Irish Botanic Gardens (ca. 1800–ca.1831)." *Archives of Natural History* 17: 49–65.

———. 2002. "The Drummonds of Forfar—Five Generations of Plantsmen in Scotland, Ireland, North America, Australia and India." *Archives of Natural History* 29, no. 1: 30.

Nelson, Gareth. 1978. "From Candolle to Croizat: Comments on the History of Biogeography." *Journal of the History of Biology* 11, no. 2 (Fall): 269–305.

Newman, Andrew. 2004. "Hutton, William (1797–1860)." In *Oxford Dictionary of National Biography*, edited by Lawrence Goldman. Oxford: Oxford University Press.

Nicolson, Dan H. 1991. "A History of Botanical Nomenclature." *Annals of the Missouri Botanical Garden* 78: 33–56.

Nicolson, Malcolm. 1987. "Alexander von Humboldt, Humboldtian Science and the Origins of the Study of Vegetation." *History of Science* 25: 167–94.

Ogilvie, Brian W. 2006. *The Science of Describing: Natural History in Renaissance Europe.* Chicago: University of Chicago Press.

O'Hara, R. J. 1991. "Representations of the Natural System in the Nineteenth Century." *Biology and Philosophy* 6, no. 2: 255–74.

Olarte, Mauricio Nieto. 1993. "Remedies for the Empire: The Eighteenth Century Spanish Botanical Expeditions to the New World." PhD diss., Imperial College, University of London.

Oldroyd, David R. 1986. *The Arch of Knowledge: An Introductory Study of the History of the Philosophy and Methodology of Science.* London: Methuen.

Osborne, Michael A. 1994. *Nature, the Exotic, and the Science of French Colonialism.* Bloomington: Indiana University Press.

Ospovat, Dov. 1995. *The Development of Darwin's Theory: Natural History, Natural Theology and Natural Selection, 1838–1859.* Cambridge: Cambridge University Press.

Outram, Dorinda. 1984. *Georges Cuvier: Vocation, Science and Authority in Post-revolutionary France.* Manchester, UK: Manchester University Press.

Owens, J. M. R. 1990. "Richard Taylor." In *The Dictionary of New Zealand Biography*, 437–38. Wellington: Allen and Unwin.

Palmer, Frederick William, and Aderji Burjorji Sahiar. 1971. *Microscopes to the End of the Nineteenth Century.* London: HMSO.

Payne, Pauline. 1994. "'Science at the Periphery': Dr Shomburgk's Garden." In *Darwin's Laboratory: Evolutionary Theory and Natural History in the Pacific*, edited by Roy MacLeod and Philip F. Rehbock, 239–59. Honolulu: University of Hawai'i Press.

Pearce, Susan, Alexandra Bounia, et al., eds. 2000–2002. *The Collector's Voice: Critical Readings in the Practice of Collecting.* Aldershot, UK: Ashgate.

Pearce, Susan Mary. 1994. *On Collecting: An Investigation into Collecting in the European Tradition.* London: Routledge.

Petersen, George Conrad. 1966. William Colenso. In *An Encyclopaedia of New Zealand*, ed. A. H. McLintock, http://www.teara.govt.nz/1966/C/ColensoWilliam/ColensoWilliam/en.

Pickering, Andrew, ed. 1992. *Science as Practice and Culture*. Chicago: University of Chicago Press.

Piper, Charles V. 1906. "Flora of the State of Washington." *Contributions from the United States National Herbarium* 11: 1–637.

Polanyi, Michael. 1967. *The Tacit Dimension*. London: Routledge and Kegan Paul.

Porter, Andrew, ed. 1999. *The Nineteenth Century*. Oxford: Oxford University Press.

Porter, Duncan. 1993. "On the Road to the Origin with Darwin, Hooker, and Gray." *Journal of the History of Biology* 26, no. 1: 1–38.

Porter, Roy. 1978. "Gentlemen and Geology: The Emergence of a Scientific Career, 1660–1920." *Historical Journal* 21, no. 4: 809–36.

Porter, Theodore M. 1986. *The Rise of Statistical Thinking, 1820–1900*. Princeton, NJ: Princeton University Press.

Pratt, Mary Louise. 1992. *Imperial Eyes: Travel Writing and Transculturation*. London: Routledge.

Raby, Peter. 2001. *Alfred Russel Wallace: A Life*. Princeton, NJ: Princeton University Press.

Rajnai, Miklos, ed. 1982. *John Sell Cotman, 1782–1842*. London: Herbert Press.

Rehbock, Philip. 1983. *The Philosophical Naturalists: Themes in Early Nineteenth-Century British Biology*. Madison: University of Wisconsin Press.

Reingold, N., and M. Rothenberg, eds. 1987. *Scientific Colonialism*. Washington: Smithsonian Institution Press.

Rendle, A. B. 1935. "A Short History of the International Botanical Congresses." *Chronica Botanica* 1: 35–40.

Richards, Evelleen. 1987. "A Question of Property Rights: Richard Owen's Evolutionism Reassessed." *British Journal of the History of Science* 20: 129–71.

Richards, Robert J. 1992. *The Meaning of Evolution: The Morphological Construction and Ideological Reconstruction of Darwin's Theory*. Chicago: University of Chicago Press.

Rignall, John. 2001. *Reader's Companion to George Eliot*. Oxford: Oxford University Press.

Ritvo, Harriet. 1997a. *The Platypus and the Mermaid: And Other Figments of the Victorian Classifying Imagination*. Cambridge, MA: Harvard University Press.

———. 1997b. "Zoological Nomenclature and the Empire of Victorian Science." In *Victorian Science in Context*, edited by Bernard Lightman, 334–53. Chicago: University of Chicago Press.

Rowland, K. T. 1970. *Steam at Sea: A History of Steam Navigation*. Newton Abbot, UK: David and Charles.

Rudwick, Martin. 1985. *The Great Devonian Controversy: The Shaping of Scientific Knowledge among Gentlemanly Specialists*. Chicago: University of Chicago Press.

Rupke, Nicolaas A. 1994. *Richard Owen: Victorian Naturalist*. New Haven, CT: Yale University Press.

Ruse, Michael. 1999. *The Darwinian Revolution: Science Red in Tooth and Claw*. Chicago: University of Chicago Press.

Saunders, Gill. 1995. *Picturing Plants: An Analytical History of Botanical Illustration*. London: Zwemmer/Victoria and Albert Museum.

Schaffer, Simon. 1984. "Priestley's Questions: An Historiographic Survey." *History of Science* 22: 151–83.

———. 1995. "Accurate Measurement Is an English Science." In *Values of Precision*, edited by M. Norton Wise, 135–72. Princeton, NJ: Princeton University Press.

———. 1997. "Metrology, Metrication and Victorian Values." In *Victorian Science in Context*, edited by Bernard Lightman, 438–74. Chicago: University of Chicago Press.

Schiebinger, Londa. 1996. "Gender and Natural History." In *Cultures of Natural History*, edited by Nicholas Jardine, James A. Secord, and Emma Spary, 163–77. Cambridge: Cambridge University Press.

Secord, Anne. 1994a. "Corresponding Interests—Artisans and Gentlemen in 19th-Century Natural History." *British Journal of the History of Science* 27, no. 95, pt. 4: 383–408.

———. 1994b. "Science in the Pub—Artisan Botanists in Early-19th-Century Lancashire." *History of Science* 32, no. 97, pt. 3: 269–315.

———. 1996. "Artisan Botany." In *Cultures of Natural History*, edited by Nicholas Jardine, James A. Secord, and Emma Spary, 378–93. Cambridge: Cambridge University Press.

———. 2002. "Botany on a Plate: Pleasure and the Power of Pictures in Promoting Early Nineteenth-Century Scientific Knowledge." *Isis* 93, no. 1: 28–57.

———. 2007. "Nature's Treasures: Dawson Turner's Botanical Collections." In *Dawson Turner*, edited by Nigel Goodman, 43–66. Chichester, UK: Phillimore.

Secord, James A. 1986a. *Controversy in Victorian Geology: The Cambrian-Silurian Dispute*. Princeton, NJ: Princeton University Press.

———. 1986b. "The Geological Survey of Great Britain as a Research School, 1839–1855." *History of Science* 24: 223–75.

———. 1989. "King of Siluria: Roderick Murchison and the Imperial Theme in Geology." In *Energy and Entropy: Science and Culture in Victorian Britain*, edited by Patrick Brantlinger, 63–92. Bloomington: Indiana University Press.

———. 1997. Introduction to *Principles of Geology*, by Charles Lyell, ix–xlvii. Harmondsworth, UK: Penguin Books.

———. 2000a. "Progress in Print." In *Books and the Sciences in History*, edited by Nicholas Jardine and Marina Frasca-Spada, 369–89. Cambridge: Cambridge University Press.

———. 2000b. *Victorian Sensation: The Extraordinary Publication, Reception, and Secret Authorship of Vestiges of the Natural History of Creation*. Chicago: University of Chicago Press.

Shapin, Steven. 1984. "Brewster and the Edinburgh Career in Science." In *"Martyr of Science": Sir David Brewster, 1781–1863*, edited by A. D. Morrison-Low and John R. R. Christie, 17–24. Edinburgh: Royal Scottish Museum.

———. 1994. *A Social History of Truth: Civility and Science in Seventeenth-Century England*. Chicago: University of Chicago Press.

Shattock, Joanne. 1989. *Politics and Reviewers: The "Edinburgh" and the "Quarterly" in the Early Victorian Age*. Leicester, UK: Leicester University Press.

Sheets-Pyenson, Susan. 1988. *Cathedrals of Science: The Development of Colonial Natural History Museums during the Late Nineteenth Century*. Kingston, ON: McGill-Queen's University Press.

Shteir, Ann B. 1989. "Botany in the Breakfast Room: Women and Early Nineteenth-Century Plant Study." In *Uneasy Careers and Intimate Lives: Women in Science,*

1789–1979, edited by Pnina G. Abir-Am and Dorinda Outram, 31–43. New Brunswick, NJ: Rutgers University Press,

———. 1996. *Cultivating Women, Cultivating Science: Flora's Daughters and Botany in England, 1760 to 1860*. Baltimore, MD: Johns Hopkins University Press.

———. 1997a. "Elegant Recreations? Configuring Science Writing for Women." In *Victorian Science in Context*, edited by Bernard Lightman, 236–55. Chicago: University of Chicago Press.

———. 1997b. "Gender and 'Modern' Botany in Victorian England." *Osiris* 12: 29–38.

———. 2003. "Bentham for 'Beginners and Amateurs' and Ladies: *Handbook of the British Flora*." *Archives of Natural History* 30, no. 2: 237–49.

Sinnema, Peter W. 2002. Introduction to *Self-Help: With Illustrations of Character, Conduct and Perseverance*, by Samuel Smiles, vii–xxxviii. Oxford: Oxford University Press.

Slinn, E. Warwick. 1999. "Poetry." In *A Companion to Victorian Literature and Culture*, edited by Herbert F. Tucker, 307–22. Oxford: Blackwell.

Sloan, Phillip R. 1996. "Natural History, 1670–1802." In *Companion to the History of Modern Science*, edited by R. C. Olby, G. N. Cantor, J. R. R. Christie, and M. J. S. Hodge, 295–313. London: Routledge.

Slotten, Ross A. 2004. *The Heretic in Darwin's Court: The Life of Alfred Russel Wallace*. New York: Columbia University Press.

Smith, Crosbie. 1998. "Nowhere but in a Great Town: William Thomson's Spiral of Classroom Credibility." In *Making Space for Science: Territorial Themes in the Shaping of Knowledge*, edited by Crosbie Smith and Jon Agar, 118–46. London: Macmillan.

Smith, Jonathan. 1994. *Fact and Feeling: Baconian Science and the Nineteenth-Century Literary Imagination*. Madison: University of Wisconsin Press.

———. 2006. *Charles Darwin and Victorian Visual Culture*. Cambridge: Cambridge University Press.

Spary, Emma C. 2000. *Utopia's Garden: French Natural History from Old Regime to Revolution*. Chicago: University of Chicago Press.

Stafford, Robert A. 1989 (2002). *Scientist of Empire: Sir Roderick Murchison, Scientific Exploration and Victorian Imperialism*. Cambridge: Cambridge University Press.

Stapf, Otto. 1926. "The Botanical Magazine: Its History and Mission." *Journal of the Royal Horticultural Society* 51: 29–43.

Star, Susan Leigh, and James R. Griesemer. 1989. "Institutional Ecology, 'Translations' and Boundary Objects: Amateurs and Professionals in Berkeley's Museum of Vertebrate Zoology, 1907–39." *Social Studies of Science* 19: 387–420.

Stearn, William T. 1971. "Linnaean Classification, Nomenclature and Method." In*The Compleat Naturalist: A Life of Linnaeus*, edited by Wilfrid Blunt, 242–48. London: Collins.

———, ed. 1999. *John Lindley, 1799–1865: Gardener–Botanist and Pioneer Orchidologist*. Woodbridge, Suffolk, UK: Antique Collector's Club.

———. "Britten, James (1846–1924)." In *Oxford Dictionary of National Biography*, Oxford University Press, http://www.oxforddnb.com/view/article/38297 (accessed 23 June 2006).

Stemerding, Dirk. 1991. *Plants, Animals and Formulae: Natural History in the Light of Latour's "Science in Action" and Foucault's "The Order of Things."* Amsterdam and Enschede: University of Twente.

Stevens, Peter F. 1984. "Haüy and A.-P. de Candolle: Crystallography, Botanical Systematics, and Comparative Morphology, 1780–1840." *Journal of the History of Biology* 17, no. 1: 49–82.

———. 1991. "George Bentham and the Kew Rule." In *Improving the Stability of Names: Needs and Options*, edited by D. L. Hawksworth, 157–68. Königstein: Koeltz Scientific Books.

———. 1994. *The Development of Biological Systematics: Antoine-Laurent de Jussieu, Nature, and the Natural System*. New York: Columbia University Press.

———. 1997a. "J. D. Hooker, George Bentham, Asa Gray and Ferdinand Mueller on Species Limits in Theory and Practice: A Mid-Nineteenth-Century Debate and Its Repercussions." *Historical Records of Australian Science* 11, no. 3: 345–70.

———. 1997b. "On Amateurs and Professionals in British Botany in 1858—J. D. Hooker on Bentham, Brown, and Lindley." *Harvard Papers in Botany* 2, no. 2: 125–32.

———. 2003. "George Bentham (1800–1884): The Life of a Botanist's Botanist." *Archives of Natural History* 30, no. 2: 189–202.

St. George, Andrew. 1993. *The Descent of Manners: Etiquette, Rules and the Victorians*. London: Chatto and Windus.

Stott, Rebecca. 2003a. *Darwin and the Barnacle*. London: Faber and Faber.

———. 2003b. *Theatres of Glass: The Woman Who Brought the Sea to the City*. London: Short Books.

Strathern, Marilyn. 1988. *The Gender of the Gift: Problems with Women and Problems with Society in Melanesia*. Berkeley and Los Angeles: University of California Press.

Strick, James E. 2000. *Sparks of Life: Darwinism and the Victorian Debates over Spontaneous Generation*. Cambridge, MA: Harvard University Press.

Thomas, Nicholas. 1991. *Entangled Objects: Exchange, Material Culture, and Colonialism in the Pacific*. Cambridge, MA: Harvard University Press.

———. 1994. *Colonialism's Culture: Anthropology, Travel and Government*. Cambridge: Polity Press.

———. 1997. *In Oceania: Visions, Artifacts, Histories*. Durham, NC: Duke University Press.

Thomas, Nicholas, and Richard Eves. 1999. *Bad Colonists: The South Seas Letters of Vernon Lee Walker and Louis Becke*. Durham, NC: Duke University Press.

Thomason, Bernard. 2004. "Dyer, Sir William Turner Thiselton (1843–1928)." In *Oxford Dictionary of National Biography*, Oxford University Press, http://www.oxforddnb.com/view/article/36467 (accessed 4 September 2006).

Thompson, E. P. 1984. *The Making of the English Working Class*. Harmondsworth, UK: Penguin Books.

———. 1991. *Customs in Common*. Harmondsworth, UK: Penguin Books.

Tosh, John. 1999. *A Man's Place: Masculinity and the Middle-Class Home in Victorian England*. New Haven, CT: Yale University Press.

Turner, Frank Miller. 1993. *Contesting Cultural Authority: Essays in Victorian Intellectual Life*. Cambridge: Cambridge University Press.

Turner, Gerard L'Estrange. 1980. *Essays on the History of the Microscope*. Oxford: Senecio.

Turrill, W. B. 1953. *Pioneer Plant Geography: The Phytogeographical Researches of Sir Joseph Dalton Hooker*. The Hague: Martinus Nijhoff.

———. 1963. *Joseph Dalton Hooker: Botanist, Explorer and Administrator*. London: Scientific Book Club.

Underwood, Matthew. 2004. "A Tool for Better Homes and Gardens: The Invention and Development of the Wardian Case." MPhil essay, Department of the History and Philosophy of Science, Cambridge University.

Vetch, R. H. 2004. "Galton, Sir Douglas Strutt (1822–1899)." Revised by David F. Channell. In *Oxford Dictionary of National Biography*, Oxford University Press, http://www.oxforddnb.com/view/article/10317 (accessed 10 November 2004).

Wakeman, Geoffrey. 1973. *Victorian Book Illustration: The Technical Revolution*. Newton Abbot: David and Charles.

Waller, John C. 2001. "Sir Francis Galton, the 'Gentlemanly Specialist' and the Professionalization of the British Life-Sciences." *Journal of the History of Biology* 34: 83–144.

Walters, S. Max, and E. Anne Stow. 2001. *Darwin's Mentor: John Stevens Henslow, 1796–1861*. Cambridge: Cambridge University Press.

Ward, Alan. 2001. "McLean, Donald, 1820–1877." In *Dictionary of New Zealand Biography*, http://www.dnzb.govt.nz/.

Webb, Joan. 1995. *George Caley: Nineteenth Century Naturalist*. Chipping Norton, NSW: S. Beatty.

White, Paul. 1999. "The Purchase of Knowledge: James Edward Smith and the Linnean Collections." *Endeavour* 23, no. 3: 126–29.

———. 2003. *Thomas Huxley: Making the "Man of Science."* Cambridge: Cambridge University Press.

Winsor, Mary P. 1976. *Starfish, Jellyfish, and the Order of Life: Issues in Nineteenth-Century Science*. New Haven, CT: Yale University Press.

Winter, Alison. 1998. *Mesmerized: Powers of Mind in Victorian Britain*. Chicago: University of Chicago Press.

Wright-St Clair, Rex. 2006. "Monro, David, 1813–1877." In *Dictionary of New Zealand Biography*, http://www.dnzb.govt.nz/ (updated 7 April 2006).

Yaldwyn, John, and Juliet Hobbs, eds. 1998. *My Dear Hector: Letters from Joseph Dalton Hooker to James Hector, 1862–1893*. Wellington: Museum of New Zealand Te Papa Tongarewa.

Yeo, Richard. 1981 (2001). "Scientific Method and the Image of Science, 1831–1891." In *Science in the Public Sphere: Natural Knowledge in British Culture, 1800–1860*, edited by Richard Yeo, 65–88. Aldershot, UK: Ashgate / Variorum.

———. 1985 (2001). "An Idol of the Marketplace: Baconianism in Nineteenth Century Science." In *Science in the Public Sphere: Natural Knowledge in British Culture, 1800–1860*, edited by Richard Yeo, 251–98. Aldershot, UK: Ashgate / Variorum.

———. 1991. "Reading Encyclopedias: Science and the Organization of Knowledge in British Dictionaries of Arts and Sciences, 1730–1850." *Isis* 82: 24–49.

———. 1993. *Defining Science: William Whewell, Natural Knowledge, and Public Debate in Early Victorian Britain*. Cambridge: Cambridge University Press.

———. 2002. "William Whewell: A Cambridge Historian and Philosopher of Science." In *Cambridge Scientific Minds*, edited by Peter Harman and Simon Mitton, 51–63. Cambridge: Cambridge University Press.

Zirkle, Conway. 1946. "The Early History of the Idea of the Inheritance of Acquired Characteristics and of Pangenesis." *Transactions of the American Philosophical Society* 35, no. 2: 91–150.

INDEX